The culture of the math

The culture of the mathematics classroom is becoming an increasingly salient topic of discussion in mathematics education. Studying and changing what happens in the mathematics classroom allow researchers and educators to recognize the social character of mathematical pedagogy and the relationship between the classroom and the culture at large. This international and interdisciplinary volume outlines the emerging paradigms of socioconstructivist and sociocultural perspectives on this topic, and it reports findings gained in both research and practice.

The volume is divided into three sections. The first presents several attempts to change classroom culture by focusing on the education of mathematics teachers and on teacher–researcher collaboration. The second section shifts to the interactive processes of the mathematics classroom and to the communal nature of learning. The third section discusses the means of constructing, filtering, and establishing mathematical knowledge that are characteristic of the classroom culture.

As an examination of the social nature of mathematical teaching and learning, the volume should appeal both to educational psychologists and to cultural and social anthropologists and sociologists. The editors have compiled a volume that explores not only the acquisition of mathematical knowledge but the communal character of such knowledge.

Dr. Falk Seeger is Lecturer in Psychology and Mathematics Education at the Institut für Didaktik der Mathematik, Universität Bielefeld, Bielefeld, Germany. Dr. Jörg Voigt is Professor of Mathematics and Mathematics Education at the Universität Münster, Münster, Germany. Ute Waschescio is Lecturer in Psychology and Mathematics Education at the Institut für Didaktik der Mathematik, Universität Bielefeld, Bielefeld, Germany.

The culture of the mathematics classroom

EDITED BY

Falk Seeger

Jörg Voigt

Ute Waschescio

CAMBRIDGE
UNIVERSITY PRESS

PUBLISHED BY THE PRESS SYNDICATE OF THE UNIVERSITY OF CAMBRIDGE
The Pitt Building, Trumpington Street, Cambridge CB2 1RP, United Kingdom

CAMBRIDGE UNIVERSITY PRESS
The Edinburgh Building, Cambridge CB2 2RU, UK   http://www.cup.cam.ac.uk
40 West 20th Street, New York, NY 10011-4211, USA   http://www.cup.org
10 Stamford Road, Oakleigh, Melbourne 3166, Australia

First published 1998
Printed in the United States of America

Typeset in Times Roman 11/13 pt., in MagnaType™ 3.52 [AG]

Library of Congress Cataloging-in-Publication Data
The culture of the mathematics classroom / edited by Falk Seeger, Jörg
Voigt, Ute Waschescio.
p.   cm.
Includes bibliographical references and indexes.
ISBN 0-521-57107-3 (hardcover). – ISBN 0-521-57798-5 (pbk.)
1. Mathematics – Study and teaching – Social aspects.   I. Seeger,
Falk.   II. Voigt, Jörg.   III. Waschescio, Ute.
QA11.C785   1998
510'.71 – dc21   97-5755
                                                                      CIP

*A catalog record for this book is available from
the British Library.*

ISBN 0-521-57107-3 hardback
ISBN 0-521-57798-5 paperback

Contents

Contributors

Albrecht Abele Fachbereich III/Mathematik, Pädagogische Hochschule, Heidelberg, Germany

Heinrich Bauersfeld Institut für Didaktik der Mathematik, Universität Bielefeld, Bielefeld, Germany

Nadine Bednarz CIRADE, WB-2220, Université de Québec à Montréal, Montréal, Canada

Maria G. Bartolini Bussi Dipartimento di Matematica, Università di Modena, Modena, Italy

Paul Cobb Peabody College, Vanderbilt University, Nashville, Tennessee, USA

Yrjö Engeström Laboratory of Comparative Human Cognition, University of California, La Jolla, California, USA, and Center for Activity Theory and Developmental Work Research, Department of Education, University of Helsinki, Helsinki, Finland

Paul Ernest School of Education, University of Exeter, Exeter, United Kingdom

Jeff Evans School of Mathematics and Statistics, Middlesex University, Queensway, United Kingdom

Lisa Hefendehl-Hebeker Lehrstuhl für Didaktik der Mathematik, Universität Augsburg, Augsburg, Germany

Stephen Lerman Centre for Mathematics Education, School of Computing, South Bank University, London, United Kingdom

Falk Seeger Institut für Didaktik der Mathematik, Universität Bielefeld, Bielefeld, Germany

Heinz Steinbring Fachbereich Mathematik, Universität Dortmund, Dortmund, Germany

Jörg Voigt   Fachbereich Mathematik, Universität Münster, Münster, Germany

Ute Waschescio   Institut für Didaktik der Mathematik, Universität Bielefeld, Bielefeld, Germany

Erna Yackel   Department of Mathematics, Purdue University Calumet, Hammond, Indiana, USA

Acknowledgments

The idea for the present volume was born during an international conference, "The culture of the mathematics classroom," held at "Haus Ohrbeck" near Osnabrück, Germany, in October 1993, supported by the Volkswagen Foundation. The preparation of this volume greatly benefited from this meeting. We gratefully acknowledge the contributions (to the conference and to discussions related to this volume) of Heinrich Bauersfeld, Gert Schubring, Heinz Steinbring, and Hans-Georg Steiner at the Institut für Didaktik der Mathematik (IDM) of the University of Bielefeld, Germany.

We also would like to thank Jonathan Harrow for reviewing the language of the chapters with scrutiny and empathy, and Renate Baute for helping with the editing and printing of the manuscripts.

Introduction

FALK SEEGER, JÖRG VOIGT, AND
UTE WASCHESCIO

The present volume explores the culture of the mathematics classroom. "Classroom culture" is a concept signaling a change of perspective on education at many levels: on those of the practice of teaching and learning as well as on research and theoretical approaches. This change of perspective on classroom processes is reflected in the attempt to capture their complexity and their wholistic character as fundamentally social processes of making sense and meaning.

While the concept of classroom culture is presently gaining some currency (see, e.g., Bruner 1996), this volume nicely reflects that, as a paradigm, it is still in the making.

Before presenting the individual chapters in this volume, we shall outline some of the fundamental questions related to this emerging paradigm.

There is some tension between "culture" as somehow general and "classroom culture" as somehow specific; between "culture" as a pre-given frame of teaching–learning processes and "classroom culture" as something to be constructed; between culture as something to be appropriated and something to be created. The question, then, is how "classroom culture" relates to "culture at large," how "sharing the culture" relates to "creating culture," and how sharing a particular culture may relate to sharing culture at large.

There is another tension in the concept of classroom culture: Are the influences that culture exerts on learning more direct or more indirect? The meaning of culture and cultivation has always been related directly to education and development. In many languages, terms referring to education also point to "cultivation" in the sense of tillage, growing of plants, and so on. The connotations of "culture" and "cultivating" direct our attention to the indirect processes of teaching and learning that are not as obvious as the direct effects of teaching acts. In the mathematics classroom, the processes of meaning making are also very much intuitive and ambiguous. For example, why is a theorem called a deep or a strong one? Why is one solution of a mathematical problem taken as more elegant than others? What is the criterion for the validity of a nonformal argument? What does an account that is taken from a domain of experience outside mathematics represent? Following this line of argument, it becomes obvious that the direct teaching of knowledge is not the whole story of mathematics education. Instead, we have to reconsider the not so new insight that some knowledge can be taught, but sense-making can only be nurtured — to put it in Leont'ev's words (cf. Leont'ev 1978). The question, then, is how the educational process in the classroom should be conceptualized as a process of cultivation resting heavily on indirect teaching–learning processes.

The connotations of sharing the culture and participating underline the social nature of the whole enterprise. Putting culture at the center of conceptions of mathematics education means a rejection of learning viewed as "solo experience." In Jerome Bruner's words: "I have come increasingly to recognize that most learning in most settings is a communal activity, a sharing of the culture" (Bruner 1986: 127). The question, then, is what a classroom that focuses on learning mathematics as communal activity culture might look like. This question is particularly challenging to mathematics education, because doing mathematics is typically considered to be a solitary game.

Taken together, the preceding three questions form a broad perspective on the culture of the mathematics classroom that may be expressed as follows: What could a culture of the mathematics classroom look like that, through its teaching content and its nurturing of sense-making, allows students to share the culture of mathematics and thereby also allows them to share the culture at large?

To contribute to this broad perspective is the explicit goal of the chapters written for this volume. It comes as no surprise that their authors share some points of departure and some common beliefs. One of these beliefs is the firm rejection of a notion that understands innovation in education largely as a matter of reorganizing mathematical content. Instead, the perspective of classroom culture underlines that new content will not be enough to bring about innovation if the social processes in the classroom are left unchanged. Another shared belief is the equally firm rejection of viewing innovation as a transformation that can be brought about by moving along the chain research → administration → teachers → students, and feeding in innovation at any given point. Instead, it is obvious that the evolution of a culture of the mathematics classroom cannot be ordered from above nor be prescribed by research.

Other common beliefs are:

- The tension between traditional, everyday mathematics classrooms and the ideal conceptions of teaching and learning mathematics is inevitable, and it forms the challenge both to understand the stability of the traditional classroom and to provide for its change. Therefore, most chapters consist of careful empirical studies and considerations on normative orientations.
- Although it is claimed that the products of the discipline of mathematics are universal, learning mathematics involves processes of meaning making that are sensitive to context. The culture of the mathematics classroom forms an important context for these processes.
- One stereotype of mathematical activity is the solitude of the problem solver, who only communicates the products of thoughts and their reasons in retrospect and in formal modes of argumentation. Nevertheless, at school, learning mathematics is organized in social situations, even in face-to-face interactions, and the mathematics learned depends on the dynamics of the social processes among the participants of the classroom.
- Studying the culture of the mathematics classroom, the researcher is confronted with highly complex processes. In order to handle this complexity, to reduce it, and to reconstruct structures and typical aspects, one is forced to develop and to apply theoretical concepts. Because of academic claims, the researcher has to explicate the theoretical concepts, and the reader cannot be saved from taking the theoretical bases into account. Also, the researcher's attempt to cope with the complexities of the classroom culture has the consequence that qualitative methods are preferred for the study of the classroom culture as a net of relations and as a meaningful context for the persons involved.

The approach presented in this volume is as multifaceted as the multicultural society of today. Sharing a culture that rests on the principled diversity of its constituent parts calls for a permanent dialogue. This dialogue is based on the assumption that one can find community in diversity. The present volume reflects this principle as a *modus vivendi* of the scientific community of mathematics education in the dialogues that are going on among interactionist, constructivist, and sociocultural approaches to classroom culture. That the dialogue is so lively is to a large extent a result of the efforts of individual persons. It is, however, also based on the very positive development of mathematics education as a field of study, as an academic discipline with both a strong international and a strong interdisciplinary character.

During the last decades, mathematics education has evolved into an integral part of academic life on an international scale. Today, teachers, mathematics educators, and researchers can benefit from theories and research data that were generated originally under diverse disciplinary perspectives: in psychology, sociology, linguistics, cultural anthropology, history, or epistemology. The chapters in this volume are based on research work that heavily draws on multi- and interdisciplinary approaches transformed into the disciplinary perspective of mathematics education. The multidisciplinary perspective makes it possible to conceive the complexity of mathematics education as the interrelationship of the divergent perspectives of learners, teachers, and educators; as an evolving system that is structured by social regularities and discursive practices; and as a domain of knowledge with a specific epistemology.

Consequently, the present volume is divided into three main sections that focus on changing classroom culture, on the analysis of classroom processes as constituting classroom culture, and on knowledge as related to classroom culture.

The first section presents attempts to change classroom culture with a focus on teacher education and teacher–researcher collaboration. Past, and mostly unsuccessful, efforts toward innovation in the mathematics classroom have shown that changes cannot be brought about through minor modifications of isolated aspects of the teaching–learning process. Instead, rather radical and complex changes seem to be necessary. These involve the role of the teacher, the role of the student, patterns of classroom interaction, underlying concepts of the learning process, major reorganizations of mathematical content, attitude change, the physical and institutional context of schooling, and many more. Such radical changes call for equally radical changes in the education of teachers and teacher students, because it would be pointless to expect teachers to break with traditions in their classrooms while continuing to be educated in the spirit of these same traditions. Consequently, teacher–researcher collaboration forms an essential element of attempts to change the culture of the mathematics classroom.

Maria Bartolini Bussi describes a project involving the change of classroom discussions and the change of the educational context as well. Interpreting transcripts of classroom interactions, she gives a vivid impression of the spirit of discussions about geometrical problems. The development of this classroom culture was based on the teachers' and researchers' involvement in the study of theoretical works. These theoretical concepts are particularly used to explain the teacher's role in a dialogue-oriented classroom culture.

Nadine Bednarz describes a promising collaborative research project that involves elementary school teachers and researchers. The project overcame the traditional model of innovation as transformation of theory into practice by focusing on the interaction between the teacher and the students and on the interaction between the teacher and the researcher. The teachers' collaboration with the researchers involved a reflection on and transformation of the meanings the teachers ascribed to classroom processes. The author brings out the parallels between the classroom culture and the culture of teacher education. She illustrates these parallels by interpreting detailed excerpts from school life and from the collaborative work.

Yrjö Engeström uses concepts from activity theory to describe and explain the motivational sphere of classroom culture. The theoretical tools are applied in the analysis of an innovative team of elementary teachers. Attempting to relate mathematical activities to the life world outside school, the teachers had to cross the traditional and institutional boundaries of school work. The author describes the teachers' team work in detail, the joint enterprises, and the conflicts, and he presents a complex network of theoretical concepts that enables us to understand the structural characteristics of the teachers' innovations at school.

Lisa Hefendehl-Hebeker concentrates on one main aspect of improving the classroom culture, that is, the change in the teacher's attitude toward the students and toward mathematics. Self-critically reflecting on her own teaching practices, she relates the students' struggle with mathematical problems at secondary school to the teacher's understanding of mathematics. The author gives reasons for why a deeper understanding of the history of mathematics improves the teacher's competence to negotiate mathematical meanings with students. Thus, understanding students' mathematical activities and understanding mathematics seem to be two sides of the same coin.

The second section, on classroom processes, focuses on how classroom culture is constituted through processes of teacher–student interaction and communication. In the ongoing interactive processes, students and teachers (re-)establish specific mathematical practices that partly differ from mathematical activity in the discipline as well as from everyday mathematical activity outside school. In these processes, a microculture is reconstructed and created. The analysis of this microculture is guided by questions addressing, for example, how mathematics learning can be seen as part of students' being introduced into culture and society at large; how classroom culture is created by the network of teachers' and students' activities; how classroom culture may lead to making personal sense of mathematics; and how an ethnographic perspective takes culture and society at large into account. The studies in this chapter are based on different theoretical approaches such as constructivism, symbolic interactionism, and activity theory. The authors also adopt different methods for the analysis of classroom life: micro-ethnographical methods, methods of cognitive analysis, and methods of discourse analysis. Generally, the results of the studies are captured in classroom documents like transcripts of discussions among students and between teacher and student.

Albrecht Abele analyzes reasoning processes among students and evaluates the quality of their arguments. Three groups of children from grades 2, 3, and 4 were confronted with an unfamiliar, complex, and open mathematical problem. Applying several category systems to the students' collaborative work, the author examines, among other topics, the winding solution paths, the persuasiveness and the validity of arguments, and the developmental differences between the grades. The study profits from establishing relations between pedagogic claims regarding cooperative learning and mathematical claims regarding reasoning.

Paul Cobb and Erna Yackel compare two different theoretical views on the mathematics classroom culture: the constructivist perspective and the sociocultural perspective. Each forms a main theoretical and normative background for studies of classroom culture. On the one hand, the authors reconstruct the theoretical assumptions of both views as presenting basically incompatible alternatives. On the other hand, following a pragmatic approach, they call for a division of labor in order to overcome the claims of hegemony. The theoretical concepts developed from the competing perspectives can be taken as instruments for studying different topics.

Jörg Voigt analyzes the interaction between the mathematics teacher and students in order to understand the emergence of intersubjectivity between differently thinking persons. The microculture of the mathematics classroom is viewed as constituted and stabilized through these very interaction processes. The study centers on classroom episodes in which the participants had interpreted a real-world phenomenon differently and negotiated its mathematical meaning. The author describes how the microculture can be improved if the teacher takes the freedom of mathematical modeling into account and has the opportunity to reflect the processes of negotiating meaning in the classroom.

Ute Waschescio continues the dispute between the socioconstructivist and sociocultural views on the mathematics classroom. She focuses on the concept of social interaction presented in the social constructivist theories of the preceding two chapters, and she offers arguments as to why these theories have not provided an explanation for the occurrence of learning and for the orientation of development in the classroom. In particular, the socioconstructivist restriction of psychological dimensions to internal processes of the individual and the neglect of cultural–historical issues must, of necessity, result in the dilemma of a missing link between learning and negotiation. The criticism leads to the preference for the sociocultural perspective based on Vygostky's concepts of internalization and cultural tool.

The third section, on epistemology and classroom culture, focuses on issues of how knowledge is generated, processed, and filtered in the mathematics classroom. The epistemological perspective is oriented toward how mathematical knowledge is created in the classroom, how it is related to application in practical situations, and how scientific knowledge of mathematics as a discipline differs from content knowledge in the classroom. Some lines of special interest are explored such as the relation between public and private knowledge, the rhetoric of the mathematics classroom, the linearization of knowledge in typical teacher–student interactions, the social construction of knowledge and subject matter, and the role of representations in the mathematics classroom.

Paul Ernest explores the relationship between personal and public knowledge of mathematics and reconstructs differences and parallels between the two cultures of the school and mathematical research. Personal and public knowledge are viewed as being linked by social processes among the members of a culture. Instead of logic, conversation, in the sense of interpersonal negotiation, is taken to be of crucial importance for the validity and the acceptance of knowledge manifested in linguistic behaviors and symbolic representations. The author's basic philosophical orientation enables him to combine a wide variety of dimensions and aspects, with the consequence that the social construction of mathematical knowledge is grasped in its entirety.

Jeff Evans studies the relationship that exists between mathematics at school and mathematics outside school. The principal focus of this study is on the problem of transfer of learning from school to practical situations. The difficulty of this problem is explained by relating different types of mathematics to different discursive practices involving specific interpersonal relations, specific language-related means, specific values, specific feelings, and so forth. Presenting a case study of an interview with an older student about trading, the author illustrates discursive practices and their specific characteristics. As a result, compared with the traditional understanding of transfer as the application of general and abstract mathematics to everyday situations, an alternative perspective is proposed that does not abstract cognition from cultural practices.

Stephen Lerman characterizes mathematics as a social construction in a fundamental way. Referring to sociologies of mathematics, he proposes a radical shift in the view on mathematics. Whereas, from the absolutist point of view, mathematics seems to be independent of history and culture, a sociology of mathematics considers mathematics, as well as the view on mathematics, as socially constructed in a given society. Using theoretical arguments, the author encourages the community of mathematics educators to make use of the potential of the sociologies of mathematics. In this way, topics of mathematics education, for example, mathematical abilities or mathematical concepts, can be studied with regard to their cultural and historical backgrounds.

Falk Seeger focuses attention on the concepts of representation. For example, the number line is taken as an external representation, and the student's mental number line as an internal representation. The author reviews several theoretical works that conceptualize especially the relationship between representation and abstraction and that are prominent in mathematics education. His cultural–historical approach to representations leads to the view of representations as exploratory artifacts. Exemplified by concrete examples, representations can function and can be interpreted in multiple ways. Although it is considered that the exploratory potential of a representation could be reduced in order to guarantee the stability of the classroom culture, this potential has to be unfolded in order to explore the richness of inner-mathematical relations as well as the often neglected relations between mathematics at school and mathematics outside school.

Heinz Steinbring considers the process of deciphering signs and symbols as the core problem facing the members of the classroom culture when trying to understand school mathematics. Because the signs and symbols have to be deciphered relative to specific contexts that are not fully recognizable for the learners, the classroom members may be tempted to follow conventionalized and ritualized procedures of using the signs and symbols, or they are tempted to reduce the meanings to concepts with which the learners are familiar. Interpreting two classroom episodes, the author describes this danger. He gives reasons for why the teacher and the students should resist the temptation to fix meanings immediately in the single classroom situation, and why they should understand the process of ascribing mathematical meanings as a relatively open one. The participants have to maintain a dynamic balance between the need to stabilize conventionalized patterns of argumentation socially and the option of constructing new conceptual understandings of the signs and symbols.

Heinrich Bauersfeld, in an epilogue, criticizes the frequent and often unreflected use of a concept like "culture," stating that it sometimes does not have a clarifying but rather an obscuring effect. He takes a look at several issues, which, within the discourse on the culture of the mathematics classroom, usually receive little attention but are nonetheless very important. These are the unity of mind and body; the crucial function of human interaction and cooperation; language, self, and reflection; the teacher's role; and the agents.

Changing and analyzing classroom culture always means to go beyond the given. In the process of change, the given may only be a starting point for the development of alternative and new ways of making mathematics classrooms more lively and more interesting places. In the process of analysis, it may also not be enough to look at what is given in the classroom. One should also transcend the confines of the classroom here and take a look at the world outside. This is definitely not at odds with the purpose of schooling that is (or should, or will be) all about opening windows and doors to the world beyond the classroom. The view that classroom culture is part of culture at large might also be strengthened through a perspective that opens the windows and doors into classrooms, thus making it possible to view what is happening there in greater detail and clarity.

References

Bruner, J.S. (1986). *Actual minds, possible worlds.* Cambridge, MA: Harvard University Press.

Bruner, J.S. (1996). *The culture of education.* Cambridge, MA: Harvard University Press.

Leont'ev, A.N. (1978). *Activity, consciousness, and personality.* Englewood Cliffs, NJ: Prentice-Hall.

Part I

Changing classroom culture

1 Joint activity in mathematics classrooms: A Vygotskian analysis

MARIA G. BARTOLINI BUSSI

A dialogue: It is a very cold and clear February morning. The scene is in a 6th-grade classroom during a mathematics lesson.

Student: Teacher, why are the shadows cast by the sun not proportioned?

Teacher (surprised and perplexed): What? What do you mean?

Student: Oh, I was coming to school this morning. While I was walking, I looked at my shadow: The body was enormously long and the head so small.

(Garuti 1994, personal communication)

What has the preceding dialogue got to do with school mathematics? Why does the student feel free to introduce an everyday problem, optimistic about the possibility of solving it in a school setting? Why does the teacher accept this dialogue? Does the student gain an answer to her request? What kind of answer?

The aim of this chapter is to answer these questions by analyzing the classroom culture that makes it possible to consider the dialogue as an emblematic representative of the standard classroom life.

The classroom in which this dialogue takes place is taught by a teacher–researcher (Rossella Garuti) who is involved in the Mathematical Discussion project; this project aims at designing, implementing, and analyzing long-term teaching experiments with a systematic recourse to classroom discussion orchestrated by the teacher. The theoretical framework of the project is mainly based on the seminal work of Vygotsky and his scholars and on the recent development of the cultural–historical school or activity theory (Bartolini Bussi 1996).

I shall divide the paper into two parts, concerning (1) the quality of classroom interaction and (2) the quality of educational contexts or fields of experience (Boero 1992; Boero et al. 1995).

This separation is functional in terms of presentation but risks being misleading. In a cultural–historical perspective, the particular modes of classroom interaction and the educational contexts for the teaching–learning activity (and for the tasks by means of which activity is realized) are closely connected to each other. In other words, the communicative and the instrumental aspects of activity cannot be separated (for a discussion see Engeström 1987). This aspect will be emphasized in the approach to mathematical discussion, as defined in our research group.

The quality of classroom interaction

Mathematical discussion: Some references to the literature

In recent years, the study of interaction in the mathematics classroom has received increasing attention in the literature on didactics of mathematics. In particular, analyses of existing situations in standard classrooms (e.g., Edwards and Mercer 1987; Maier and Voigt 1992) have shown that verbal interaction gives rise to patterns and routines. Maybe the most quoted example is the basic "I–R–F" exchange structure (initiation by a teacher, which elicits a response from a student, followed by an evaluative comment or feedback from the teacher; see Edwards and Mercer 1987: 9) that is observable in most classroom dialogues between the teacher and students.

Several correctives have been proposed to encourage the learner's responsibility for learning: for example, small group work (e.g., Laborde 1991); one-to-one tutoring (e.g., Schoenfeld et al. 1992; other articles in the same issue); or whole class discussion. In this chapter, I shall elaborate more on the last topic.

Instances of discussion in the classroom can be found in the verbal part of the teaching–learning activity. The verbal part does not exhaust the whole of activity (e.g., it does not describe gestures), yet it gives us a relevant perspective on the teaching–learning process because of the importance of verbal communication in the school setting.

The term "mathematical discussion" has been introduced by Pirie and Schwarzenberger (1988), who define it as a purposeful talk on a mathematical subject in which there are genuine student contributions and interaction. Each part of the definition is related by the authors to a specific aim in order to distinguish it from social chat on whatever subject and to emphasize students as active and interactive participants.

Examples of such discussions may be found in the international literature (see Bartolini Bussi 1991, 1992, for reviews). In order to distinguish better discussion from teacher-led I–R–F exchange, the role of
the teacher is not emphasized, to the extent that most of the examples
studied by the authors concern mathematical discussion in small group
work, that is, peer interaction.

Mathematical discussion is also analyzed by Richards (1991) from an
explicitly stated constructivist perspective. In order for students to become mathematically literate adults, they must learn to speak "inquiry
math," that is, to ask mathematical questions; to solve mathematical
problems that are new for them; to propose conjectures, to listen to
mathematical arguments; and to read and challenge popular articles
containing mathematical content. The teacher's role is explicitly addressed, with extensive reference to Schoenfeld's work: The essence of
good teaching is stated as the fact that both the teacher and the student
are prepared to learn from their mutual participation in a discussion. The
mathematical literacy of teachers is explicitly required for good teaching in order to design a context in which the student can construct
mathematics.

The conceptualizations of discussion quoted are representative of the
most common perspective in the literature: Mathematical discussion is
conceived as a good way for establishing, through negotiation in the
classroom, consensual domains in which communication about a mathematical topic takes place. The need to encourage the learner's responsibility for learning is clear.

It is beyond the scope of this chapter to analyze the reasons underlying this conceptualization of mathematical discussion. They may be
many and from different origins: the need to contrast the traditional
ideology of teaching; the need to reduce the number of variables in the
study of classroom processes (Artigue 1992); the need to meet the myth
of the self-made man or woman or to remain faithful to the underlying
individualist theories of learning (Wertsch 1991). However, the infre-
quent focus, if any, on the teacher's role as a representative of the
existing culture and his or her responsibility for introducing it to the
classroom is a constant in the literature on mathematical discussion (and
on the whole of didactics of mathematics too).

Emphasizing the need to construct consensual domains is surely important but tells only a part of the story. It would be really too naive to
believe that students can reconstruct the centuries of products of mathematics by themselves, and, in any case, this would not fit the present
organization of the school system. Yet, as long as research focuses only

on the individual learning pole and ignores the issues of teaching, teachers are offered few suitable instruments for managing the complexity of classroom life. A comprehensive approach to mathematical discussion that takes account of the learner's responsibility for learning as well as the teacher's responsibility for teaching requires a different foundation for the whole field (further elements for discussion can be found in Waschescio this volume).

Mathematical discussion: A Vygotskian (and Bachtinian) perspective

The number of references to Vygotsky is increasing in the literature on classroom interaction. What seems to be the most popular citation concerns internalization or interiorization (understood differently from Piaget's sense) as the transformation of object-oriented (collective) activity into the structure of internal consciousness. The famous genetic law of cultural development states:

Any function in the child's cultural development appears twice, or on two planes. First it appears on the social plane, and then on the psychological plane. First it appears between people as an interpsychological category, and then within the child as an intrapsychological category. (Vygotsky 1981: 163)

This is the almost universally quoted part that may be related very easily to the interpersonal activity that takes part in every discussion, from social chats to debates at scientific symposia. Actually, it is quoted in many papers on classroom interaction (and even on peer interaction) to mean the shift from taking part in a dialogue to being able to enter a dialogue with oneself (e.g., posing self-questions in problem-solving). But the full quotation is longer and more meaningful:

Social relations or relations among people genetically underlie all higher functions and their relationships. Hence, one of the basic principles of our volition is that of the division of functions among people, the new division into two parts of what is now combined into one. It is the development of a higher mental process in the drama that takes place among people. (Vygotsky 1981: 163)

In the school setting too, what has to be internalized is related to the division of functions. Even if both teachers and students are engaged in the same indivisible activity (the teaching–learning activity in the school setting), their functions are not the same, so the roles they have to play in the drama are different. The teacher is not one among many peers, but rather the guide in the metaphorical zone of proximal development (Vygotsky 1978).

What makes a mathematical discussion an effective part of school experience, in a Vygotskian perspective, is its suitability to represent some part of the drama played among persons that is related to the individual mathematical concept or procedure under scrutiny. The more voices are represented, the more meaningful the discussion is for the participants. Here we are using the term "voice" after Wertsch (1991), following Bakhtin, to mean a form of speaking and thinking that represents an individual perspective and the membership of a particular cultural and social category. Yet, in order to be acknowledged by all the participants, different voices need time enough to articulate themselves: Hence, classroom dialogues led by the teacher by means of a very strong I–R–F exchange structure are not meaningful in this sense, because the voices diverging from the expected flow of the discourse are immediately evaluated as not pertinent or even wrong by the teacher and cancelled from the collective memory. In spite of the dialogical form, most classroom discourse in traditional classrooms is only a teacher monologue. On the contrary, the most distinguishing feature in those kinds of mathematical discussions we wish to design and to study is the presence of articulated different voices as in a polyphony (Bakhtin, as quoted in Engeström 1987: 310–316).

The musical metaphor can help us to a better description of a mathematical discussion as a polyphony of articulated voices on a mathematical object (either concept or problem or procedure or belief) that is one of the motives of the teaching–learning activity. The term motive is used after Leont'ev (1977) to mean the object of the collective activity.

It is beyond the scope of this chapter to analyze in more depth the limits and the potentials of this metaphor that actually introduces a cultural perspective on discussion that is very different from the constructivist one (see, e.g., the relevance of different kinds of imitation in the theory and practice of counterpoint, Schönberg 1963).

Our approach to mathematical discussion can be contrasted to the approach of Pirie and Schwarzenberger (1988). Some elements of their approach are present (the focus on an explicit theme that concerns mathematical activity; the focus on interaction that we include in polyphony), but we introduce additional elements: the positive value given to imitation; the necessary presence of different voices, among which the teacher's voice acts as a representative of culture; and the reference to the motives of the teaching–learning activity that recalls, on the one side, collective long-term processes and, on the other side, the material or ideal objects that determine the direction of activity.

A form of mathematical discussion is the debate orchestrated and organized by the teacher in the whole classroom on a common mathematical object. In this case, the teacher has the important role of introducing a different perspective on the object different from the ones introduced by the students (e.g., the speech genre of formal education, Wertsch 1991: 127). It has to be emphasized that the motives of activity in teaching–learning mathematics (or a special part of it) determine very strongly the cultural voice of the teacher on the basis of the careful analysis of the epistemological foundations of the concepts and introduce strategies that are dependent on the fact that the discussion is about mathematics and are quite different from the ones that would be observed in a discussion about social relations, justice, or history (see, e.g., the paraphrasing on a general level that is described in the following or Bartolini Bussi and Boni 1995).

In more expert classrooms, after some months or years of experience in the kind of debate described, different voices come from students themselves, and, in the extreme case of the individual task that incorporates previous collective activity, different voices are internalized and can be uttered by an individual actor who is speaking to himself or herself (some examples of discussion with oneself are quoted in this section and in the following one). In fact, the voices of a mathematical discussion are not necessarily identified with verbal utterances pronounced by different individual participants (as stressed, in contrast, by Pirie and Schwarzenberger 1988); indirect utterances of other concrete voices that have been part of the speaker's experience also act as distinguished voices in the dialogue. It is the case of the students' reference to everyday practical experience or of the teacher's reference to pieces of established knowledge that are part of his or her education as a mathematically literate adult and of the things he or she is supposed to teach in order to introduce students into the same culture. For instance, a direct quotation from a book is an explicit way to introduce a voice from outside into the classroom (a detailed study of the status of voices can be found in Bartolini Bussi 1996; Bartolini Bussi, Boni, and Ferri 1995).

A direct help in a joint activity with a single student can also be transformed into a voice in the collective discussion, as the following transcript provided by Mara Boni shows:

The students (5th graders) have made a real-life drawing of a corner of their own classroom. The drawing is one step in a complex teaching experiment on the representation of the visible world by means of perspective drawing (Bartolini Bussi 1996). An excerpt of the initial part of the later discussion follows. All the students are sitting in a circle. The teacher is walking around among them. The situation is relaxed.

1. Teacher: Yesterday we finished our drawings. Now everybody should tell me what they have done. How did you organize your own work?

(some students explain their difficulties in drawing)

11. Teacher: I'm not asking everybody about their mistakes. I'm asking you what helped you with your drawing? What created difficulties for you? Not only fine, ugly crooked lines, but what did you use to help yourselves while working . . . this is what interests me and what can help you yourselves.

12. Francesca: I had trouble in the upper corner, and I asked myself: Are the walls made this very way? They do not seem to give the impression that you are inside the room. There's no depth there. I've got the lines wrong. I don't know. . . .

13. Greta: Er, I've organized myself for drawing in this way. I looked at the ceiling to make the shelf; not exactly the ceiling, but the border between the wall and the ceiling: It was straight and the shelf was straight like the ceiling.

14. Teacher: Straight? What do you mean?

(The teacher is interested. She approaches Greta's desk and bends down to listen better)

15. Greta: It was not crooked, it was perfectly . . . wait . . . it was parallel to the ceiling and I've drawn it parallel; but it didn't work very well, and my difficulty was that I've drawn the top of both the shelf and the table, even if I saw only the thickness of the shelf and only a piece of the top of the table. I didn't succeed, and you have. . . .

(Greta breaks off her talk. The teacher turns toward the whole class)

16. Teacher: I've told you: Are you afraid? Why? *(to the classroom)* She saw the open side and an obtuse angle, nearly a straight line. She pointed it out on the sheet but she was afraid of tracing the line. *(to Greta):* Why were you afraid?

17. Greta: I continued to be stubborn about the top *(of the table)*, but I didn't see it. Later, when you traced it on my sheet I understood, and I've changed the shelf too.

18. Teacher: Let's stop for a while. I shall recall the episode. *(to the classroom)* I don't know if you have understood. Greta sees the table and tries to draw it with an obtuse angle (but not very obtuse). This is an old problem for most of you too. She looked at reality and at the sheet,

without drawing anything; she pointed out the right line, that is, the line as she saw it, but she didn't draw it. She shook her head; she stopped. At a certain point, I asked her: "Why are you so afraid? I can help you," and I traced the line she was pointing at. Then Greta unblocked herself and was reassured; she drew the other parts of the table; at the beginning she outlined them, she controlled them, and then she took away the other surfaces. I didn't understand (*laughing*) what happened in your head. Maybe the problem was not just yours alone. Let's try to discuss it a little. What can help us in this kind of problem? How can the problem be solved with the tools we have?

The transcript cannot convey completely the warm and encouraging attitude of the teacher. As usual, she has defined some norms of the contract (1 and 11) but she is ready to make the most of Greta's talk, for Greta and for the class as well. Greta is helped to recall a difficult moment of her process of drawing. On the one side, the teacher helps Greta to tell the whole story; and, on the other side, she uses the story for all the students. The teacher is also constructing the collective memory of the classroom with emphasis on the special role she has played as a guide.

Types of mathematical discussions

The preceding description of mathematical discussion identifies it with the verbal part of the continuous flow of activity going on in the classroom. This description may be suggestive, but it does not provide a way of designing, analyzing, or modeling classroom processes.

A research project on mathematical discussion has been in progress in our research group since 1987, as a form of cooperation between a professional researcher in mathematics education (the author of this chapter) and a group of schoolteachers (teacher–researchers Ferri 1992). Up to now, we have studied some models of discussions (and identified more in the flow of activity) and have used them to design, try out, and analyze some teaching experiments in classrooms from grade 1 to grade 8 (Bartolini Bussi 1991, 1992).

Every teaching experiment is long term (1 year at least, and up to 3 years) and consists of a complex alternation of individual and collective tasks. Examples of the former are open problem situations to be solved and commented on in writing by students; examples of the latter are mathematical discussions in the form of either classroom dialogues or guided collective readings of sources (from textbooks, documents of classroom history, documents of human history). A particular model of mathematical discussion is described in the following.

A balance discussion is a whole class discussion whose goals are the socialization and collective evaluation of strategies already set up by students in individual problem solving. It takes place some days after every individual problem-solving session. The delay fosters the distancing of the students from their own products to the extent that they may be considered in a detached way. The teacher does not propose comparison of all the strategies, but selects only the prototypes of different strategies or maybe different problem representations; following Bakhtin, we can say that the delay allows the teacher to individualize different voices in the classroom and to organize their entrance into the scene.

The standard plot of a balance discussion is the following:

> First act. *True balance: Individual presentation and collective evaluation of different solutions before the classroom.* (Some authors of prototypical solutions are requested by the teacher to explain their solution; fellow students are invited to join the authors of similar solutions, adding their own explanation; both right and wrong solutions are discussed, and consensus solutions are sought.)
>
> Second act. *Process: Individual reconstruction of the process of problem solving.* (Students are requested to elicit the individual process of solution, mentioning either effective or abandoned strategies; the difficulties they have met, and how they have overcome them.)
>
> Third act. *New learning: Collective elicitation of new learning.* (After the teacher states that something new has been learned through activity, students are requested to state what this is, by comparing their performances and/or beliefs before the problem-solving session and after the discussion.)
>
> Fourth act. *Institutionalization: Statement of the official status of the new learning.* (If consensus has been reached on one or more solutions coherent with accepted mathematical behavior, they are institutionalized by the teacher and stored in the memory of the classroom, e.g., by means of some written statement in personal notebooks; if no consensus has been reached or if consensus seems to be converging on unacceptable solutions, the fact that it is not . yet possible to have a shared strategy and that the problem will have to be considered again after some time is institutionalized; institutionalization can also concern a definition of mathematical objects shared by the classroom or an attitude toward mathematics.)

The short-term effect of a balance discussion is the socialization of individual strategies up to the collective statement (if this is attained) of a method of solution to be applied to a whole class of problems and to be stored in the memory. However, long-term effects also include (1) the nonliteral interpretation of teacher's questions and (2) the internalization of a model of the metacognitive process to be activated in problem solving.

The following transcript (provided by Mara Boni) contains the first exchanges of a balance discussion carried out in a second-grade class on the real-life drawing of a cube with an edge in the foreground. Every student has his or her product on the desk. Before selecting prototypes to be discussed, the teacher invites the students to intervene freely:

1. Teacher: Have you succeeded in drawing the cube? Let's go.
2. Ingrid: I . . . the thing that I have . . . yes . . . I've asked myself . . . I've said: What don't I see? And then I've drawn what I saw and not what I didn't see, and it worked out.
3. Elisa: At the beginning it was so difficult. I did the same as Ingrid and I've succeeded. But very very slowly . . . it was so difficult.
4. Greta (*gesturing with the finger*): Er . . . I followed the angles so I was able . . . then . . . . at first I said, I asked myself: Do I also have to draw what I do not see or what I see? Then I read what was written and I realized that we had to draw only what we saw.
5. Marco: Me too, like Ingrid, I asked myself: What is it that I don't see? Then after I'd asked myself, I first drew the upper side and then I began the bottom side.

The first observation concerns the multivocality of individual utterances. Besides, it is evident that the students are not answering the teacher's explicit question, but the implicit questions in the second act of every balance discussion: How did you succeed in drawing? Which difficulties did you meet?

The following excerpt from a transcript of Franca Ferri is the introduction to the written comment of a fourth grader, concerning her real-life drawing of the classroom.

SCHEME THAT CAN HELP ME
1. Which problems did I have?
2. How have I solved them?
3. We learn by looking!
4. Which techniques have to be used besides looking?
5. Is this drawing equal to others already done?
6. Did I like it? Why?
7. Is it similar to reality?
8. How do our eyes see?
9. What is similar and what is not similar [to reality] in my drawing?

The quantity of self-posed questions to direct the comment is evidence of the internalization of the plot of balance discussion.

The preceding standard plot is now used systematically by the teachers in our group to design balance discussions after the phase of individual problem solving. Knowing the plot in advance is a great help, but the teacher has to make a lot of "on-the-spot" decisions in the flow of debate. The teacher's position is similar to that of an actor in the ancient commedia dell'arte, in which improvisation on the plot played a major role: Yet improvisation was also not governed by chance but by very refined actor education; it was rather a science, based on a personal repertory of variants (which kind of jokes, which kind of provocation for the audience, which kind of reactions to some words from the audience, etc.). A fine-grain analysis of discussions is needed to offer teachers a repertory of communicative strategies to be used.

Examples of communicative strategies

Each communicative strategy has a short-term impact on the classroom, but the system of communicative strategies determines the discussion itself and, through it, influences the overall activity.

An example of communicative strategy that is used especially in the third act of the balance discussion is paraphrasing on a general level (Bartolini Bussi and Boni 1995; Bartolini Bussi in press). During the course of discussion, students very often offer contributions concerning individual instances of objects or situations or problem solutions. "For instance" is a typical opening of students' speech. When, in the specific situation, the student grasps the relationships that make the specific an instance of a general procedure or of a theoretical concept of mathematics, the teacher repeats the student's utterance, imitating the syntactic structure, but replacing individual with general (e.g., substituting the word "object" or "thing" for the individual object or introducing a universal quantifier such as "every" or "whenever," or changing the numerical data of an arithmetic problem, etc.).

The short-term effects of this strategy is the splitting of the classroom into two subgroups: students who follow the teacher and produce general statements; and students who still need the reference to concrete and specific situations; however, the communication between them is maintained because of the existence of the concrete case in the shared experience.

Long-term effects of the systematic paraphrases on a theoretical level are observed in the classroom tendency (among low attainers as well) to favor general and rational methods to the detriment of individual and empirical ways to solve problems. The following excerpt from a transcript of Franca Ferri, while starting from a wrong statement, provides evidence of this attitude:

What is an angle? Try and explain the concept of angle; give some example and help yourself by means of drawing too. (Some polygons are traced in black ink with red points in the vertices.)
The red ones are angles.
An angle is that kind of tip that does not contain space, always has the same size, is in squares, rectangles, triangles. . . . By means of my reasoning, I explain that all the angles are equal all over the world.

This shift from individual to general is only a part of a double movement to be attained in the process of theoretical thinking in mathematics: The other, from general to individual (see the ascent from the abstract to the concrete, Davydov 1979), defines another case of communicative strategy (a meaningful example is in Mariotti 1994).

> *Changing the culture of the mathematics*
> *classroom: How teachers learn to orchestrate a*
> *discussion*

When the research project started in 1987, we tried to collect some examples of discussions in the classrooms. They had not been designed very carefully, but rather managed by teachers on the basis of their previous experience and on some deep beliefs about schooling. The project was carried out with a systematic alternation of a phenomenological analysis of data from classrooms and of the study of the literature on interaction in mathematics classrooms.

At the very beginning, teachers found it very difficult to free themselves from the desire to talk nonstop and to teach something at all costs. The following excerpt from a transcript by Franca Ferri is taken from a balance discussion on infinity (1988: fourth grade) and concerns the "number" of points in a line segment; the reference is to the classical problem of comparing the sets of points of two line segments of different length that had been posed individually some days before.

1. Teacher (*drawing on the blackboard*): I have a line and two points on the line (A and B). You have told me that there are infinitely many points between A and B. Matteo has another line and two more points on the line (A' and B'). Their distance is greater than the one between A and B. How many points are there between A' and B'?
2. Many students (*buzz*): Infinitely many.

3. Some students: It all depends . . . we'll see.
4. Beppe: Many points.
5. Riccardo: It depends on the distance.
6. Matteo: No, if there are infinitely many points between two points on a line, there are infinitely many points between two points on another line too, and so on . . . they are infinitely many. . . .
7. Christian (*angry*): You haven't asked to speak. I'd like to say it myself!
8. Riccardo (*warding off*): No . . . no, or, rather, I'm telling it. In my opinion, it depends. If you had specified and given the measure . . . it depends on the millimeters. For instance, if a line is 5 cm long and the other is 10 cm long, the points on the former are half the points on the latter.
9. Teacher: Richi, pay attention, we are talking about points and not about measure.

When the transcript of this discussion was analyzed collectively by the research group (including the teacher), everybody agreed that the last intervention was inconvenient, because it could have resulted in cutting off Riccardo's explanation, which was representative of a voice, that is, a true epistemological obstacle in understanding the concept of mathematical infinity. Luckily, in that specific case, the classroom atmosphere was very open and nonauthoritarian, to the extent that some minutes later he once more defended his theory.

In order to break the stereotyped routines of classroom discourse, we considered very seriously the literature on mathematical discussion shaped by the (more or less radical) constructivist approach. In the following experiments, the teachers tried to avoid direct teaching of strategies or evaluation of students' utterances such as the one quoted that could have inhibited students' involvement in classroom interaction. An example from a transcript by Franca Ferri follows.

Fifth graders are discussing integer numbers. As usual, they are sitting in a circle. The atmosphere is relaxed. As an exercise in addition with integers, a student has proposed $4 + (-8)$, and they have agreed that it gives -4.

1. Fabio: When we were young, we didn't know negative numbers; zero was supposed to be the smallest, yet it's true only for natural numbers. Now we know decimals, negatives, and we can do more operations than just addition.
2. Teacher: Well, Christian (*he asks to speak*), can we do any operation?
3. Christian: Minus 100 times 3 is plus 300 or minus 300?
4. Teacher: Minus 100 times 3.
5. Christian: In my mind, it is plus 200, 'cause, minus times 300, that you take away from 100; but if you make (*plus*) times 300, plus 300 is more. . . .

6. Fabio: No, there is an easier way, you make 100 times 3, that is 300, and you see that you have exceeded; it's 300; you had minus 100, and with the hundred you have a piece, you arrive at zero; the other 200 are in the other piece *(he's gesturing and moving his finger on an imaginary line, i.e., the number line)*.
7. Teacher: Plus 200.
8. Simone: Minus 100 times 3 *(with astonishment)*?
9. Teacher: Yes.
10. Dora: Bah! *(with perplexity)*.
11. Dora: I agree, 'cause as he said before . . .
12. Matteo: No, we need division *(much laughter because Matteo always has very original ideas)*.
13. Teacher: Let's listen to Matteo: why do we need division?
14. Matteo: If you make 100 *(he stops)*.
15. Teacher: But the number is minus 100.
16. Matteo: . . . divided by 100 is minus 1. That is I go toward *(he stops)*.
17. Teacher: OK, but this is another example. Christian's problem was minus 100 times 3, what is it?
18. Simona: I agree, to my mind it's plus 200, as minus 10 plus 3 is minus 7.
19. Teacher: OK, but this is a multiplication isn't it?
20. Simone: To my mind, it's not plus 200, 'cause when we prove with division, 200 divided by 3 is not minus 100. You have a remainder.
21. Matteo: Exactly what I said. It's minus 300.
22. Teacher: OK, Matteo, did you mean that?
23. Matteo: I did.
24. Teacher *(smiling)*: You have to learn to express yourself better. OK, I see that there is somebody with a pocket calculator. Have you done it? Yes? OK. We can do it, but you have to reason before. Matteo says that it's minus 300 . . . and one group says that it is plus 200.
25. Stefano: It's minus 300, 'cause if you make 100 times 3, you have minus times 3 *(There are some debates among different students who defend different positions)*.
26. Serena: I agree with Stefano. Why? 'cause, in the first grade, in the second grade, how did we discover multiplication? We made, 100 times 3: 100 plus 100 plus 100. It's the same with negative numbers.

Even a sketchy analysis of this excerpt shows a lot of things:

1. Even if the teacher intervenes several times, most of her utterances do not alter the spontaneity of students' interaction. For instance, the second part of n. 2 is supposed to be not relevant to Christian because he probably has already decided to pose this specific problem before asking to speak, and he interprets the teacher's utterance as permission to speak only; n. 19 also does not break the flow of the students' agreement or disagreement.

2. The teacher is patient: She gives the students time to develop their own arguments without evaluating their proposals too quickly; so she accepts (n. 7 and n. 9) a "wrong" solution, even if it is disputed by a fellow student.

3. Different communicative strategies are used to regulate the flow of debate: She grants the right to speak (n. 2); she invites students to listen (n. 13); she sums up the status of the debate without taking a position (n. 24); she refocuses on the problem (e.g., n. 15, n. 17, n. 19); she mirrors students' utterances to reassign them with the responsibility of going on (e.g., n. 4, n. 7). The preceding communicative strategies are part of the repertory collected by the research group.

Surely the teacher is neither teaching any strategy directly nor evaluating directly students' utterances as right or wrong. Rather, she is supposed to accept either right or wrong strategies in the same way. Only once does she forbid the use of a strategy (the recourse to the pocket calculator, n. 24), defining the specific contract of the task (to reason beforehand). However, she is teaching something concerning classroom culture (or, better, changing traditional classroom culture). We have much evidence that something concerning classroom culture is emerging:

1. The students are learning that they are allowed/encouraged to pose their own problems (n. 3);

2. The students are learning that they can manage agreement/disagreement without being explicitly requested by the teacher (e.g., n. 5/6; n. 18/20);

3. The students are learning that a teacher's game is hiding her knowledge, to go on (e.g., n. 24);

4. The students are appropriating a metacognitive strategy—remembering together (n. 1 and n. 26), to locate them in the history of the classroom;

5. Some students are interiorizing dialogue with the teacher (e.g., n. 26).

Up to now, we have considered the shift from a more "traditional" to a "constructivist" management of discussion. This management proved to be very effective in establishing a fruitful classroom climate. But we soon met the limits of this "constructivist" phase expressed by a contradiction with teacher expectations. The consideration of two extreme cases, which are judged to be very important in Italy (i.e., remedial education of low attainers and introduction of advanced mathematical instruments, see Bartolini Bussi 1994a), had decisive effects on further work.

Through deep and inflamed discussions, the members of the research group became aware of three very important facts: (1) the "constructivist" phase has to be meant as a physiological answer to the need to contrast some traditional routines of classroom interaction; (2) the emphasis on the responsibility of the learner in his or her own learning, typical of the constructivist perspective, minimizes the teacher's teaching responsibility to act as a more expert guide in the introduction into a complex society; (3) the contradiction between learning and teaching responsibilities cannot be solved within a constructivist framework. The "Vygotskian" phase of research was starting. Actually, the preceding crisis depended on the personalities of the teachers (who shared a problematic rather than an ideological attitude toward the process of teaching–learning in school); however, the process was facilitated and accelerated by the introduction into the research team of contrasting voices from outside: At that moment, theoretical constructs defined by Vygotsky and the scholars of the cultural–historical school entered as constitutive elements of the theoretical framework. This choice did not mean the rejection of the previous phase: a "constructivist" phase seems to be physiological for both novice teachers (with experience of traditional management of classroom discourse) and novice students (with no experience of direct and genuine engagement in classroom discourse). However, this choice allowed us to frame the teaching function of the teacher in a general theory of learning (or, better, of teaching–learning in the school setting) from which we could borrow instruments for design and analysis.

The choice of the cultural–historical perspective deeply affected the following work. The theoretical constructs of internalization and of semiotic mediation entered as tools in designing, implementing, and analyzing teaching experiments in the classroom: The most problematic issue concerned the analysis of the teacher's role as a guide in the learning process. This analysis cannot be limited to a single lesson, as the guide's role concerns a long-term process. The meaning of some teachers' utterances in the dialogue can be reconstructed only by recalling the whole process.

An example

To give an example, we can analyze the following excerpt from a transcript by Franca Ferri. The excerpt is taken from the balance discussion of the following problem: Draw the tablecloth on the table; explain your reasoning (fourth grade, see Figure 1.1).

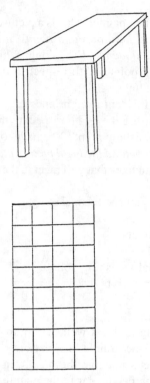

Draw the tablecloth on the table.
Explain your reasoning.

Figure 1.1. Perspective drawing (from Bartolini Bussi, 1996).

The students are trying to state how to draw the "vertical" lines of the table-cloth ("vertical" means parallel to the right and left borders of the sheet, according to an explicit convention of the classroom). All the students are sitting in a circle, as is usual in the discussions.

276. Teacher: . . . How can we state the rule well for the vertical lines?
277. Marco: First it's necessary to extend the vertical sides of the table.
278. Teacher: Darling, of the table?
279. Marco: Of anything. You extend them and when they meet, you fix a point where all the vertical lines of the drawing go.
280. Lisa: We extend the sides, "vertical" for us, of whichever thing until they meet in the vanishing point.
281. Teacher: Now, what's the vanishing point?
282. Lisa: It's the meeting point of the extension of the "vertical" sides of whichever figure.

283. Teacher: How are these lines?
284. Marco: They are incident.
285. Lisa (*ironically*): It doesn't take much effort: They meet.
286. Teacher: How are they in reality?
287. Voices: Parallel.
288. Teacher: Let's take an example. I'm a painter. This is a picture I wish to paint (*she goes to the middle of the group and points at a vertical sheet in front of both her and the students*). I'm asking: Which are the lines that meet in the vanishing point? With respect to the picture.
289. Francesco: First, they are straight lines.
290. Lisa: That will become incident, incident straight lines.
291. Teacher: Which are the lines; how are they with respect to the picture, these lines that will go to the vanishing point? You have to draw this table (*she points at a table that is behind the ideal picture plane*): With respect to the picture, how are the lines that will meet in the vanishing point?
292. Chiara: They change measure as you see them sloping.
293. Marco: The horizontal lines are parallel.
294. Federico: The others are perpendicular.
295. Luna: It's true: I lacked the word.
296. Teacher: Now, what's the principal vanishing point?
297. Luna: Can you sum it up well for everybody?
298. Teacher: Yes, that's the point . . .
299. Giulia: Luna, we already know it.
300. Teacher: I hope so. Yet I'm trying to tell it well, with a more precise language. It's the point where all the straight lines that are perpendicular to the picture will meet. Let's imagine the picture in a vertical position. All the straight lines perpendicular to the picture will go to the principal vanishing point (*she gestures in the meantime*).
301. Marco: Oh, we said "vertical" but we knew that it was a conventional expression.
302. Chiara: We have made the sheet with all the conventions (*she refers to a previous activity*). To say "up" in the sheet is a convention too.
303. Teacher: I know that you sometimes say it in other ways, but I've summed up for you, as Luna says, in a more mathematical language. You say "the vertical lines"; we, oh not me as Franca and you as Class B, but mathematicians, say in a more correct way "all the lines perpendicular to the picture."
304. Costanza: So there is no longer any problem of understanding: "vertical" is ambiguous, it's a convention, but perpendicular is clear. Think of the drawings by Dürer.
305. Marco: Yes, there is a vertical picture in front of Dürer.
306. Giulia: Can we repeat it all together? The vanishing point is obtained by . . .

307. Voices: . . . representing the lines perpendicular to the picture.
308. Costanza: And everybody knows that perpendicular means forming a right angle.
309. Giulia: We look at perpendicular lines.
310. Teacher: Is it clear for everybody? We have finished with lines perpendicular to the picture. And now, let's go to the famous horizontal lines, the ones parallel to the picture (*she alludes to the original problem*).

It is impossible to understand the style of the dialogue without knowing what happened before. We are looking at a classroom in the middle of a long experiment (3 years) on the representation of the visible world by means of perspective drawing (Bartolini Bussi 1996). The students have learned to draw objects of the microspace (Berthelot and Salin 1992); they have stated some rules for drawing (such as straight lines in reality remain straight lines in the drawing); they have clashed with the difficult problem of representing the mesospace (Berthelot and Salin 1992) of the whole classroom; they have read collectively an excerpt from Piero della Francesca (see Figure 1.3 from chapter XIV in Piero della Francesca 1460) in which the principal vanishing point (or, as Piero says, the eye) is used to solve the problem of drawing a square grid on a "degraded" square (i.e., the perspective image of a square) in the picture plane; they have consciously referred Piero's rule to their efforts in drawing the floor and the walls of the classroom; they have reconstructed the three-dimensional meaning of the two-dimensional rule stated by Piero by observing and coloring the copies of some famous engravings by Dürer that show the perspectographs; then they have been given the problem of the tablecloth: The first part of the problem (i.e., to draw the "vertical" lines) is an application of Piero's rule: All but three students have succeeded in solving the problem individually. In the first part of the discussion (1–275), the other three students have been called to the overhead projector and helped by the teacher and by the classroom to contrast and to modify their (wrong) solution. The excerpt under scrutiny (276–310) concerns the institutionalization of the solution.

The teacher aims to state the general rule for the image of the lines that are perpendicular to the picture plane. This is not a repetition of Piero's discourse, but rather an interpretation: to state explicitly the relationships between an object and its image.

The first part of the excerpt (276–295) concerns the collective statement of the general rule on the basis of the particular solution that has been agreed. In this case, the teacher uses the standard pattern of repeated questions to collect all the elements she needs from the previous part of the discussion. At the beginning, she takes attention away from the table (278), but, later, she comes back to a real table (291) and repeats gestures that embody the geometrical relationships between the objects (the hands skim through the ideal vertical plane of the picture; the forefinger points at the contour of a real table). This continuous movement from the real (or ideal) object to anything (and vice versa) is a fundamental part of the construction of the general rule that has to refer to any kind of situation but has to be verified in concrete situations. Federico's contribution adds what was lacking (294). All the elements are present now.

The second part of the excerpt (296–303) concerns the teacher's role as a representative of the culture. She asks a direct question (296), and Luna (297) asks the teacher to play her role. The teacher gives a general statement and relates it by pointing to the concrete situation. The two different tools of semiotic mediation (words and gestures) have different, yet strictly related, functions: Whereas words are referring to the general, gestures are referring to the particular. So the meaning of the general statement is built by referring to the particular situation. The teacher asks the classroom to identify her role when she speaks on behalf of the community of mathematicians (303). Usually, she can also accept classroom conventions, but when she is playing the role of the mathematician, the tendency toward a "correct" mathematical language is explicit.

The last part of the excerpt concerns the students' reactions: Costanza recalls the work on Dürer's engravings (another example of semiotic mediation) and adds meaning to that work; Giulia (who seemed uninterested before – 299) wishes all to repeat the statement together and emphasizes the value of imitation. The teacher (310) closes this part of the discussion by returning to the initial problem in a new form, in which "perpendicular to the picture plane" substitutes for "vertical." The new problem is introduced by referring to classroom conventions and to mathematical conventions as well, by recalling Marco's proposal (293).

In this excerpt, the teacher plays the guide's role in a very evident way: This role is acknowledged by the students (see 297) but does not inhibit their dialogue. Further examples can be found in Bartolini Bussi and Boni (1995).

The quality of educational contexts (or fields of experience)

Taking the distance from school tradition

The second part of this chapter is devoted to the quality of educational contexts.

The contexts in which we make the teaching experiments are manifold: perspective drawing, number fields, shadows of the sun, and so on. However, one feature has to be emphasized and contrasted with traditional teaching.

In traditional classrooms, following textbooks, fragmented and isolated problems are provided. The most common clustering of problems is determined by a "higher knowledge" (e.g., problems for addition or problems on divisors; or problems to be solved by means of the quadratic equation) that refers to the internal organization of mathematics and is neither known by nor available to students. This feature also seems to affect a lot of research that, in order to keep the length of the study reasonable, focuses on decontextualized individual problems, whose function in the overall activity is not clear at all. We could agree that a context exists in the form of this higher knowledge but is not acknowledged as such by the classroom actors. In other words, the higher knowledge context is not a field of experience for the classroom. We use the term "field of experience" after Boero (1992, Boero et al. 1995) to mean a sector of human culture that the teacher and students can recognize and consider as unitary and homogeneous.

The attitude of focusing on fragmented problems contrasts with a part of the constructivist and with the Vygotskian approach to mathematical discussion. On the one side, Richards (1991: 40) claims that inquiry math is distinguished by settings (provided by the teacher) that persist across several class sessions, with an entire class session often devoted to resolving a single question; a setting that is explored over time gives a context for discussion in which the consensual domain can be established. On the other side, Leont'ev (1976, 1977) considers it to be senseless and unjustified to focus on single actions, defined by specific goals but not included in an activity that develops over time to the extent that not even the unit of analysis can be reduced to a single task.

Long-term exploration of a particular context is consistent with the activity of working mathematicians. Usually, mathematicians work in the same field for several years: Problems are not isolated and fragmented, but are contextualized and generated by the activity of exploring the field. Sometimes, mathematicians solve isolated problems, but the core of their professional work is contextualized activity within a research field. Their own research field will be described vividly by mathematicians as a field of experience with a lot of internal connections. Academic mathematics is determined by the activity of mathematicians, and probably some features of this activity should be maintained in school mathematics. We suggest that one feature to be maintained is the long-term exploration of a field of experience; isolated problems can have their space (with a local change of the didactical contract) as problem exercises in order to become more expert in managing some tools, as games, as application of mathematics to meaningful problems from other activity fields, and so on, but they do not constitute the core of the curriculum.

The contexts of working mathematicians are usually generated from within mathematics because of internal problems: Thus, a mathematician can work for years in the context of "boundary value problems for an ordinary differential equation" or "topological structure of differentiable 4-manifolds" and become an acknowledged expert in the field. The aim of schooling is not preparing expert mathematicians: Hence, the analogy is not perfect. On the one side, the choice of suitable fields of experience is very delicate, as the long exploration of them deeply shapes the image of school mathematics. On the other side, it cannot be done only on the basis of the experience of mathematicians for the reasons mentioned. It is necessary to find fields of experience in which the solution of problem situations requires a "nonartificial" recourse to the mathematical tools (either concepts or procedures) that teachers are asked to teach on the basis of school programs.

Fields of experience can be taken from either inside mathematics or outside mathematics. For instance, the arithmetic of natural numbers (addition, multiplication, multiples, divisors, etc.) can be assumed. This sounds familiar, but we have other examples. Fields of experience from outside mathematics address the epistemological complexity (Hanna and Jahnke 1993) of relationships between mathematics and reality. We are speaking of fields of experience rich in true modeling problems and not about those collections of curious problems in which mathematical tasks are disguised as applied mathematics by introducing everyday words (see Pollack 1969 for a discussion). We could even wonder whether the complication that exists within such fields is acceptable for school mathematics. A special class of them, suitable for school, is suggested by careful investigations in the history of mathematics (Boero 1989).

In the following sections, three different fields of experience, suitable for the teaching—learning of geometry in Grades 1–8, will be sketched and related to each other: Activity in them has been designed, implemented, and analyzed independently over the last 10 years by three different Italian research groups. An experiment on coordinating them all is now in progress (Garuti personal communication).

We have chosen (randomly) an order of presentation in this chapter that is not representative of a fixed order of presentation in the classroom: Rather, the experiences in all three fields are interlaced with each other, and only specific invariants of each field are focused and institutionalized from time to time by the teacher.

All the fields deal with teaching—learning geometry by means of modeling either natural or cultural phenomena. The size of the space taken into account is wide-ranging: from the individual body scheme (Lurçat 1980) to external spaces; microspace (close to subject, to be explored by handling and seeing); mesospace (accessible through movement and global vision); macrospace (not accessible through global vision) (Berthelot and Salin 1992); and a special case of macrospace not even accessible through movement that we can call the megaspace of cosmos.

A first example: Representation of the visible world by means of perspective drawing

The first example concerns the field of experience of the representation of the visible world by means of perspective drawing. A teaching experiment focusing on the exploration of this field is in progress in our research group in a limited number of classrooms ranging from grade 3 to grade 8 (Bartolini Bussi 1996; Bartolini Bussi, Boni, and Ferri 1995; Costa, Ferri, and Garuti 1994; Ferri 1993).

The function of the theory of linear perspective in the development of modern mathematics is well known as the root of projective geometry and of the study of geometric transformations. The theory of linear perspective lies on the border between geometry and art, because its development is strictly intertwined with the development of a mathematical model of vision and with the theory of infinite and homogeneous space that is the basis for modern conceptions of space.

We operationalized the exploration of this field of experience by means of a sequence of problems in real-life drawing and the reconstruction or the appropriation of instruments for perspective drawing produced during human history in order to master perspective. An ex-

Draw a small ball in the centre of the table.
You can use instruments.
Explain your reasoning.

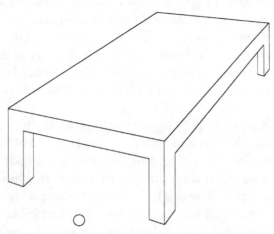

Figure 1.2. Perspective drawing (from Bartolini Bussi, 1996)

ample of the former is the problem in Figure 1.2; an example of the latter is the method given in Figure 1.3.

The drawing problems concern representation of either microspace (e.g., a set of partially interposed boxes) or mesospace (e.g., the classroom) on a microspace (the sheet of paper) (Berthelot and Salin 1992). Relationships between different frames of reference are involved (e.g., the frame of reference of mesospace and of the microspace given by the sheet). The basic invariant is alignment, according to the modeling of *prospectiva artificialis* (i.e., artificial or mathematical perspective), which is different from *prospectiva naturalis* or *communis* (i.e., natural or common perspective), because it is based on the projection on a plane and not on a spherical surface (Panofsky 1961).

A second example: Shadows

The second example concerns the field of experience of the shadows cast by the sun. Teaching experiments ranging from grade 3 to grade 8 are in progress in the research group at Genoa (Boero 1989, 1992; Boero and Ferrero 1994; Boero, Garuti, and Mariotti 1996; Boero et al. 1996).

La superficie quadrata deminuita
in più parti equali divisa,
quelle devisioni in quadrati producere
(Piero della Francesca, De Prospectiva Pingendi, fig. XVa)

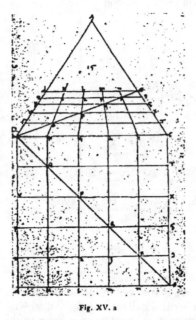

Fig. XV. a

Figure 1.3. Perspective drawing (from Piero della Francesca, *De Prospectiva Pingendi,* fig. XVa).

Observation and study of shadows are documented from ancient times, in relation to either time measurement (sundials) or the development of theoretical geometry (Serres 1993).

The exploration of this field is carried out with problems concerning the production of hypotheses (either predictive, justificative, interpretative, or planning and heuristic ones) or statements and proofs on the whole mathematized phenomenon of shadows. Examples of problems are given in Figures 1.4, 1.5, and 1.6.

The problems concern relationships between either a number set (measures) or a microspace (e.g., a nail on a tablet) or a mesospace (e.g., a courtyard) and the megaspace containing the sun. When representation is required, the microspace of the sheet of paper is also involved. The basic invariant is parallelism, according to the modeling that assumes sunrays as coming from an infinitely distant point and represented by parallel lines.

First problem:
Find the height of a tower,
knowing its shadow on the ground

Second problem:
Complete the hourly shadow fan (if possible)
traced by means of a nail on a tablet

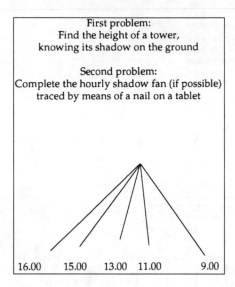

Figure 1.4. Sun shadows (from Boero, 1992).

The situation:
In the classroom experience fans of sun shadows
(produced by a vertical nail on a horizontal tablet)
are drawn at the same hour (maybe on different days)
in two different courtyards:
the one in front of the school and
the one behind the school (higher than the previous one).

The problem:
Are the two fans equal on the same day or not? Why?

Figure 1.5. Sun shadow problem (from Scali, 1994).

We have seen in the past years
that the shadows of two vertical sticks on the horizontal ground
are parallel.
What can you say about parallelism of two sticks,
if the first is vertical and the second is not vertical?
(Can the shadows be parallel? Sometimes? When? Always? Never?)
Produce your conjecture as a general statement.

Figure 1.6. Sun shadow problem (from Boero et al. 1996).

A third example: Anthropometry

The third example concerns the field of experience of anthropometry. A teaching experiment focusing on the exploration of this context was carried out in a limited number of classrooms ranging from grade 6 to grade 8 (Lanciano 1986).

Anthropometry is documented from ancient times in statements on the standard for the representation of the human body (Panofsky 1962).

Draw your own shape (first step)
To draw your own shape from actual measures
of some parts of your body,
make the following measures with the help
of your schoolfellows:

AB = total height	AB = ...cm
CD = arm span	CD = ...cm
EF = shoulder width	EF = ...cm
GH = pelvis width	GH = ...cm
IB = navel height above ground	IB = ...cm
AN = height when seated	AN = ...cm
LC = palm length	LC = ...cm
AO = head height	AO = ...cm

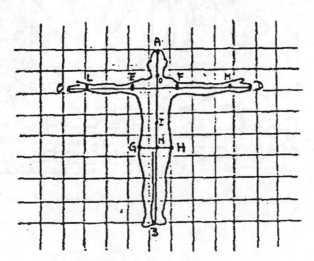

Figure 1.7. Anthropometry (from Lanciano, 1986).

*Se apri quanto le gambe che cali da capo 1/14 die tua altezza
e apri e alza tanto le braccia
che con le lunghe dita tu tocchi la linea della sommità del capo,
sappi che'l centro delle estremità delle aperte membra
sia il bellico e lo spazio che si trova infra le gambe
sia triangolo equilateralo.
Tanto apre l'omo ne le braccia quanto è la sua altezza.*
(Leonardo da Vinci)

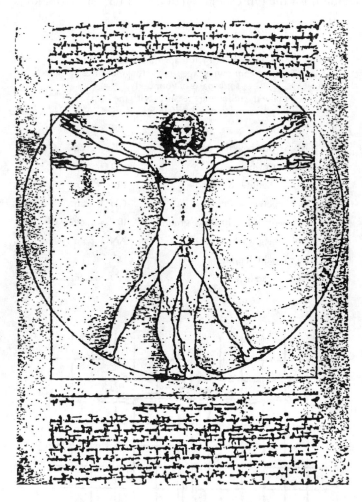

Figure 1.8. Anthropometry according to Leonardo da Vinci (from Lanciano, 1986).

This field is explored with problems concerning the statistical analysis of data from measures of student bodies, for example, or the appropriation of standard representations of the human body (e.g., Figures 1.7 and 1.8). They concern relationships between body schema and either sets of numbers or the microspace of the sheet when graphical representation is involved. The basic invariant is shape (involving angles and length ratios).

Coordination and conflicts

The preceding fields of experience are not the only ones suitable for this kind of work: Other examples related to geometry are offered by map making and cartography (Tizzani 1994), by astronomy (Lanciano 1994), by unfolding polyhedra (Mariotti 1994), and by machines and curve drawing devices (Bartolini Bussi 1993; Bartolini Bussi and Pergola 1996; Dennis 1995).

In the preceding fields, students cope with a whole network of concepts such as line segments; parallel, incident, and perpendicular lines; horizontal and vertical lines; planes; angles; length; and ratios. However, the path in the network is not determined by the structure of mathematics as a scientific discipline (as in the axiomatic approach), but by the need to solve more and more complex problems generated by the exploration of the field itself (see Boero et al. 1995 for a discussion).

In addition, each field offers a case of geometrical modeling, which makes explicit the gap between what is known about reality and what is assumed in the mathematical model: In the three cases under scrutiny, these are the difference between *prospectiva artificialis* and *naturalis;* the difference between infinitely and finitely distant points; and the difference between the measuring results in standard positions (in which the skeleton is identified with an articulated system with only right or straight angles) and in moving bodies.

The three fields are not isolated from each other. For instance, Table 1.1 shows conflicting problems generated by the interplay between the field of shadows, in which parallelism is invariant, and the field of perspective drawing, when particular standpoints are assumed. Figure 1.10 (either a drawing or a photo) shows the conflicting situation generated by the interplay among all the contexts when the shadow of a human body is observed from two different standpoints. Other examples suggested by Table 1.1 are left to the reader (see, e.g., Figure 1.9).

All the examples concern problems that are close to students' experience: The recourse to mathematical concepts as tools is "natural" in the

Table 1.1. *Three contexts for geometry in Grades 3--8*

CONTEXTS FOR GEOMETRY IN
COMPULSORY EDUCATION (GRADES 3--8)

CONTEXT	Representation of the visible world by means of perspective drawing	Sunshadows	Anthropometry
GEOMETRY OF	Vision and painting	Sunrays and sunshadows	Bodies and shapes
CONSTRAINTS OF MODELLING	Prospectiva artificialis	Sunrays as parallel lines	Standard body positions
CONCEPT INVARIANTS	Alignment	Parallelism	Geometric shapes
SIZE	From micro-space and mesospace to microspace (sheet)	From macro-space (mega-space) to micro-space (sheet)	From body schema to microspace (sheet)
INTERPLAY	Looking at (or drawing or photographing) sunshadows		
OF		Shadows of human bodies (on the wall)	
CONTEXTS	Looking at (or drawing or photographing)		human bodies
	Looking at (or drawing or photographing) shadows of human bodies		

Comment on the following pictures
(taken on Feb 14th in the schoolyard)
of the same three sticks and their shadows
(original pictures have been transformed into
ink drawings for the purpose of reproduction):

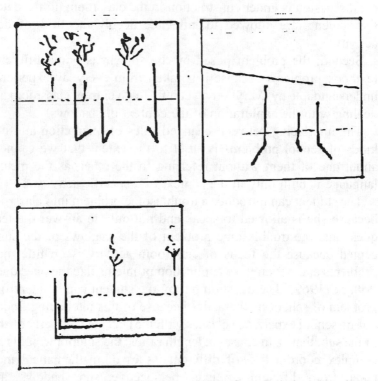

Figure 1.9. Interplay between contexts (from Garuti, personal communication).

sense that it repeats the process used in history when the borders between practice and theory, between technology and science, were crossed, with reciprocal influences on each other.

The initial dialogue: A second look

If we now return to the initial dialogue, we detect in it the signs of a changing culture of the mathematics classroom.

First, the dialogue clashes with the ground rules of communication in standard classrooms stated by Edwards and Mercer (1987: 45) as follows:

1. It is the teacher who asks the questions.
2. The teacher knows the answers.
3. Repeated questions imply wrong answers.

Actually, the problem of the dialogue is posed by a student and accepted by the teacher. It is not by chance that a tradition of mathematical discussion is under construction in the classroom, to the extent that the students are allowed to influence deeply the flow of designed activity.

Second, the problem posed by the student is quite different from textbook problems because it is taken from everyday experience and understandable by everybody; yet it fits with current classroom activity, dealing with the exploration of the context of shadows.

The interlacing between changed rules of interaction and changed kinds of school problems is strict to the extent that we cannot think about one of them without referring to the other, as the meaning of language is built only in the context of a specific activity.

The student can introduce a troublesome problem into classroom life because she is allowed to speak and not only to answer the teacher's questions; the troublesome problem of the shadows of the sun is accepted because the focus of classroom activity is on true modeling problems and not on false application problems like the ones quoted by Pollack (1969). The question put by the student can be identified as a problem of school mathematics because of this interlacing. Individual, out-of-school experience clashes with the accepted property of shadows of the sun, that is, to conserve length ratios; the solution would require a complex coordination of different issues of mathematical modeling taken from different contexts: perspective, sun shadows, and anthropometry. That coordination has not yet been worked out by the teacher, who actually profits from the problem to develop the interplay between contexts: The problem in Figure 1.10 is the institutionalized version of the verbal problem posed by the student.

The emergence and the evaluation of such a dialogue in a sixth-grade classroom are outcomes of the productive conjunction of new models of interaction as well as of educational contexts, and this makes it a living example of changing culture in the mathematics classroom.

Acknowledgments

The author has the financial support of MURST and CNR. The Mathematical Discussion project is being developed in cooperation with a group of teachers: Stefania Anderlini, Bianca Betti, Mara Boni, Claudia Costa, Maria Franca

On the basis of all your knowledge
comment upon the following picture
(the original picture has been transformed
into an ink drawing for the purpose
of reproduction)

Figure 1.10. Interplay between contexts (from Garuti, personal communication).

Ferraroni, Franca Ferri, Cinzia Fortini, Rossella Garuti, Marialaura Lapucci, Federica Monari, Anna Mucci, Silvia Salvini.

References

Artigue, M. (1992). Didactic engineering. In R. Douady and A. Mercier (Eds.), *Research in didactic of mathematics* (pp. 41–65). Grenoble: La Pensée Sauvage.

Bartolini Bussi, M. (1991). Social interaction and mathematical knowledge. In F. Furinghetti (Ed.), *Proceedings of the Fifteenth International Conference For the Psychology of Mathematics Education* (Vol. 1, pp. 1–16). Assisi (Italy): The Program Committee of the 15th PME Conference.

Bartolini Bussi, M. (1992). Mathematics knowledge as a collective enterprise. In F. Seeger and H. Steinbring (Eds.), *The dialogue between theory and practice in mathematics education: Overcoming the Broadcast Metaphor* (pp. 121–151). Materialien und Studien Band 38, Bielefeld: IDM University of Bielefeld.

Bartolini Bussi, M. (1993). Geometrical proofs and mathematical machines: An exploratory study. In I. Hirabayashi, N. Nohda, K. Shigematsu, and F.L. Lin (Eds.), *Proceedings of the Seventeenth International Conference for the Psychology of Mathematics Education* (Vol. II, pp. 97–104). Tsukuba, Japan: The Program Committee of the 17th PME Conference.

Bartolini Bussi, M. (1994a). Theoretical and empirical approaches to classroom interaction. In R. Biehler, R.W. Scholz, R. Strässer, and B. Winkelmann (Eds.), *Mathematics didactics as a scientific discipline* (pp. 121–132). Dordrecht: Kluwer Academic Publishers.

Bartolini Bussi, M. (1994 b). The mathematical discussion in primary school project: Analysis of long term processes. In L. Bazzini and H.G. Steiner (Eds.), *Proceedings of the Second Italian–German Bilateral Symposium on Didactics of Mathematics.* Materialien und Studien, Band 39 (pp. 3–16). Bielefeld: IDM, University of Bielefeld.

Bartolini Bussi, M. (1996). Mathematical discussion and perspective drawing in primary school. *Educational Studies in Mathematics, 31*, 11–41.

Bartolini Bussi, M. (in press). Verbal interaction in the mathematics classroom: A Vygotskian analysis. In M. Bartolini Bussi, A. Sierpinska and H. Steinbring (Eds.), *Language and communication in the mathematics classroom* (pp. 75–97). Washington, DC: National Council of Teachers of Mathematics.

Bartolini Bussi, M., and Boni, M. (1995). Analisi dell'interazione verbale nella discussione matematica: un approccio vygotskkiano, *L'Insegnamento della Matematica e delle Scienze Integrate, 18* (5), 221–256.

Bartolini Bussi, M., Boni, M., and Ferri, F. (1995). *Interazione sociale e conoscenza a scuola: la discussione matematica.* Modena: Centro Documentazione Educativa.

Bartolini Bussi, M., and Pergola, M. (1996). History in the mathematics class-room: Linkages and kinematic geometry. In H.N. Jahnke, N. Knoche, and M. Otte (Eds.), *Geschichte der Mathematik in der Lehre* (pp. 39–67). Göttingen: Vandenhoeck & Ruprecht.

Berthelot, R., and Salin, M.H. (1992). *L'enseignement de l'espace et de la géometrie dans la scolarité obligatoire.* Thèse, Université Bordeaux I.

Boero, P. (1989). Semantic fields suggested by history. *Zentralblatt für Didaktik der Mathematik,* 21 (4), 128–132.

Boero, P. (1992). The crucial role of semantic fields in the development of problem solving skills in the school environment. In J.P. Ponte, J.F. Matos, J.M. Matos and D. Fernandes (Eds.), *Mathematical problem solving and new information technologies* (pp. 77–91). Berlin: Springer-Verlag.

Boero, P., and Ferrero, E. (1994) Production and management of the hypotheses in elementary problem solving. In N.A. Malara and L. Rico (Eds.), *Proceedings of the First Italian–Spanish Research Symposium in Mathematics Education* (pp. 83–90). Modena: Consiglio Nazionale delle Ricerche.

Boero, P., Garuti, R., Lemut, E., and Mariotti, M.A. (1996). Challenging the traditional school approach to theorems: A hypothesis about the cognitive unity of theorems. In *Proceedings of the Twentieth International Conference for the Psychology of Mathematics Education* (Vol. 2, pp. 121–128). Valencia (Spain).

Boero, P., Garuti, R., and Mariotti, M.A. (1996). Some dynamic mental processes underlying producing and proving conjectures. In *Proceedings of the Twentieth International Conference for the Psychology of Mathematics Education* (Vol. 2, pp. 113–120). Valencia (Spain).

Boero, P., Dapueto, C., Ferrari, P., Ferrero, E., Garuti, R., Lemut, E., Parenti, L. and Scali, E. (1995). Aspects of the mathematics–culture relationship in mathematics teaching–learning in compulsory school. In L. Meira and D. Carraher (Eds.), *Proceedings of the Nineteenth International Conference for the Psychology of Mathematics Education* (Vol. 1, pp. 151–166). Recife (Brazil): The Program Committee of the 19th PME Conference.

Costa, C., Ferri, F., and Garuti, R. (1994). Perspective drawing as a semiotic tool towards the statements of geometry for young pupils (grades 3–6). In *Proceedings of the 46th CIAEM* (pp. 156–162; 213–217). Toulouse: CIAEM.

Davydov, V.V. (1979). *Gli aspetti della generalizzazione nell'insegnamento.* Firenze: Giunti Barbèra (original Russian edition 1972).

Edwards, D., and Mercer, N. (1987). *Common knowledge: The development of understanding in the classroom.* London: Routledge.

Engeström, Y. (1987). *Learning by expanding: An activity theoretical approach to developmental research.* Helsinki: Orienta-Konsultit Oy.

Ferri, F. (1992). What is the teacher's role in research on didactics of mathematics? Paper read at the Topic Group: Cooperation between Theory and Practice in Mathematics Education. *7th International Congress on Mathematics Education,* August 17–23.

Ferri, F. (1993) "Perspective is a solvable problem" or "To be born five hundred years before." In H. Milan and J. Novotna (Eds.), *Proceedings of the SEMT 93 – International Symposium on Elementary Math Teaching* (pp. 33–35). Prague: Charles University Publications.

Hanna, G., and Jahnke, H.N. (1993). Proof and application. *Educational Studies in Mathematics,* 24, 421–438.

Laborde, C. (1991). Deux usages complementaires de la dimension sociale dans les situations d'apprentissage en mathématiques. In C. Garnier, N. Bednarz and I. Ulanovskaya (Eds.), *Aprés Vygotsky et Piaget: Perspectives sociale et constructiviste. écoles russe et occidentale* (pp. 29–50). Bruxelles: De Boeck-Wesmael

Lanciano, N. (1986). Antropometria e arte. In L. Lazotti-Fontana (Ed.), *Arte e scienza: Riflessioni teoriche e prospettive didattiche* (pp. 221–238). Atti del Seminario CEDE (5–7 dicembre 1984). Frascati: Quaderni di Villa Falconieri n. 8.

Lanciano, N. (1994). Aspects of teaching learning geometry by means of astronomy. In N.A. Malara and L. Rico (Eds.), *Proceedings of the First Italian–Spanish Research Symposium in Mathematics Education* (pp. 43–49). Modena: Consiglio Nazionale delle Ricerche.

Leont'ev, A.N. (1976). *Problemi dello sviluppo psichico.* Roma: Editori Riuniti (original Russian editions 1959–1964).

Leont'ev, A.N. (1977). *Attività, coscienza, personalità.* Firenze: Giunti Barbèra (original Russian edition 1975).

Lurçat, L. (1980). *Il bambino e lo spazio: Il ruolo del corpo.* Firenze: La Nuova Italia Editrice (orig. ed. 1976).

Maier, H., and Voigt, J. (1992). Teaching styles in mathematics education. In H. Schupp et al. (Eds.) Mathematics Education in the Federal Republic of Germany. *Zentralblatt fuer Didaktik der Mathematik,* 24 (7), 248–252.

Mariotti, M.A. (1994). Figural and conceptual aspects in a defining process. In J.P. da Ponte and J.F. Matos J.F. (Eds.), *Proceedings of the Eighteenth International Conference for the Psychology of Mathematics Education* (vol. III, pp. 232–238). Lisbon, Portugal: The Program Committee of the 18th PME Conference.

Panofsky, E. (1961). *La prospettiva come forma simbolica.* Milano: Feltrinelli (original German edition 1924).

Panofsky, E. (1962). Storia della teoria delle proporzioni. In *Il significato delle arti visive.* Torino: Einaudi (original German edition 1921).

Piero della Francesca (1460). *De Prospectiva Pingendi* (critical edition by G. Nicco Fasola, 1942). Firenze: Sansoni.

Pirie, S.E.B., and Schwarzenberger, L.E. (1988). Mathematical discussion and mathematical understanding, *Educational Studies in Mathematics,* 19, 459–470.

Pollack, H.O. (1969). How can we teach applications of mathematics? *Educational Studies in Mathematics,* 2 (2/3), 393–404.

Richards, J. (1991). Mathematical discussions. In E. von Glasersfeld (Ed.), *Radical constructivism in mathematics education* (pp. 13–51). Dordrecht: Kluwer Academic Publishers.

Scali, E. (1994). Le role du dessin dans la modélisation géométrique elémentaire des phénomènes astronomiques. In *Proceedings of the 46th CIAEM* (pp. 168–176; 240–243). Toulouse: CIAEM.

Schoenfeld, A.H., Gamoran, M., Kessel C. and Leonard M. (1992). Toward a comprehensive model of human tutoring in complex subject matter domains. *The Journal of Mathematical Behaviour,* 11, 293–320.

Schönberg, A. (1963). *Preliminary exercises in counterpoint.* London: Faber & Faber.

Serres, M. (1993). *Les origines de la Géométrie.* Paris: Flammarion.

Tizzani, P. (1994). Quelques hypothéses pour expliquer les difficultés surgissant dans la représentation plane de situations dans l'espace et leurs implications pour ce qui est de l'enseignement. In *Proceedings of the 46th CIAEM* (pp. 185–192; 244–247).

Vygotsky, L.S. (1978). *Mind in society: The development of higher psychological processes.* Cambridge, MA: Harvard University Press.

Vygotsky, L.S. (1981). The genesis of higher mental functions. In J.V. Wertsch (Ed.), *The concept of activity in Soviet psychology* (pp. 144–188). Armonk, NY: M.E. Sharpe.

Wertsch, J.V. (1991). *Voices of the mind: A sociocultural approach to mediated action.* London: Harvester Wheatsheaf.

2   Evolution of classroom culture in mathematics, teacher education, and reflection on action

NADINE BEDNARZ

The projects implemented over the last 4 years within the framework provided by the CIRADE-associated research schools (CIRADE is the Centre Interdisciplinaire de recherche sur l'Apprentissage et le Développement en éducation at the Université du Québec à Montréal) are in keeping with a certain and, in my opinion necessary, relationship between research and teachers' professional development.

I shall concentrate on one collaborative research project that is currently being conducted in one of the schools mentioned. It involves elementary school teachers and a team of researchers at CIRADE in a process of reflection concerning their teaching practices. This project centers on the development of mathematical reasoning in young children and is based on a socioconstructivist approach to teaching. A set of actions was developed through which the participating teachers would be able to appropriate this approach.

This set of actions is patterned on the dynamic of reflective analysis described by Schön (1983, 1987) in *The reflective practitioner* that has provided abundant matter for reflection and comment within the sphere of teacher training in recent years. In essence, what is being proposed here is regular, planned alternation of in-class experimentation with the approach and group or subgroup reflection on this process. Emphasis is placed on establishing a shared process of reflection on action and interaction in mathematics classes.

Part of the project is currently investigating the process of appropriation through which teachers proceed. It is possible to retrace this through the following steps.

1. *The teaching practices occurring in class:* A series of investigations are focusing their analyses on the interactions between teacher and children in relation to the various activities developed and discussed

by the team (Bauersfeld 1980; Voigt 1985). Such analyses make it possible to trace eventual changes that emerge in classroom mathematical culture.

2. *Interactions between university researchers and the practitioners themselves during reflective appraisal of the approach in a classroom setting.*

But before actually discussing this appropriation of the approach by teachers and the evolution of classroom culture, I shall discuss our choice of collaborative research as an avenue of research on continuing education for teachers. In that connection, it is our intention to show the consistency of such a choice from an epistemological point of view, in light of the socioconstructivist perspective that has informed our research for quite some time now (Bednarz et al. 1993; Bednarz and Garnier 1989; Garnier, Bednarz and Ulanovskaya 1991). Before proceeding to explicate this choice, however, we wish to outline the framework used to deal with questions relating to mathematics teaching.

A socioconstructivist approach to teaching

In recent years, the alternatives underlying our research have centered on a conception of mathematics education that is, in large part, informed by the perspectives of the socioconstructivist school (on this subject, see Bednarz and Garnier 1989; Garnier, Bednarz and Ulanovskaya 1991) and more especially by Brousseau's theory of didactic situations (1986).

The foundational core, as it were, is to be located in an epistemological analysis of the development of mathematical knowledge (Bachelard 1983) and the works of Piaget. Against such a theoretical backdrop, the process by which mathematical knowledge is constructed is in no way viewed as linear but as a discontinuous process that often requires the children to engage in major conceptual restructuring. Such is the case, for example, in the transition from problem solving in arithmetic to problem solving in algebra (Bednarz and Janvier 1996) or in the conceptual shifts required of children in order to solve certain classes of problems in arithmetic (Poirier 1991) or even in algebra (Bednarz, Radford, and Janvier 1995). The processes involved in the construction of mathematical knowledge have provided an analytical focus on which to elaborate teaching strategies emphasizing the knowledge that children develop in context while also aiming toward increasing the conceptual complexity of this knowledge (Bednarz and Dufour-Janvier 1988; Bednarz et al. 1993). It is within this problematic of conceptual shifts that classroom action is conducted along two lines, involving, on the one hand, a certain social structuring of the class, and, on the other, the introduction of constraints at the very center of the situations being proposed, with the objective of encouraging the child to develop new problem-solving procedures and new organizational or notational modes. Thus, to take the example of number work, to which the remainder of this chapter is devoted, the introduction of constraints governing the size of collections or the way in which these are presented (real vs. drawn collections) into the tasks of counting, comparing, and transforming collections requires children to work out new problem-solving procedures and new modes of organization. Classroom action is thus organized around a whole series of situational problems in which the child is called to develop a set of problem-solving procedures and to implement a number of implicit models or conceptions that direct his or her action (i.e., a kind of personal framework allowing the child to interpret data, to act, etc.). Now, these implicit models, which refer to a certain mental representation by the child of data and the locally functioning relationships among the data, take on their full meaning in connection with the situations at hand and the actions of subjects in these situations. The learning process will, at this point, draw on these modes of problem solving as well as on subjects' locally constructed knowledge; to that end, a social structuring of the class will take shape in such a way as to leave room for various types of confrontation and hence bring out the various positions of the children.

In situations of communication, involving a dialectic of formulation between two subjects or two groups of subjects, children can be induced to construct a description, an explicit model of the situations, and the actions to be applied to these situations. Here, the social situation of communication is used to trigger a certain process of external representation that, from the perspective we have adopted, is designed to play a fundamental role in the construction of knowledge. It is often the case that knowledge shows up in connection with implicit models; that is why communication situations are used to make the different models used by children explicit. Thus, the representations developed have a dual function: communication with others, on the one hand, and conceptualization, on the other. Together, they provide evidence of the ongoing development of mathematical activity. It is this dialectic of formulation, taken with that of validation (with which it is intimately connected), that brings out what is crucial in interaction among children and, again, between children and their teacher. Thus, throughout this dialectic, and continually in relationship to others, the child must explicate the rules, symbols, and conventions that he or she uses and justify his or her choices. In this way, the student will develop a certain reflexive competency vis-à-vis not only the content but also the symbolism applied in the classroom. The various interactions will bring out a number of explicit models, thus giving rise to confrontations between student points of view over the course of a debate.

Doing mathematics does not consist simply in transmitting and receiving information in mathematical language, even when this information has been understood. The child-cum-mathematician must adopt a critical attitude toward the models which he or she has constructed. (Brousseau 1972: 64)

A perspective such as this one requires teachers to perceive the scope and limits of the situations they propose, and to remain attuned to the strategies, conceptions, and notations developed by children in order to work out their own action accordingly. The issue is no longer one of teaching mathematics as though it formed a field of knowledge that has been delimited a priori, having a form and organization that have been defined once and for all, with the attendant risk that its use will remain a strictly formal affair, devoid of meaning for children. Instead, the issue is one of implementing a new didactic contract within the classroom situation that fosters a better-informed, reflective relationship to mathematical knowledge.

Teaching describes the attempt to organize an interactive and reflective process, with the teacher engaging in a constantly continuing and mutually differentiating and actualizing of activities with the students, and the establishing and maintaining of a classroom culture, rather than the transmission, introduction, or even rediscovery of pre-given and objectively codified knowledge. (Bauersfeld 1994: 195)

It is from this perspective, then, that a socioconstructivist conception of teaching presupposes a number of major breaks with the usual norms of teaching (Larochelle and Bednarz 1994). Such a conception of teaching implies similar changeovers for teacher training and continuing education that are consistent with the objectives one hopes will be pursued in the classroom.

The socioconstructivist perspective and collaborative research: A conception of teaching knowledge–ability (savoir–enseigner) to be constructed on a situational basis

The following postulates serve to specify the kinds of relationships among practice, training, and research that receive special consideration in connection with the collaborative project in question here.

If there is to be any recognition of the constructed, reflective, and contextual nature of the knowledge of students, there must also be a recognition of the constructed, reflective, and contextual nature of teaching knowledge–ability (*savoir–enseigner*).[1] The development and appropriation of teaching knowledge–ability would appear, then, as a construction in a specific situation rather than as a preconceived kind of training suitable for all situations. In epistemological terms, this means that knowledge relating to a set of particular professional practices is not constructed in any setting other than the one in which this practice is actualized (Lave 1988) and is conditional upon the actor/practitioner's understanding of these practice settings. As far as continuing education is concerned, such a conception of teaching knowledge–ability, which advocates greater contextualization and differentiation, entails taking as point of departure the meanings that the teacher has developed in context and that will frame each situation or action proposed by the training supervisor.

In order to give meaning to the activities suggested to them, teachers make use of a background of knowledge, postulates, and constructs that enable them to interpret these activities. What implicit meanings do teachers give the activities proposed by researchers in context? What didactic contract is developed in the classroom (Brousseau 1986; Schubauer–Leoni 1986)? These meanings will structure the learning tasks that are proposed to students and the way in which teachers and students will collaborate or not on the construction of knowledge in a class setting.

In this perspective, the collaborative project conducted here constantly evolved in connection with the meanings worked out by teachers in context and resulted in a relationship between training and practice that differed considerably from that found in the classic models of continuing education for teachers.

Of an overly mechanistic cast, these classical training models maintain a standardized conception of practice in which the tools provided to practitioners are designed to offer ready-made solutions to foreseeable problems. However, many situations arising in practice are to some extent indeterminate, thus, in turn, requiring that the problem be constructed in context and that the solution be developed in relation to the problem.

It is the indeterminate nature of situations that casts a critical light on the metaphor of the teacher as an efficient user of the tools that others have fabricated for the purpose of running an apparatus. What indeterminateness brings to the fore instead is the metaphor of the reflective practitioner who must have a certain margin for action enabling him or her to develop a capacity for exercising judgment in context.

This change in perspective completely transforms the relationship between the practitioner and the researcher, and indeed the training supervisor. In effect, the point of departure no longer consists of the researcher's understanding of the apparatus that is supposed to be operated in all contexts, but rather in the practitioner's understanding of the specific situation that has to be dealt with. It is on the basis of this understanding that the practitioner can transform his or her ability to act. In this view, then, the researcher serves as the one who makes this understanding explicit.

Would that mean, then, that the role of the researcher in the construction of teaching knowledge–ability consists only in explaining and making explicit the knowledge that the practitioner has of his or her ability to act? Such an assertion would, by the same token, deny the potential contribution of the researcher in the construction of this teaching knowledge–ability.

But how is this influence to be exerted without reverting to a mechanistic relationship in which the researcher dictates how the practitioner is to act? Here we arrive at the fundamental challenge facing collaborative research and the researcher who initiates it. The situations developed by researchers will, on that score, serve as a springboard for an approach in which action in class and reflection on action will play important roles while also inducing teachers, on the one hand, to place situations at a distance, reflectively, and, on the other hand, to reappropriate and reconstruct them progressively.

In the approach adopted in the class in mathematics, as will be seen further on in this chapter, teachers must constantly interpret the cognitive activity of children (procedures used, reasoning approaches, difficulties) to provide an analysis of how students function and evolve in connection with the implemented activities. We are dealing here with a fundamental, crucial component of the approach, which, moreover, is difficult to pull off. What holds for the teacher applies to the researcher as well: The latter must develop an interpretation of the activity of teachers, of how they function and evolve in connection with their practices (as these are actually being enacted), as well as of their reflection on action (as occurs during reflective stocktaking and teacher–researcher interactions).

This analysis makes it incumbent upon researchers to elaborate a reinterpretation of their own perspective in light of what teachers themselves do and say (Confrey 1994).

The elaboration of teaching knowledge–ability (i.e., knowledge, pedagogical strategies related to the approach proposed) thus appears in the context of this collaborative process as a coconstruction, to which researchers and teachers contribute in a form of interactive interplay. What, in other words, is a multifarious process of argumentation, confrontation, spelling-out of points of view, negotiation, and so forth, thus helps to enrich a certain experiential reality and brings forth a whole range of possibilities. It is in this way that collaborative research takes on meaning.

Description of the collaborative research project

Forming a backdrop to the project at the time of its inception were the teaching/learning situation and the development of meaningful learning experiences for children.

The project arose from critical preoccupations over the way mathematics has been taught in a traditional context. In essence, the project was aimed at developing critical reflection among teachers and contributing to the restructuring of their teaching knowledge–ability in order to bring about a new definition of the roles of teachers and children in the appropriation of knowledge, that is, in order to form a "research community."

A socioconstructivist theoretical perspective provided researchers with a frame of reference with which to elaborate the approach that would be proposed to the teachers (in mathematics) for experimentation in class.

A range of activities that were consistent with this approach to learning (here, various elements were accorded greater or lesser importance by the approach) would then be subjected to experimentation in class (see Table 2.1). These situations would provide the basis on which an unplanned evolutionary dynamic could be set in motion, and would

Table 2.1. *Highlights of the approach, as used in mathematics*

Importance put on strategies, organizational procedures, ideas, reasoning approaches developed by children: In each classroom situation, children must show evidence of pattern, be able to link or contrast ideas in addition to defending them or presenting arguments for them.

Social interactions: the spelling-out of points of view, argumentation, confrontation, and negotiation is emphasized.

Choices by teachers of strategies or emphasis in class worked out in accordance with the student's mode of reasoning, through attempts at identifying the strategies, conceptions, and ideas developed by children.

Increasing the complexity of children's knowledge and not the transmission of a pre-given form of knowledge is emphasized.

themselves take shape at the outset in accordance with the specific needs expressed by the teachers (for an example of activities elaborated in a classroom setting, see Bednarz et al. 1993).

Thus, the project of collaborative research (begun in 1990–1991) stemmed from the needs of a team of first-grade teachers concerning problem solving. Their questioning process led to a project (at the time, the team included three teachers, one remedial educator, and two researchers) that was pursued in a more systematic fashion over the next 2 years. The underlying question was this: Is it possible to adopt a problem-solving approach for first-grade children (6- to 7-year-olds)? What would be the meaning of such an approach with young children? And how can this approach be worked out?

More specifically, in 1991–1992 and in 1992–1993, the teachers' project led to the elaboration of and experimentation with learning situations devised in connection with children's mathematical reasoning approaches, ideas, organizational procedures, and first learning experiences (number, operations, measurement, geometry, etc.).

In addition, the project was expanded to include three second-grade (7- to 8-year-olds) and two third-grade (8- to 9-year-olds) children at the specific request of their teachers, who were now handling students who had benefited from a different approach to mathematics in the preceding 1 or 2 years.

During the meetings with the participating teachers held at various stages of the process of collaborative research, the approach served as a springboard for reflection on action. Thus, discussions touched on the situations that were elaborated, the strategies used by the children, their patterns of reasoning, the management of the activity in context, and the underlying educational approach. This provided matter for reflection on the teaching and learning of mathematics and served to call habits of doing and conceiving of teaching and learning into question.

In a kind of continual round robin of elaboration, experimentation, and observation of situations in class, alternating with stocktaking and analysis in teams, the teachers and researchers found themselves confronted with questions relating to different aspects of teaching activity. Thus, questioning of the theoretical foundations underlying the pedagogical strategy or emphasis adopted alternated with reflection on the mathematical knowledge serving as focus of the situation at hand. Likewise, interrogation on the subject of children's knowledge and strategies used by them alternated with reflection on those situations that fostered the development of mathematical reasoning in young children, all of which necessarily entailed preoccupation with how to bring these situations to realization in class and engendered concern over the attendant tasks of management and adaptation. The reflections of teachers (as expressed during exchanges between researchers and teachers) and their classroom teaching practices (as recorded on a regular basis over a 2-year period) provided an analytical basis for reconstructing the different phases that teachers pass through in appropriating the approach, in addition to illustrating the doubts and breaks with habits of thinking and doing confronting them as they attempted to create a new didactic contract within the classroom.

Predetermined knowledge and constructed knowledge: A relationship to teachers' knowledge with which researchers must come to terms

The choice of a collaborative type of research presupposes a kind of knowledge that is constructed on the basis of the various partners' respective capacities, that is, knowledge constructed in the interactions of researchers and practitioners, and again, in the interactions between teachers themselves during reflective stocktaking on action in groups.

One must not be tricked into believing that all teachers immediately subscribe to this relationship to knowledge, especially when it is considered that the predominant relationship to knowledge in Western culture is more "technicist" (an offshoot of positivism) than constructivist. This implies that the researcher must deal with some teachers who are more at ease in a technicist type of relationship and who refuse to adopt a relationship of a different kind, which, among other things makes greater demands on them in terms of their involvement in the entire process of coconstructing knowledge, including the underlying process of reflective stocktaking on action.

All the same, at the outset, all teachers entertained a technical relationship to the approach. The following exchange, which took place between the researcher and a teacher, is illustrative of this technicist relation. It followed discussion of a number of first-grade activities that had been tried out by the teacher at the beginning of the school year. With their focus on counting collections or comparing the number of elements contained within different collections, these activities brought a range of variations into play within the situations presented to the children (the nature of the elements, the distance or proximity of elements, the size of the collection, etc.).

Researcher: What comes after these activities (in reference to actions conducted in connection with real collections)? What are they followed up with?

Teacher: (She examines her plan of activities.) We move on to the drawn collection. But it would be better if I were to continue to work with real collections.

R: (On that score, the researcher suggests another activity that would make it possible to see how far the children have got with counting a real collection): Put a collection of objects on the overhead projector, ask how many objects there are, move it around in front of the students so they can clearly see the action being performed, and then ask them the question again, "how many objects are there?" (Another member of the team suggests an alternative activity that is derived from a context in which an elf plays pranks and pushes around a collection of objects that the children have on their desks.)

T: (The teacher wants to know in which order the suggested activities should come.) First, there's the activity with the overhead projector comparing two collections that are equal but are spread out differently? Then there's the elf activity? And then there's the activity using the over-

Table 2.2. *Management of activities in the classroom*

Management of the activity is focused on the material aspects -- the "how-to" rather than on the contents (how, in concrete terms, I lead the activity in class, while nonetheless losing sight of the purpose).

The type of question put to the children does little to encourage a reflective relationship (the activity of the child is strongly influenced by the instructions given).

The way in which the children's answers are picked up on focuses more on the results than on the "how did you do it?"; makes do with nodding observance of what the child has done rather than making explicit what it is he or she has done; does not seek the reasons why.

head for the collection that's moved around? When all these activities are over, the children should be able to tell me it's the same thing.

The more technical relationship to the approach may be seen in the teacher's questions, which are focused more on the material aspects (i.e., the how-to) than on the contents of the approach. At that point, the researcher took a tack in a conceptual direction by forcing reflection on the underlying intentions of the activity, on the one hand, and the introduction of a number of constraints, on the other.

R: Well, what do you think? Is the order of activities important? What are you attempting to achieve with these activities? Is one more important than the other?

This technical-learning type of relationship to the approach will also appear in connection with the management of activities in class as outlined in Table 2.2.

At this stage, although researchers mimic or simulate activities with teachers so as to respond to these technical preoccupations, they nevertheless attempt to involve teachers in a reflective process by questioning the underlying intentions of an activity and also by anticipating different possible strategies and patterns of reasoning that children may use. The teachers will progressively adopt an entirely different relationship to the approach.

Practical reflection: Restructuring of a repertory of actions

When a new approach such as the one used in conjunction with the project in mathematics is proposed to teachers, it goes without saying that the routine way of doing things is upset. The process of appropriating a new way of teaching necessarily confronts teachers with the indeterminate and the unknown. They will have to develop a new repertory of actions or, at least, alter or restructure any repertory that they already possess.

In the phase labeled practical reflection, which corresponds to the teacher's concern with personalizing the approach, efforts are directed at attuning the underlying intentions of the approach to the context (children's needs, class requirements). The following exchange, taken from a meeting of teachers and researchers, provides an illustration of issues surrounding the process at this stage.

A first-grade teacher recounts what happened during an activity that she had tried out in class. The activity refers to children's counting of collections, in which the collections are composed of different objects. (Who has the same number of elements as I? The teacher presents a collection on the overhead projector located at the front of the class.)

Teacher: *Very few children made a mistake,* in contrast with last year. I put the first collection on the projector, 9 or 11 objects. The children counted their own collections, and *I saw* a number of them counting the collection on the projector by pointing at the screen.

I separated my display with a string so that on one side there was my collection and on the other side the child could set out his or her own collection.

I tried something that hadn't been planned for: I moved the objects around on the projector (put them closer to one another) and asked the question again.

Then I noticed that the children counted the collection on the projector all over again. *I think that* that's because I stood in front of the overhead as I was moving the objects around and the children couldn't see what I was doing.

After that, one of the children (Bertrand) came up front (one who said he had the same number of objects) and he set out his objects, but scattered around; what's more, his objects were bigger than mine. *Now there I had a great opportunity to see* if the children could be taken in by appearances.

(I said to the other children) "Do you think Bertrand has the same number of
  elements as me?" Some children said yes, others no. *Their arguments
  went like this:*
"Look, it's because he hid more of the white ones than you did." "It's not
  important; that's because the objects are bigger."
I also noticed that when the children counted objects, they counted objects that
  were alike: 3 buttons, 4 cubes, and so forth.
Researcher: And the ones who trusted appearances, how did they react to the
  arguments of the others? Could they be convinced?

In this excerpt, it is possible to observe how the teacher integrates a
number of different elements into her teaching approach:

Description of classroom management ("I separated my display") is
interspersed with *careful observation* of the children's strategies ("I saw
a number of them counting the collection on the projector by pointing at
the screen"; "I noticed the children counted the collection on the projector all over again"; "Their arguments went like this." With reference
to the action taking place, *she produces an analysis* of a number of
possible reasons for this strategy ("I think that that's because I stood in
front of the overhead."). *Personal initiatives are taken while teaching
and are based on what the teacher has observed* (this is where the
strategic actor may be glimpsed): "I tried something that hadn't been
planned for"; and, further on, "Now there I had a great opportunity to
see if the children could be taken in by appearances."

In this example of personal reappropriation of the activity, the teacher
registers what is occurring in the classroom with this group of children
and yet does not lose sight of the underlying intention of the activity,
which she has followed attentively: "Now there I had a great opportunity to see if the children could be taken in by appearances."

This form of reflectivity leads to a personal repertory of learning
strategies that have to be developed as the situation with students requires. At this stage, interactions between researchers and teachers play
a key role and undergo articulation in connection with a variety of
situations (narration of the activities tried out in the classroom; problems encountered by the teachers; observation and analysis of children's
work; viewing of videos taken in the classroom, etc.). In addition, a
combined process of analysis, argumentation, and negotiation is embarked upon, thus helping to clarify a situation and generating potential
ways for delving further into it.

During reflective stocktaking sessions concerning the approach in
class, tensions, which, from a constructivist perspective, are part and
parcel of teaching and learning mathematics, will come to the fore of the

debate. Such tensions derive particularly from the fact that the teaching practices employed in mathematics must be both meaningful for students (meaning here that emphasis is given to personal construction, the elaboration of strategies, children's own knowledge, and their habits of argumentation) and yet relevant from the standpoint of established practice (e.g., teachers are required to teach conventional ways of writing numbers, computational algorithms, etc.). Confronted with a dilemma such as this, a number of teachers continue to take the path that they judge to be the best.

The emancipatory relationship to the approach

In a crucial phase corresponding to the teacher's concern with taking control of the approach, a teacher proceeds from a personal mode of appropriation to one that is more institutional in character. One can picture the dilemmas arising between the values and ends conveyed by the approach (increasing complexity of the student's knowledge, entertaining of various possible ways of doing things, etc.) and those conveyed by the institution (programs that emphasize meeting a number of specific requirements with respect to computational algorithms, for example; evaluations; etc.) and, then again, by society (parents). It is at this stage that the teacher establishes an emancipated relationship to the approach, as may be seen in the following two excerpts taken from field notes.

21 January 1993

L. (the researcher) went into the class (Grade 2) to film the activity called "jeu du marchand" (a game of buying and selling using multibases, which, *within a context,* situates activities involving comparison, a range of transformations performed on the collections themselves, addition, subtraction, dividing up, etc.). In this particular context, children have to ascribe meaning to the actions performed in terms of groupings, or groups of groupings. It was an open day at the school. The teacher told the researcher that she had had several parents that morning. There were 30 minutes left before the lunch break, so she decided to do mathematics. She works on the "jeu du marchand" with the children using the Multibases Dienes blocks, and takes care to bring out the different aspects of the approach (i.e., spells out the various procedures used by children to arrive at solutions; spells out the reasons that lead children to take a position in the comparison of various collections; etc.) so that parents can see what the children are capable of. She also uses the opportunity to get the message across to parents: *don't offer the quick and easy route to children* (i.e., don't give them some computational procedure that they only have to apply to problems – the children already have many means at their disposal that are just as effective).

4 February 1993

Discussion of the test results (the tests were reexamined in the course of the year with the researchers, at the request of this particular teacher. Adopting a critical point of view toward the traditional examinations provided by the school board, she came forward with the desire to have an approach to evaluation that was more in keeping with the type of learning approach being developed with the children). She is satisfied; it seems to her that the scores are better this year than in previous years. Nevertheless, she has received some complaints from parents who do not agree with the test, nor, for that very reason, with the approach. In the case of one student whose parents are worried, she reminds them that the child is having difficulty in all subjects. She can tell that it is not because this student is being taught with a new approach that things are harder for him. In the case of another child, she thinks that it is because the parents do not understand what is being presented in the test. *She has a meeting with them next Monday, at which she is going to present the material and have them work with it and have them bring out the elements being worked on* (in other words, she is going to pursue the same idea that she spontaneously put into effect during the open day when she attempted to explain to parents the intentions underlying what she was doing).

Some time later, she mentions in connection with a report that the principal requested from her that she has spent 25 years looking for an approach that leads children to reason, and at last she has found it.

What may be grasped in the preceding excerpts is how, by means of a number of personal initiatives undertaken by this mathematics teacher vis-à-vis parents, the interplay of institutional constraints (in the present case, represented by the parents and the values they wish to bring to bear on the math teacher) in fact becomes a springboard that can be used to broaden one's repertory of teaching practices – notably, the simulation of activities in front of parents so as to bring out the underlying intentions of this approach, in addition to the learning experiences that children gain through it. This mainly institutional reappropriation of the approach will, in the case of this teacher, result in her taking greater charge of evaluation (developing new forms of evaluation that are more in keeping with the approach) and in greater involvement with other teachers (running workshops, discussions based on students' work, etc.).

Classroom mathematics culture and collaborative research culture involving researchers and teachers

What stands out in the collaborative model, via the interaction occurring between researchers and teachers during reflective stocktaking on the approach, is a certain type of culture that inescapably suggests parallels with the type of culture that teachers develop with children in class (see Table 2.3). The two excerpts provide an illustration of how the classroom learning situation mirrors, and is mirrored in, the training situation in which researchers and teachers interact with one another so as to restructure teaching knowledge–ability.

The first of the following two excerpts is taken from a verbatim transcription of a class discussion between a third-grade teacher involved in the collaborative research project and her students.

It brings out the mathematics culture that has been gradually developed by the teachers involved in the collaborative research project. Each child was assigned a problem to solve ahead of time (in this case, children were given different problems with a multiplicational structure of varying complexity), using an illustration and question that he or she had been provided with (of the problems put to the children, no two were alike). Then, taking turns, each child went up to the front of the class to present his or her problem to the others, so that they could try to solve it. The idea behind this activity, in which the focus consists in the formulation of problems by the children themselves so that others may arrive at a solution, has been dealt with previously (Bednarz 1996).

Jean-Philippe was given the following problem to solve: How many rows are hidden? He was given an illustration of a spice rack in which each row can accommodate 4 bottles and in which 3 rows of bottles are visible; he was also provided with the information that the rack contains 16 bottles altogether. In front of all the others, Jean-Philippe stated the problem this way:

1. J.-P: There are 16 spice bottles and 3 rows. How many rows have been
2. hidden?
3. Teacher: Can somebody else reformulate Jean-Philippe's question to see if
4. we've all understood correctly? Can somebody tell us what he or she has
5. understood? Emmanuel, what did you understand?
6. Emmanuel: I think, well, I think he forgot to tell us something; I mean, how
7. many rows are in each . . . in each bottle . . .

Table 2.3. *Comparison of classroom culture and research culture*

Classroom culture group: teachers / children	Collaborative research culture group: researchers / teachers
Point of departure: children and their learning; work out teaching strategies in accordance with children's patterns of reasoning	Focus action on the teaching situation, on learning and the meaning that teachers attach to this process: analyze what students produce, the problems that arise, teaching strategies, etc.
A different approach: a genuine process of explanation, argumentation, and negotiation in class	Construction in context of potential repertory of knowledge and student-directed teaching practices through discussion, interaction with colleagues and researchers
A model favoring participation in a certain kind of culture in which questioning and the explanation of solutions play an important role	A model that favors questioning, the "making explicit of" various points of view
Perceive the student in terms of the diversity and wealth of his or her resources/abilities; in terms of the various solutions and arguments that he or she is able to advance	Perceive the teacher in terms of his or her potential for action (strategies, solutions, routines, etc.)
Perceive teaching as an interactive and reflective process that develops in accordance with the solutions advanced	Perceive collaborative research as an interactive and reflective process that develops in accordance with the various strategies evolved
Encourage evolution: identify conditions that will provide the child with the opportunity to change his or her way of doing things	Develop critical reflection toward one's own practice, so as to reorganize it
Assumes that the teacher forms a conceptual model of what the child does (so as to better understand this, and act accordingly)	The researcher identifies bases that serve to promote development

8. Teacher: What do you mean? (All the students begin talking at the same
9. time, and no one can hear what's being said. The teacher asks for another
10. reformulation of the problem.)
11. Jacques-Olivier, could you reformulate the problem nice and loud so
12. everyone can hear?

13. Jacques-Olivier: Well, there are three rows. You've got 16 spice bottles
      and
14. 3 rows. How many rows are hidden?
15. Teacher: (Turning to a child who has raised his hand) Hugo has got
16. something to say. Say it out loud, Hugo.
17. Hugo: I don't understand why the rows are hidden.
18. J.-P.: (Reacting to what Hugo has just said) Well, there are 16 spice
19. bottles and there are 3 rows.
20. Teacher: (Turning to another child who wants to say something)
21. Guillaume, have you understood?
22. Guillaume: There are several ways to solve it.
23. Teacher: Well now! Listen. Guillaume says there are several ways to
      solve
24. the problem. Can you explain this to us, Guillaume?
25. Guillaume: Because, if you put one bottle for each row, then you have 16
26. rows. Two bottles for each row, then you have fewer rows. The more
27. bottles you put on each row, the fewer, the fewer rows there'll be.
28. Teacher: What do you think, Jean-Philippe?
29. J.-P.: Well, that's not quite what my picture shows . . .
30. Teacher: So what do you suggest so that your buddies here can solve the
31. problem you were given?
32. (Turning to the other children) What could Jean-Philippe also say so that
33. we can solve the problem?
34. Guillaume: Well, he could say h . . . how many bottles there are for each
35. row. You didn't say, how many bottles there are for each row?
36. J.-P.: There are 4 bottles per row.
37. The other students, in chorus: Oh ho!
38. Teacher: Can we solve his problem now?
39. Students: Yes.
40. Teacher: Before solving the problem, could you reformulate your
      problem
41. for us, Jean-Philippe?
42. J.-P.: There are 16 spice bottles. There are 4 spice bottles in each row, and
43. there are 3 rows. How many rows are hidden? (Several students tell each
44. other that they can find the answer)

During this particular excerpt devoted to the formulation of the problem
by Jean-Philippe so that the others can understand it and solve it, the
teacher has brought out a reformulation of the problem by the children
themselves through a number of different stages in the process (l. 3 to
l. 5; l. 8 to l. 12; l. 40 to l. 41). From the outset, this teacher has attempted
to use this reformulation activity to understand what it is that children
themselves have understood of the problem and to stimulate the initial
interaction between them (l. 13 to l. 19). In the usual didactic contract,

the attitude adopted toward Jean-Philippe's incorrectly formulated problem would have undoubtedly consisted in pointing out the error so as to induce the child to make the necessary corrections. In this instance, however, the teacher has, from the start, avoided taking a position with regard to the problem that Jean-Philippe formulated inadequately (l. 1 to l. 2: Jean-Philippe forgets to specify that there are 4 bottles per row). Instead, she has attempted to bring the children to the realization that the presentation of the problem was incomplete (l. 6 to l. 7, and l. 17). Had she not encouraged the children's interactions, Guillaume's interpretation of that presentation (l. 22 to l. 27), in which he conceives of different solutions using made-up data, would never have come to the fore. Now, Guillaume's solution will not really be taken up by the teacher. She refers the decision concerning G.'s solution back to Jean-Philippe, who will thus be required to reexamine his own formulation of the problem. The interaction that the teacher has set in motion is aimed in this case at encouraging the child who formulates the problem to do so in a way that enables the other children to solve it (l. 28 to l. 33). It follows, too, that awareness of the incompleteness of the presentation emerged (l. 34 to l. 36).

In the remaining portion of this excerpt, devoted to a review of the proposed solutions, we will see how the teacher fosters a process of spelling out, argumentation, and negotiation among the children themselves.

1. Arnaud: If you've got 16, and then you have 4 . . . all right, if you add
2. 16 and 16, that makes 32; 32 plus 32 gives you 64.
3. Guillaume: Huh!? He said there were 3 rows and 16 bottles. Like,
4. there's no way, because . . .
5. Teacher: All right, all right. If you don't agree, but one at a time.
6. Guillaume started to say something.
7. Guillaume: I don't agree, because if you take 16, there are 4 bottles per
8. row. Look here (he motions with his hands to show Arnaud that his
9. solution can't work): here's your spice rack, and here are your spice
10. bottles. When there are four bottles to each row, then you've got 16
11. altogether. Now you've already got 3 rows, and he wants to know how
12. many others are hidden.
13. Arnaud: (Unconvinced) Look, Guillaume. Like so, you've got 16
14. bottles. And then you've got 4: 1, 2, 3, 4; 1, 2, 3, 4. (he counts out
15. several times).
16. Guillaume: No, because . . .
17. Teacher: (Pointing to another student who wants to speak) So, now,
18. over here, what do you want to say?
19. Michel: Like, you won't get 16. Look, I noticed that he said he put 4

20. in each row (showing his illustration) and then he asked how many
21. (rows) were hidden. I saw that 3 rows of 4, that gives you 12, and
22. when you add on another one . . .
23. I think there's one that's hidden because 12 + 4 = 16.
24. Teacher: (Showing his illustration to the others) Would you like to
25. explain once again why you don't agree with Arnaud's solution?
26. Michel: Well then you have a bigger number; you don't get 16.
27. (Addressing Arnaud) You know, Jean-Philippe didn't say that he had
28. more than 16. He said, "How many hidden rows are there?" If you add
29. up everything, you don't get 16.
30. Teacher: All right, then, Joannie has another way. Let's turn around
31. and see how Joannie does it. We'll see if it's really a different way.
32. Joannie: I was thinking, 3 times 4 is 12, but you have to come up with
33. 16 bottles. Well, 16 is greater than 12. Four times four is 16, I mean, I
34. continued on to 16, because there were 16 bottles.
35. Magalie: No, but look, she took 4 times 4; you can't tell that it's 3.
36. Joannie: Yes you can. That there (the last row) it would not be there?
37. Michel: But we think it's 4 times 4. You'd better put it somewhere
38. else if you really want to see 3 times 4.

Arnaud models the problem at hand as one of multiplication – 16 bottles
per row, with 4 rows (l. 1 to l. 2) – prompting Guillaume to react at once
(l. 3 to l. 4). The teacher (l. 5 to l. 6) then encourages interaction between
both children, who present their respective arguments (l. 7 to l. 16). In
his context-bound verbalization, Guillaume (l. 7 to l. 12) attempts to
make Arnaud understand that there cannot be 64 bottles. Arnaud remains totally unconvinced of this (l. 13 to l. 15) and goes back over his
own explication of the solution to the problem. At this point, to prevent
limiting the discussion to the two children involved, the teacher encourages the other children to take part, too (l. 17 to l. 18). Then, another
child tries to explain to Arnaud why his line of reasoning does not work
(l. 19 to l. 23). To do so, the child shows the solution she has worked out
on her board (each child had a squared-off surface on which he or she
can place self-sticking markers in order to try out various solutions): she
shows 3 rows of 4 markers, for a total of 12 markers, and then shows
what has to be added – another row of 4 markers – to produce 16
markers, which was the total given at the beginning. After showing this
illustration to the other children, the teacher asks the child to spell out
where she disagrees with Arnaud for the others just one more time (l. 24
to l. 25). Then she encourages another child (Joannie) to explicate her
solution (l. 30 to l. 31), which gives rise to a working out of the respective arguments of these two children (l. 32 to l. 38) concerning the
illustration on the board, that is, Joannie does not spell out the difference

between the 3 rows (which are a given in this problem) and what has to
be added in order to come up with 16 bottles.

Argumentation among the children continues and will help bring out
the type of organization that works best in order to help grasp the end
result most quickly.

The second excerpt, which was taken from a meeting of researchers
and third-grade teachers involved in the collaborative research process,
provides an illustration of the same interplay of spelling-out, argumentation, and negotiation among the participants in this project.

The following series of reflections focused on the solution offered by
a student to a problem presented in class by two third-grade teachers.
The oppositions that arise provide a window not only on the different
ways the project participants have of interpreting but also on the way in
which the meaning given to the solution in question is capable of
evolving.

1.   Line (a teacher): Take the SaintValentine problem:
2.   "The teachers for the two third-grade classes (30 children in each
3.   classroom) buy two chocolates for each student. How many chocolates
4.   did they buy?" (When put to the children, the question went as follows:
5.   The third-grade teachers brought two chocolate hearts for each student in
6.   each class. How many chocolates did they buy for both classes?) They
7.   knew right off it was 120. Some of them wrote down 4 times 30.
8.   I said, "Where did you come up with 4 times 30? Why did you give me 4
9.   times 30?"
10. He says, "There are 120 chocolates, so that gives you 4 times 30." "But
11. how do you get 4? How many chocolates does each child get?" "Two, he
12. says." "But then you tell me that each child gets 4! Four times 30." "No!
13. There are 30 in one class and 30 in the other, it gives 4."
14. Anne (a teacher) It might be o.k.
15. Researcher: (To Anne): What makes you think so?
16. Anne: Because 2 chocolates in . . . ; 2 and 2 is 4, 4 with 30. Rather than 2
17. times 60 – 2 chocolates and 60 students – then, 4 chocolates for 30
18. children. Now, that student of yours, he just pictures the group of 30 . . .
19. But there are two classes, so he doubles it, and it works out as though it
20. were 4 chocolates.
21. Researcher: You mean, it's as though there were one child with four
22. chocolates if you put both classes together . . . ?
23. Line: He put two children together.
24. Researcher: It's as though one child in one class got 4 chocolates.
25. Line: Sure, but that's not the reality of the situation.
26. Researcher: You could say: "In each class, you have one child who gets 2
27. chocolates. It's as if, if I gave everything to one class, I'd get 4 chocolates

28. per child."
29. Line: If I use the class as a basis.
30. Anne: Well he used the class as a basis. He doubled the number of
31. chocolates for the class.
32. Researcher: That might be it.
33. Line: But it's not logical.
34. Researcher: Oh ho! It's not logical, you say. There's a certain logic. . . .
    But
35. it's a reconstruction that he's carried out.
36. Line: It doesn't stick to reality, does it? It doesn't stick to the problem!
37. Researcher: It's a reconstruction.
38. Line: He knew he got 120; he said, "4 times 30."
39. Anne: If I work it out using 60 students, 2 chocolates (each) gives me (a
40. total of) 120.
41. But if I just use 30 – there are 30 children in a class – then it's as though
42. each one got 4, and that gives me 120. Isn't that reality!
43. Line: There's a certain logic. He put the two classes together, and
44. acted as though. . . . But in reality, that's not how it works. Each child
45. didn't receive 4.
46. Researcher: But was he able to explain how he went about getting his
    result?
47. Line: Yes. It's like both classes – that gives you 4. There's nothing in
48. the problem which says you can do that. He came up with that logic in
49. order to make it work out.
50. Researcher: He came up with a means of calculation.
51. Anne: To find out how many chocolates there were, you take 2, you take
52. your 60 children – it's like 4 times 30. That gives you 120. To his way of
53. thinking, the number of chocolates was a sure thing.
54. Line: I know, I know! But it's not the most logical way. Forget for a
55. moment . . .
56. Anne: I think it's really bright to have hit on a way of calculating like
    that!
57. Line: . . . the children who said that, that doesn't prove that he
58. really understood what the problem involved!
59. Anne: Well, yes! Because he found an equivalence.
60. Line: As you like, yes, there is an equivalence from that point of view.
61. He wasn't the only one – there were two or three who did it that way.
62. Anne: Right, because 2 per child here and 2 per child there, if it were all
63. in the same class, you'd get 4, and 4 times 30, that gives you 120
64. chocolates.
65. Line: Looked on as a kind of logic, it works.

During this interaction between the teachers and the researcher, certain basic conceptions involving mathematics and problem solving are brought into play in the ensuing debate. The solution put forward by several students, and which was narrated by one of the teachers (l. 1 to l. 13), immediately triggers a reaction on the part of the other teacher (Anne) (l. 14). At this point, the researcher (l. 15) makes Anne spell out the reasons that lead her to think this particular child's reasoning may be correct (l. 16 to l. 20). It is this explication that the researcher reformulates in order to bring it home to the first teacher (Line) (l. 21 to l. 24). This, in turn, leads the first teacher to spell out the underlying conception she has of problem-solving in mathematics, via a kind of the child's reply, "Sure, but that's not the reality of the situation" (l. 25); in other words, the solution to a problem must stick to the context, reflect the reality of the situation, and not diverge from this.

The first episode of this excerpt ends in this way, and, thus, a new type of interaction begins (l. 26 to l. 45). Here, the researcher takes up a position within the terms laid out by Line, attempting to ascribe meaning to the solution proposed by the child within the context (l. 26 to l. 28); the other teacher backs up this effort with an explication of her own (l. 30 to l. 31). At that point, Line's argumentation takes a different tack: "but it's not logical" (l. 33). The researcher takes up a position in line with these new terms and attempts to show that, on the contrary, there is a certain logic, if of a different kind, that is, the child has come up with a reconstruction, has developed a new representation, in which he has strayed from the context in a way, however, that can be grasped in terms of arithmetical operations (l. 34 to l. 37); Line's reaction is not long in coming. Anne relates the meaning ascribed by the researcher to the context; as for Line, she comes around to the logicality of the solution and yet maintains that the reality of the context is a different story altogether (l. 43 to l. 45). The explanation provided by the child is briefly returned to (l. 46 to l. 53) in order to demonstrate its greater effectiveness in terms of computation; Line again refuses this (l. 54 to l. 55). Anne states that she sees this procedure as a brilliant one (l. 56); Line adopts a different tack: She admits that it is an interesting way of figuring out an answer but wants to know whether the child really understands the problem (l. 57 to l. 58). To counter this reservation, the other teacher indicates that there is indeed understanding by the child, since he was able to establish an equivalence. This episode ends with Line's recognizing the logic that the child has applied but still considering that it remains at a remove from the reality of the problem (l. 59 to l. 65).

As is apparent in this excerpt, we are in fact dealing with deep-rooted conceptions of what it means to do mathematics and solve problems here; with the prompting of the researcher, the two teachers are, as a result, led to express different points of view.

This analysis sheds some light on the role of social interactions in the evolution of teachers' thinking. During this shared process of reflection on action and interaction in mathematics classes, we can actually see the beneficial effects of forms of cooperative action as well as explicit confrontations in points of view between teachers and teachers as well as teachers and researchers. The interactions bring the various meanings that teachers ascribe to situations to bear on one another, thus serving to illustrate the different underlying models, to clarify ideas, to bring forth reformulations, and so forth.

Conclusion

The preceding examples provide a brief illustration of the dynamic that is set in motion between, on the one hand, teachers and students, and, on the other hand, teachers and researchers. They also serve to shed light on the way in which this dynamic contributes to the restructuring of a certain kind of teaching knowledge—ability.

This approach is nonetheless quite a slow one and requires a profound change in attitude and practice of the teacher in order to break with the usual didactic contract serving to define the roles of students and teacher in the appropriation of a certain kind of mathematical knowledge.

The approach also requires a profound change in attitude and practice of the researcher/training supervisor involved in collaborative research in order to break with the usual didactic contract as defined by classical training models. The desire to change classroom culture and contribute to its evolution certainly requires us to dwell at some length on the meanings that the teacher gives to the situations that have been put to him or her in their context, and, equally, that the teacher should be given many opportunities to ponder the actions and interactions as they arise.

Note

1. Translator's note: this term is offered as an equivalent for a neologism in French, which indicates an active relationship to the understanding of teaching: "knowledge that" instead of "knowledge of" or, indeed, mere technical know-how (i.e., troubleshooting).

References

Bachelard, G. (1983). *La formation de l'esprit scientifique* (12ᵉ édition). Paris: Librairie philosophique, J. Vrin.

Bauersfeld, H. (1994). Réflexions sur la formation des maîtres et sur l'enseignement des mathématiques au primaire [Remarks on the education of elementary teachers, pre-service and in-service]. *Revue des Sciences de l'éducation, 20*(1), 175–198.

Bauersfeld, H. (1980). Hidden dimensions in the so-called reality of a mathematics classroom. *Educational Studies in Mathematics, 11*, 23–41.

Bednarz, N. (1991). Interactions sociales et construction d'un système d'écriture des nombres en classe primaire [Social interactions and the construction of a system of numerical notation at the primary level]. In C. Garnier, N. Bednarz, and I. Ulanovskaya (Eds.), *Après Vygotski et Piaget. Perspectives sociale et constructiviste. Écoles russe et occidentale* [After Vygotsky and Piaget: Social and constructivist perspectives: Russian and Western schools], (pp. 51–67). Bruxelles: éditions de Boeck.

Bednarz, N. (1996). Language activities, conceptualization and problem solving: The role played by verbalization in the development of mathematical thought in young children. In H.M. Mansfield, N.A. Pateman, and N. Bednarz (Eds.), *Mathematics for tomorrow's young children: International perspectives on curriculum* (pp. 228–239). Dordrecht: Kluwer.

Bednarz, N., and Dufour-Janvier, B. (1988). A constructivist approach to numeration in primary school: Results of a three year intervention with the same group of children. *Educational Studies in Mathematics, 19*, 299–331.

Bednarz, N., Dufour-Janvier, B., Poirier, L., and Bacon, L. (1993, March). Socio-constructivist viewpoint on the use of symbolism in mathematics education. *The Alberta Journal of Educational Research, 49*(1), 41–58.

Bednarz, N., and Garnier, C. (1989). *Construction des savoirs: obstacles et conflits.* Montréal: Éditions Agence d'Arc.

Bednarz, N., and Janvier, B. (1996). Emergence and development of algebra as a problem solving tool: Continuities and discontinuities with arithmetic. In N. Bednarz, C. Kieran, and L. Lee (Eds.), *Approaches to algebra: perspectives for research and teaching* (pp. 115–136). Dordrecht: Kluwer.

Bednarz, N., Radford, L., and Janvier, B. (1995). Algebra as a problem solving tool: One unknown or several unknowns? In. L. Meira and D. Carraher (Eds.), *Proceedings of the 19th annual conference of the international group for the psychology of mathematics education (PME).* (Vol. 3, pp. 160–167). Recife, Brasil.

Brousseau, G. (1986). Théorisation des phénomènes d'enseignement des mathématiques [The theorization of the phenomena of mathematics teaching]. Thèse de doctorat d'état. Université de Bordeaux 1.

Brousseau, G. (1972). Processus de mathématisation. *Bulletin de l'association des professeurs de mathématiques de l'enseignement public (APMEP),* Février, 282, 57–84.

Confrey, J. (1994). Voix et perspective: à l'écoute des innovations épistémologiques des étudiants et des étudiantes [Voice and perspective: Hearing epistemological innovation in students' words]. *Revue des Sciences de l'éducation,* 20(1), 115–134.

Dufour-Janvier, B., and Bednarz, N. (1989). Situations conflictuelles expérimentées pour faire face à quelques obstacles dans une approche constructiviste de l'arithmétique au primaire [The essaying of conflict situations for use in dealing with obstacles within a constructivist approach to elementary-school arithmetic]. In N. Bednarz and C. Garnier (Eds.), *Construction des savoirs: Obstacles et conflits* [The construction of knowledge: obstacles and conflicts] (pp. 315–333). Montréal: Éditions Agence d'Arc.

Garnier, C., Bednarz, N. and Ulanovskaya, I. (1991). *Après Vygotsky et Piaget. Perspectives sociale et constructiviste. écoles russe et occidentale.* Brussels: éditions De Boeck.

Larochelle, M., and Bednarz, N. (1994). À propos du constructivisme et de l'éducation [Constructivism and education: A propos of some debates and practices]. *Revue des Sciences de l'Éducation,* 20(1), 5–20.

Lave, J. (1988). *Cognition in practice.* Cambridge: Cambridge University Press.

Poirier, L. (1991). *Étude des modèles implicites mis en oeuvre par les enfants lors de la réslution de problèmes arithmétiques complexes mettant en jeu la reconstruction d'une transformation.* Thèse de doctorat, Université du Québec à Montréal.

Schön, D.A. (1983). *The reflective practitioner. How professionals think in action.* New York: Basic Books.

Schön, D.A. (1987). *Educating the reflective practitioner.* San Francisco: Jossey-Bass.

Schubauer-Leoni, M.L. (1986). *Maître–élève–savoir: Analyse psychosociale du jeu et des enjeux de la relation didactique* [Teacher–student–knowledge: A psychosocial analysis of the game, and stakes, of the teaching relationship]. Ph.D. dissertation, Université de Genève, Switzerland.

Voigt, J. (1985) Patterns and routines in classroom interaction. *Recherches en Didactique des Mathématiques,* 6(1), 69–118.

3   Reorganizing the motivational sphere of classroom culture: An activity-theoretical analysis of planning in a teacher team

YRJÖ ENGESTRÖM

The culture of the classroom has been found to be an extraordinarily uniform and persistently stable formation (Cuban 1984). Numerous attempts at school reform seem to have produced relatively few lasting effects (Sarason 1990). Tharp and Gallimore (1988) suggest that this is because school reforms have remained at the level of systems and structures, not reaching the daily practices of teaching and learning in classrooms. On the other hand, attempts to change the daily instructional practices, such as the program designed by Tharp and Gallimore themselves, have not been particularly successful in the long run either.

The dichotomy of systems and structures, on the one hand, and daily classroom practices, on the other hand, may itself be an important reason for the difficulties. These two levels are explicit – one is codified on laws, regulations, and budgets, the other is codified in curricula, textbooks, and study materials. There is, however, a middle level between the formal structure of school systems and the content and methods of teaching. The middle level consists of relatively inconspicuous, recurrent, and taken-for-granted aspects of school life. These include grading and testing practices, patterning and punctuation of time, uses (not contents) of textbooks, bounding and use of the physical space, grouping of students, patterns of discipline and control, connections to the world outside the school, and interaction among teachers as well as between teachers and parents. This middle level has been called "hidden curriculum" (Snyder 1971; see also Henry 1963; Holt 1964; for a recent powerful account of one version of "hidden curriculum," see Høeg 1994).

Much like going to work in any complex institution, going to school is an exercise in trying to make sense of what is going on. According to Leont'ev (1978: 171), "sense expresses the relation of motive of activity to the immediate goal of action." The middle-level features and processes characterized previously are fundamental in that they largely determine the sense of schoolwork, and thus the experience of what it means to be a student or a teacher. As sense- and identity-building features, they are of decisive importance in the formation of motivation among students and teachers.

In attempts at school reform, the middle-level phenomena of the motivational sphere have been largely neglected. In other words, motives, the crucial driving force behind the actions of students and teachers, have been neglected. As Leont'ev (1978) put it, you cannot teach motives; you can only cultivate them by means of organizing people's lives.

Motives are formed in the life activity of the child; to the uniqueness of life corresponds the uniqueness of the motivational sphere of the personality; for this reason motives cannot be developed along isolated lines unconnected one with another. Consequently, we must speak of the problems of nurturing the motives of learning in connection with the development of life, with the development of the content of the actual vital relations of the child. (Leont'ev 1978: 185–186)

Recent studies on everyday or street mathematics (Lave 1988; Nunes, Schliemann, and Carraher 1993; Saxe 1991) point to the same conclusion: the organization of the collective life activities in which the mathematical tasks are embedded is of crucial importance for learning and successful performance. While this realization is an important step forward, the same studies sometimes tend to evoke a less fruitful opposition between school and everyday practices. If everyday practices are idealized and school is condemned, the internally contradictory nature of both is overlooked. Activity theory calls attention to those very contradictions and the developmental potentials they entail.

In this chapter, I will use conceptual tools from cultural-historical activity theory (see Leont'ev 1978, 1981; Wertsch 1981; Engeström 1987, 1990; Cole and Engeström 1993) to analyze and delineate the motivational middle level as a strategic focus of change in classroom cultures. I will analyze an attempt by an innovative team of primary school teachers to change the middle-level features, or the motivational sphere, of their schoolwork and classroom cultures. My data are from a longitudinal ethnographic study of an innovative teacher team aiming at the creation and implementation of an integrated Global Education curriculum for grades K through 6 in a public primary school in southern California.

Object and motive of schoolwork

Activity is here seen as a collective, systemic formation that has a complex mediational structure. Activities are not short-lived events or actions that have a temporally clear-cut beginning and end. They are systems that *produce* events and actions and evolve over lengthy periods of sociohistorical time. I use the schematic diagram of Figure 3.1 to represent the mediational structure of an activity system.

Leont'ev (1978: 52) pointed out that the concept of object is already implicitly contained in the very concept of activity; there is no such thing as objectless activity. An object is both something given and something projected or anticipated. An entity of the outside world becomes an object of activity as it meets a human need. This meeting is "an extraordinary act" (Leont'ev 1978: 54). The subject constructs the object, "singles out those properties that prove to be essential for developing social practice" (Lektorsky 1984: 137). In this constructed, need-related capacity, the object gains motivating force that gives shape and direction to activity. The object determines the horizon of possible goals and actions.

The model reveals the decisive feature of *multiple mediations* in activity. The subject and the object, or the actor and the environment, are mediated by instruments, including symbols and representations of various kinds. This triangle, however, is but "the tip of an iceberg." The less visible social mediators of activity — rules, community, and division of labor — are depicted at the bottom of the model. Among the components of the system, there are continuous transformations. The activity system incessantly reconstructs itself.

An activity system contains a variety of different viewpoints or "voices," as well as layers of historically accumulated artifacts, rules, and patterns of division of labor. This multivoiced and multilayered nature of activity systems is both a resource for collective achievement and a source of compartmentalization and conflict. A conceptual model of the activity system is particularly useful when one wants to make sense of systemic factors behind seemingly individual and accidental disturbances, deviations, and innovations occurring in the daily practice of workplaces.

The work activity of schoolteachers is called teaching. The activity of school students may be called "schoolgoing."

The essential peculiarity of school-going as the activity of pupils is the strange "reversal" of object and instrument. In societal practice text (including the text

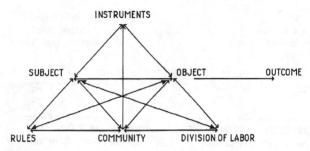

Figure 3.1. The mediational structure of an activity system (from Engeström, 1987: 78).

of arithmetic algorithms) appears as a general secondary instrument. In school-going, text takes the role of the object. This object is molded by the pupils in a curious manner: the outcome of their activity is above all the same text reproduced and modified orally or in written form. . . . In other words, text becomes a closed world, a dead object cut off from its living context. (Engeström 1987: 101)

The text itself as object carries an internal contradiction:

First of all, it is a dead object to be reproduced for the purpose of gaining grades or other "success markers" which cumulatively determine the future value of the pupil himself in the labor market. On the other hand, text tendentially also appears as a living instrument of mastering one's own relation to society outside the school. . . . As the object of the activity is also its true motive, the inherently dual nature of the motive of school-going is now visible. (Engeström 1987: 102)

In Figure 3.2, the traditional structures of teaching and schoolgoing are depicted as a pair of activity systems. Like any complex activity, schoolwork resembles an iceberg. Goal-oriented, publicly scripted in-strumental actions are the easily discernible tip of the iceberg; the deep social structure of the activity is underneath the surface but provides stability and inertia for the system. Accordingly, the topmost sub-triangles (subject–instruments–object) of Figure 3.2 represent the vis-ible instrumental actions of teachers and students. The "hidden curricu-lum" is largely located in the bottom parts of the diagram: in the nature of the rules, the community, and the division of labor of the activity. The activity structure depicted in Figure 3.2 is widely considered alienating for both students and teachers. It leads to encapsulation of school learn-ing from experience and cognition outside the school (Engeström 1991). But the traditional structure is robust. Activity theory suggests

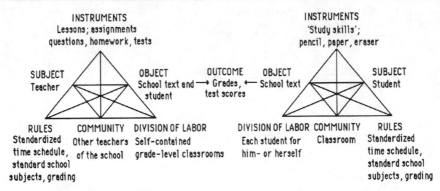

Figure 3.2. Traditional teaching and schoolgoing as interconnected activity systems.

that significant and sustainable change in the nature of schooling may not be attainable by means of manipulating any single component or isolated group of components of the activity systems. Change requires construction of a new object and cultivation of corresponding new motives. This in turn is attainable only by transforming all the components of the activity systems in concert, including and strategically emphasizing the bottom part of the iceberg.

What could replace the text as object of schoolwork? And how could such a transformation take place in practice? I will address these questions by examining the process of planning a curriculum unit in the Global Education teacher team.

The Global Education team

The teacher team studied by my research group was a vertically integrated one, consisting of five teachers responsible for five classrooms covering grades K through 6. The team was initiated and formed by the teachers themselves, with backing from the principal and a group of parents. The team abandoned regular textbooks. It aimed at creating a new curriculum, integrated around the idea of Global Education. This was a team with a mission and a lot of teacher agency, as Paris (1993) would call it.

The Global Education team was preceded by some 10 years of an Alternative Education program in the same school. The two teachers of the alternative program were instrumental in launching the Global Education program in fall 1992, after nearly a year of discussion and prepa-

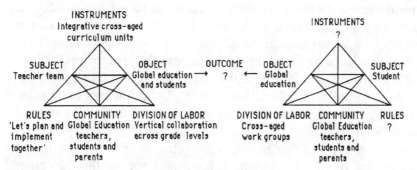

Figure 3.3. Structure of teaching and school going activities as proposed by the Global Education teacher team.

ration. The idea of a team was an integral part of the new program from the beginning. The Global Education program occupied a hexagonal building of its own on the school campus. We began to observe and record the team meetings in October 1992.

On the basis of interviews we conducted with the teachers of the team, the intended structure of Global Education activity systems may be sketched as follows (Figure 3.3). I emphasize the word "intended"; Figure 3.3 is not yet a picture based on observations of the actual practice of the team. The teachers wanted to create a new kind of educational environment and experience, called Global Education. In the United States, Global Education has become a broad movement in recent years, and there is a rapidly growing literature on it (e.g., Kobus 1983; Tye 1990; Tye and Tye 1992). The prominence of the movement is reflected in the fact that the foreword to the recent book by Barbara and Kenneth Tye (1992) was written by Bill Clinton, then the governor of Arkansas.

In spite of this prominence and literature, the teachers we studied did not want to adopt or imitate any of the available programs. In their interviews, they insisted on the local hands-on creation of their own program. They made a point of not having arrived at the idea through literature or courses but through their own varied practical experiences and personal beliefs. They shared a belief in global responsibility and multiculturalism. But they did not offer prepared definitions of what they were trying to do.

In activity-theoretical terms, the object of the team's activity was deliberately fuzzy and emergent. The term "Global Education" was an emerging notion for the team, almost a convenient gloss or proxy for something that the team members could not quite put into words in the interviews we conducted with them. It is my interpretation that the teachers wanted to make the *complexity and increasing boundary crossing* evident in the surrounding world the new object of their schoolwork (on complexity, see Perrow 1986; on boundary crossing, see Engeström, Engeström, and Kärkkäinen 1995). Instead of only telling about the complexity and boundary crossing, they wanted to organize the work itself to correspond to and reflect them. They seemed to base their work on the premise that the object of schoolwork is necessarily constructed by living through the total experience of the educational process.

The team sought to cultivate the new object by making visible and redefining, if only temporarily, certain key aspects of the conventional "hidden curriculum." I will focus on the team's efforts in the following aspects:

- To make visible and redefine the given division of labor for teachers and students by breaking down or crossing the traditional boundaries between self-contained grade-level classrooms; this effort was aimed at creating new ways of *grouping* the students and new ways of bounding and using school *space*.
- To make visible and redefine the given rules for teachers and students by breaking down the traditional standardized *time schedule* and by crossing the traditional boundaries between *school subjects*.

These redefinitions were in themselves deliberate measures to invite complexity and unpredictability into schoolwork. The team seemed determined to thrive on chaos. It constructed the curriculum units as complex experiences of crossing multiple boundaries. Interestingly enough, the very planning of those curriculum units seemed to follow the same logic. The collaborative planning discussions in the meetings of the teacher team were characterized by features that deviated from the standards of efficient goal-directed planning.

Features of talk in the planning meetings

In the Global Education team, instruction was loosely structured around integrated topics or curriculum units which included shorter intensive sequences of teaching in mixed-age groups, called "cross-aging" by the teachers. In the summer of 1992, the team drew up a preliminary one-page outline of the topics and activities for the coming school year. The more detailed content and forms of work in each topic were planned in weekly team meetings as the instruction went on.

In November 1992, the integrated topic was "Harvest Celebration." The team began to discuss and plan the details of the curriculum unit on October 6. They discussed the topic in five successive meetings (October 6, 13, 15, and 20 and November 5). In November, teachers began to handle issues related to the topic in their own classes. On November 18 and 19, they implemented the intensive core of their plan in cross-aged groups, each teacher working with students from all grade levels from K through 6. On November 23, the team had a sixth planning meeting. On November 24, cross-aged instruction continued. On November 25, all 150 students gathered in a joint celebration where they presented their work and ate various foods they had prepared. On December 8, the team discussed the topic one more time in its meeting.

All in all, the planning, implementation, and evaluation of the curriculum unit lasted 2 months and included seven team meetings. Such a relatively complex and multifaceted topic was the organizing unit of work of the Global Education team. Such a unit is a cycle or a gradually unfolding web of collaborative problem solving, planning, execution, and assessment.

We videotaped and audiotaped all seven planning meetings of the team. We also videotaped the events in and around the classrooms during the 4 days of cross-aged instruction. In addition, we interviewed the teachers and the principal of the school. In the remainder of this chapter, I use only transcripts of discourse in the planning meetings as data.

A preliminary analysis of the transcripts (Buchwald 1993; Engeström 1994) revealed certain interesting features of the way the teachers talked in their meetings. The first of these features was *a lack of pauses and conventional turn-taking*. There was a tremendous degree of overlap and immediate response in the teachers' utterances. The teachers frequently spoke simultaneously, expressed affirmation or disaffirmation during one another's speech, or broke in to respond or continue the thought in their own words. Table 3.1, based on a large sample of discourse from three meetings, indicates the frequency of the different types of turn exchange in the teachers' talk.

During these planning sessions, fewer than half of all turns at talk (40%) were followed by a pause in the discourse. This means that conversational breaks did not occur after an individual utterance or pair of utterances so much as between clusters of statements, or flurries of talk, from multiple participants. A turn at talk in this team may thus be characterized as a collaborative achievement more than an individual chance to speak.

Table 3.1. *Frequency of types of speaker exchange (Buchwald, 1993: 10)*

Exchange occurrence		
Statement followed by a pause	245	40%
Simultaneous speech	218	35%
Interrupted speech	40	6%
Speech w/o preceding pause	116	19%
Total	619	100%

The second feature in the teachers' planning talk was a *lack of imperative mood and prevalence of conditional statements*. There are three moods in the English language: the imperative, the subjunctive (or conditional), and the indicative. Interestingly enough, Angelika Wagner (1987: 165) in her study of "knots" in teachers' thinking identifies only the imperative and the indicative moods; she does not even consider the role of conditionals in teachers' dilemmatic thinking. In fact, Wagner reduces the "knots" in thinking to the imperative mood alone.

A "knot" arises in consciousness if a discrepancy is detected between "what is" and "what *must* be," with consciousness reacting to this with the self-imposed injuncton "this discrepancy *must* not exist." Because the discrepancy is already there, as part of consciousness itself, thinking does indeed go around in circles without finding an exit. . . . Hence, tension arises, with thinking quite futilely attempting to solve a problem it has created itself by continuing to imperate itself that "This *must* not be!" (Wagner 1987: 168; italics in the original)

According to Wagner's findings, such self-imperated "knots" are tremendously common in teachers' thinking, as expressed in in-depth interviews and stimulated recall protocols. While interesting and imaginative, Wagner's work displays an individualist and Cartesian bias. The "knots" were identified solely on the basis of monological responses from individual teachers. No data were collected from collaborative action and dialogue among teachers.

Findings from the analysis of our discourse data are practically opposite those of Wagner's (Table 3.2). There was a practically total absence of imperatives in the talk of the teacher team. And there was a striking frequency of conditionals. Conditional phrases represent more

Table 3.2. *Relative frequencies of moods in the planning meetings*
(Buchwald, 1993: 14)

Verb form occurrence		
Imperative	4	<1%
Conditional	485	43%
Indicative	516	46%
Indicative working as Conditional	119	10%
Conditional working as Indicative	7	<1%
Total	1131	100%

than half of the total verb forms (53%) if one counts phrases in which indicatives act as part of a conditional string. Questions and statements were stated as possibilities. Statements and recommendations were softened by conditional verbs. Most of each meeting was moved forward by conditionals: "perhaps we could . . . ."

The heavy use of conditional strings suggests a shared sense of process which includes the tracing of ideas together rather than acceptance of givens from outside or reliance on yes/no decisions. The coconstructed strings of conditionals are instances of joint attention to and extension of ideas, of teachers imagining together. The team worked through potential choices and their possible outcomes. Ideas did not seem to be owned by the teacher who put them forth – it was not her obligation to defend them. Examining the transcripts after the execution of the curriculum unit, one sees that there are very few parts of the overall plan that can be traced back to a single teacher's initiative.

The third feature of the teachers' planning talk was a prevalence of *circling back and repetition* of issues. An idea or a problem was typically taken up over and over again. However, this was not a case where "thinking does indeed go around in circles without finding an exit," as Wagner put it. The issue did not return in the same form and with the same content. Repeated raising of the issue allowed it to be considered from various angles and by different teachers. Food preservation, for example, was discussed so as to express and weigh various concerns that may be relevant in the shaping of the topic: from conceptualization of the entire unit to division of children into groups and selection of appropriate materials.

The entire process of planning and talking in the team meetings took the shape of a spiral, consisting of various parallel smaller spirals. It was a far cry from models of rational goal-oriented planning that proceeds in a linear order toward a predetermined destination. Rather, the teacher team's planning progressed like a vessel on a giant potter's wheel, emerging gradually as each teacher shapes it and adds to it. The result was robust and sturdy in that the teachers seemed confident and coordinated in the heat of the cross-aged implementation, in spite of having no written plan at all.

These features of the teachers' planning talk — lack of pauses and conventional turn-taking, lack of imperative mood and prevalence of conditional statements, and prevalence of circling back and repetition — suggest that planning sessions themselves were already part and parcel of the emerging new object, characterized by complexity and boundary crossing.

Redefining the division of labor by cross-aged grouping

A crucial issue in the formation of the curriculum unit of Harvest Celebration was how to accomplish the cross-aging of the 150 children involved in the program. This issue was more crucial than any specific contents of the unit. In cross-aging, the teacher was to face children from the age of 6 to the age of 12 in her classroom, charged with providing meaningful tasks that enable the children of various ages to work together. Cross-aging is a leap beyond the confines of the self-contained single-grade classroom, into the realm of unpredictable interactions. This was at the heart of what the Global Education team wanted to achieve.

In the literature on everyday or street mathematics, it has often been pointed out how much more easily mathematical tasks are accomplished when they are embedded in meaningful practices outside the school, as opposed to their difficulty when framed as abstract assignments of schoollike reasoning. The teacher team's discussion on the formation of cross-aged groups is in some sense an opposite example. In abstract terms, dividing 150 students into equally large groups, each consisting of equal numbers of students from all participating grade levels, requires elementary operations. In the concrete and complex context of creating a viable cross-aged curriculum unit, these operations became curiously difficult and awkward.

[GLOBAL EDUCATION TEAM MEETING 2; OCT. 13, 1992]

JW: Could we split up the rooms like, per food type?

BH: Yeah.

LL: Oh, yeah. We could do that. There's five rooms. What if we gave every, like one of the grains to each of the groups.

BH: Classrooms, right.

LL: But then, then you see, then your groups are gigantic. That means you've got thirty kids in a group!

The problem of "gigantic groups" was approached with the possibility of dividing each class into three subgroups with slightly different subactivities.

BH: But even if you were doing one grain, maybe you could come up with, you could have three different groups doing the same grain. And maybe come up with three different recipes. One could make a cereal.

LL: A cereal, uh huh, uh huh.

BH: One could make a pancake or you know a flatbread or something. And the other one could I don't know, (laughs) come up with something.

LL: Make sushi (laughs)

BH: (laughs) Yeah right, come up with something.

This idea evoked an attempt to recollect an earlier experience of dividing the students into small subgroups. This, however, led to a momentary confusion and dead end.

BH: Right. Well, if each of us had a grain and we work it that maybe we had three groups, how many groups would we have to have (inaudible) do you remember, maybe ten?

LL: What groups?

BH: When we did the timeline, we had . . .

LL: Groups of four people, working to . . .

BH: What did we have? We had twenty-some groups?

LL: Well, there's about, let's see, there's five times, there's about a hundred and fifty kids.

BH: A hundred and fifty kids?

LL: More or less.

BH: And we put four kids in a group, or five. We had five kids in a group. That's five kids in a group; that's thirty groups. Oh God!

LL: Okay. That's unrealistic. If you had, if you have, if we have five teachers.

Why did the mathematical operations of dividing 150 students into five cross-aged groups look so awkward in the teachers' conversation? Certainly the teachers were familiar with how many students were in the program; with the fact that 150 students divided among the 5 teachers makes groups of 30; and with the division of 30 into either 5 or 6. In fact, all the operations needed should have been automatic and smooth. However, automatic operations are dependent on the conditions of performing the actions in which they are embedded (Leont'ev 1978: 65–67). When the conditions become very problematic – for instance, when we try to drive a car with a different stick shift from that which we are used to – the automatic operations are blocked and slowed down. For a passing moment, they become deliberate actions again. In the case of the teachers, the elementary arithmetic operations were embedded in planning and problem-solving actions on a novel and problematic terrain of various contingencies and unknown consequences.

The notion of 30 subgroups was declared unrealistic. However, one of the teachers immediately realized that the seemingly large number is not necessarily unrealistic after all. A realistic division was constructed jointly.

BH: Well, wait a minute! Thirty kids in a group. I had six in my room. It wasn't unrealistic. It was okay.
LL: Thirty kids in one group. But I'm talking about, if we had five teachers, let's say we had five teachers participating.
BH: 'Kay.
LL: If we divide the kids subsequently into five groups, which gives us *thirty?*
BH: Right.
LL: And inside the thirty we could divide it into three groups and . . .
BH: Right. Well, even smaller. Ten is too many.
LL: Maybe. You could divide it into six groups . . .
JW: So that one teacher six groups of five. That teacher, 'kay one person in that group, then somebody, you could split that thirty, the class.
?: Right. Yes. Yes, right.

Soon the teachers returned to the issue of dividing the students into groups, to draw a summary of the provisional solution they had achieved.

LL: Listen, if we had, if we had five teachers, and each teacher takes thirty kids' responsibility. And then within that thirty you've got, like, five groups of six, or whatever.
BH: Five groups of six, six groups of five.
LL: And let's say I have the theme wheat. Wheat.
JW: Okay.
LL: Okay. I need to come up with five activities that we can do with wheat. In other words. Or six, or whatever the number is.
BH: Right, right.

However, this solution was still feeble and fragile. It was like a first draft, with little or no mutually constructed concreteness.

Some minutes later, the solution was challenged. The teachers envisioned that giving each student in a classroom basically the same topic would become too uniform and rigid. This led to the second major step in the formation of the solution. The second draft solution was a radical deviation from the first one.

LL: But see, once *again* if I've got groups of six, am I gonna ask each of those six to do a little bit of research to find out what civilization used wheat? *What* am I gonna have these kids do?

BH: Kids do?

LL: That's right.

BH: 'Cause usually we have a different topic for each kid.

LL: Right.

JW: Yea.

BH: And this way it would be the same.

This worry of "sameness" led to a search for an alternative. A different model was gradually worked out.

JW: That's why it almost seems like to me and I don't know how we would do that exactly, but it seems like, like, I don't know. . . .

LL: Well, we might want to do it this way and that is, I still have responsibility for thirty and I still have responsibility for six groups, but I'll have a rice group, a wheat group.

JW: That's what I was thinking.

LL: A whatever group, a whatever group, but *maybe?*

JW: In within those thirty?

LL: Within the thirty.

The alternative model began to gain momentum when the idea of each teacher's doing exactly the same set of different activities emerged as a labor-saving solution.

JW: That would work. That way you're saying, like in your classroom, you'd have all different grades, but in the sa-, in the other classroom, they'd be doing the same thing, all different recipes or whatever?

BH: Yeah, we'd all be doing the same thing but it would be just mixed cross age.

JW: Yeah, that would work.

LL: But I think, I think what we'd be smarter to do is to have all five be exactly the same.

BH: The same, exactly the same.

LL: So I'll have the same stuff going on that you have going on and you have going on.

BH: Right. We'll all do the same thing for rice. We'll all do the same thing for wheat.

LL: But we'll each have different *kids!*
BH: Except for different rooms will have different *kids* doing it.
JW: I see.
BH: 'Cause see then we could (inaudible) amount of work.
LL: But then see we could save ourselves an incredible amount of work.
JW: I see. Right.

The formulation of the alternative model was drafted by three teachers; one teacher had to leave the meeting early and another one was sick. Two days later, the entire team met again. The three teachers presented their alternative model to the two others. The discussion led to the third major step in the formation of the solution. It was a return to the first one, this time with more concreteness.

The alternative model of the previous meeting was first presented with enthusiasm and added detail.

[GLOBAL EDUCATION TEAM MEETING 3; OCT. 15, 1992]

JW: And then like LL would be in charge of just corn, and she would get all the information for *all* of us.
LL: About corn. And then one person would get all the information about wheat. And one person would get all the information about whatever. So wh- *if* we decided to do it cross-age, what we'd do is we'd, we'd fan the kids out – so what, there's five of us, right? – yeah, we'd have each each one of us in the five would have a cross-age section. So I'd have thirty kids in my room, but they wouldn't all be my own kids. Some of yours. And then inside that group, I would have (inaudible) and one about wheat and one about rice and one about corn or whatever it is. And so would you and so would you and so would you.
JW: But one teacher would only have to do all the stuff for corn.
LL: Mmhmm.
JW: And they'd give all their information to LL, to TS, to me, to you, to BH. And then you would maybe do wheat, and you'd give all your information, so we'd each have . . .
LL: It would cut down on the amount of work.

The alternative model was first questioned by TS, one of the teachers who were absent from the previous meeting.

TS: So you're like expert group, you're from experts and you would teach first the information to your own class, and then they disperse then and do it with the other kids?
LL: Well see, we could do it that way, too. And that's just another way to do it. I, I . . .
TS: One teacher does the planning and just hands it and you'd say, "Oh, you're

gonna have five kids in your class doing corn and this is the activities they're doing."

JW: Right. That's what we were thinking of. That way, you could do all the research and gather everything for corn, I could do wheat, and you could do rice, you could do – I don't know, whatever – and then you we'd all give the other information to each of us. So we'd end up with four, like *I* would end up with *your* information on wheat, TS's information on rice, and then . . .

Next, JL, the other teacher who was absent, added a further question. She first received an immediate answer explaining the virtues of the alternative model.

JL: Right. So then why would you need all four, if . . . ?

JW: Because we want to cross-age.

LL: Because inside one group. If we did cross-age, inside my classroom, I would have a group for rice, a group for wheat, a group for corn, a group for whatever the grains are? So I might have four or five groups that *inside* themselves were cross-age. So the oldest kids would be, be able to read the information that I had managed to collect about corn. The oldest kids would become the experts in the group, and then help the other people do the activities about corn.

JW: See what I mean? (quietly)

JL then continued questioning the alternative model somewhat vaguely. Her doubts were now picked up and clearly formulated by JW, one of the teachers who created the alternative model.

JL: I was thinking yeah. If there is like, cross-age groups but like all the kids for who are doing the wheat are in one room (inaudible) like gathering (inaudible).

(BH comes back in. There is some indistinguishable talk.)

We'll just have, just go ahead and go on. I'm not thinking exactly; that's all right.

JW: Well, you're just thinking why, why should we have the kids, why shouldn't we just have all the wheat in one room? That's what you're saying?

JL: That's because the expert is in that room. But . . .

LL, another originator of the alternative model, immediately adopted JL's implicit suggestion.

LL: Yeah, that makes sense. We could do that way, too.

JL: And then still have cross-age groups.

JW: Why are we thinking corn before?

LL: Yeah. So then you'd have, so we'd have a wheat person. And you'd have, you know, one fifth of the kids, that would be wheat?
JW: All . . .
JL: Right.

It was realized by the teachers that in spite of the original intention, the alternative model would not save labor, because of the burden of sharing one's special expertise across all the classrooms. A return to the first model now seemed natural.

LL: That would be fine. Be easier.
JW: Yeah, that would be fine. Yeah.
JL: How does that sound?
TS: And you wouldn't have to share your information with someone else.
JL: You see, I think it would be easier if you're in the one room with the one teacher.
TS: Yeah.

After this, the teachers drew up a list of the five different themes: rice, wheat, corn, hunting, and gathering. They conducted a lottery to decide who had which theme. This allocation of substantive themes represented a major consolidation of the model.

A particularly interesting feature of the third step was the ease with which the originators of the second, alternative model gave up their idea and adopted the implicitly suggested return to elaborating the first model. In fact, the originators were instrumental in formulating a critique of their own model – a critique only very vaguely implied in the utterances of the two other teachers. It seems that in spite of their enthusiasm, the originators of the second model were operating very much in a tentative, conditional mode which kept doors wide open to doubts and further alternatives. The behavior of the originators cannot be explained away on grounds of a desire to please senior colleagues; one of the originators (LL) was herself the most senior teacher of the whole team.

Crossing spatial boundaries

The creation of cross-aged groups necessarily involved a new kind of spatial movement. On the mornings of the cross-aged days, the students arrived in the home classrooms only to leave for the various thematic cross-aged classrooms to which they were assigned. This meant that on those mornings, a flow of 150 students rotated around the hexagonal building of the Global Education program, looking for their assigned rooms. At those moments, the scene was that of a social whirlpool.

The fact that the teachers had abandoned textbooks as the object forced them to go out to find objects and instruments in a pragmatic way, wherever they might be available.

LL: I know but I thought about walking them to my house, and just letting them go through my garden, which is a total mess. But they might be able to find some corn and stuff in there. But I can't imagine what you'd do instead. I mean I just can't imagine that much about gathering.

JW: And then you know what um one thing may be just to assign the the certain kids in your room who will bring. And you can get like pi—on nuts.

Others: Mmhmm. Mmhmm.

JW: And you can go, and you can just buy the stuff.

What emerges is an image of a school as an open base, or a laboratory, to which various kinds of "stuff" is introduced from the outside world for investigation and experimentation.

Redefining the rules by negotiating time

The succession of days was an endless line, grey. They ran past you. . . . When all the days were the same, when they recurred and recurred, and were planned out ten years into the future, why did you feel that time was passing, that it was linear, that your school days were a kind of countdown, that time was a train that you must and ought to be fit enough to hang on to? I think it was because of the insistence on achievement. Otherwise it is impossible to explain. . . . It only seemed as though the same subjects and the same classrooms and the same teachers and the same pupils came around again and again. In reality, the requirement was that you should, with every day, be transformed. Every day you should be better, you should have developed, all the repetition in the life of the school was there only so that, against an unchanging background, you could show that you had improved. (Høeg 1994: 226)

This is the image of standardized, linear time schedules which permeate and dominate every aspect of traditional schoolwork. Time is not negotiable. It is the universal "unchanging background" of linear achievement.

The Global Education teachers' team refused to live by this notion of time. Time was turned into multiple negotiable time spans and rhythms. For one thing, they had to select and prepare for the cross-aged days.

JL: When are we thinking, the sixteenth next Monday, right?

BH: We had we had the sixteenth, seventeenth, and eighteenth. We had it all, right? Three full days?

JW: I had it the sixteenth, seventeenth, eighteenth, nineteenth, twentieth.

BH: You had it all?

JW: I don't know.

JL: I remember that we reserved the spot in case we didn't get it done. I remember now. In case we didn't get it done.

The teachers also considered how much time it would take for each teacher to prepare for the unit, how her preparation would fit with the other work she needed to do, and how that time could be reduced.

BH: I mean 'cause gathering all this material. Usually like we've done this taken *weeks,* I mean a week ahead of time and gotten all the materials. And each group had their packet, so when the kids sat down, you had all they had all the information totally directed.

JL: Well, and we *will.* I mean I know I'm gonna have stuff by the day that we start this but that's a (inaudible).

BH: We're starting it on Monday. And I got report cards, and I'm not I'm not gonna have much (inaudible).

JL: Well, that's. Today's Tuesday. I mean to me that's that's light years away.

TS: (laughing) It's not Tuesday *yet.*

The teachers also had to consider how much time should be budgeted for each subactivity.

LL: Because food preservation takes a lot of time. It'll *take* a month to make raisins.

BH: Right.

LL: Especially if the weather is like what it is.

BH: Right, what it is. And even the apples, I mean you can't do anything.

LL: Oh yeah, you can't do that overnight.

Finally, the teachers planned the presentations and the schedule for the actual harvest festival celebration.

JW: Otherwise, I don't, I don't shorter than ten minutes, they're not really gonna be able to present too much. I mean. Especially if . . .

BH: Well maybe if we do it in *stages.* Instead of having you know having it all in one block, having presentations maybe the first and then we go and do some eating?

?: Have two (inaudible)?

BH: Have two presentations and some eating and come back. I don't know if they, you know my kids sitting for fifteen minutes, even if they're getting up and down.

LL: Well how about if we did *three* presentations and then eat, and then come back for two presentations?

BH: Okay.

Crossing boundaries between curricular subjects

As noted, the teachers divided the harvest festival topic into five sub-topics, each of which was assigned to a teacher by means of a lottery. In other words, the teachers did not divide the work into mathematics, English, history, or other standard subjects in the cross-aged groups. Even the boundaries of the assigned subtopics were open to renego-tiation.

This was manifested when the least experienced teacher, JW, wanted to make sure that it was okay to deviate from the plan for her hunting group by incorporating discussion of animal domestication as well as hunting of animals.

JW: So do you mind if I do animal domestication as like the next thing to hunting? Do you guys mind if I do that?
Others: Yeah. No. We don't care what you do. Everything sounds good to me.
JW: Because that's basically, if we're talking about grains and things like that, they're getting past the hunting. They're going into domestication.
Others: Right. Mmmhmm. Oh yeah. That's great.

Contradiction and argumentation

Thus far, my analysis may look as if transforming the motivational sphere, or the "hidden curriculum," of classroom cultures is simply a question of goodwill, effort, and unanimity. Yet, earlier in this chapter I argued that activity theory sees contradictions as the source of change and development. Indeed, it is time to look into the argumentative and contradictory aspect of the teacher team's planning discourse.

In the fifth planning meeting, the model of the curriculum unit was questioned once more. This time, the teacher, BH, who was assigned corn in the lottery presented her own package for teaching the cross-aged unit and suggested that JW, and eventually the other teachers, might follow her model. BH's package was based on preparation of poster reports in small groups.

[GLOBAL EDUCATION TEAM MEETING 5, NOV. 10, 1992]

BH: Now see, if you, now like I got the thing and you could get in cooperative groups and she could do that with cooperative groups, assign an animal to each group. You could xerox it off and then you would make a poster report, and do some kind of a mural. That's what you do.
JW: Right, but I don't, I wanna do something more, I mean I . . .

BH: More?! Do you want to know how soon we have to have this ready? Plus your report cards ready? Are you nuts! More! (laughs)

JW: I wanna do something more than just a poster though. I mean if everyone else is going to be cooking things and grinding and . . .

JW rejected BH's suggestion and indicated that she wanted to do "something more" than poster reports. This kind of a nonconditional tone of discussion was unusual in the transcripts. BH responded by invoking time constraints and then by making a virtual sales pitch for her package, based on an earlier experience of teaching corn.

BH: No, no, no, no, no. Here's a list of things that we do. Now, this is what we did.

JL: But now this a corn group.

BH: Yeah, but this would be the (inaudible). We had the cooperative groups and each. They were cross-age groups and they got together and they actually like, for instance, corn. They met each other and they made a big corn husk. For corn, a piece of corn. And they *all* put their names on it.

JW: Oh.

BH: So that was the name of the group and that would be over their station the day of the thing, but we didn't have a hundred and fifty kids. So that was how they got to come in groups. They talked about it and they all, they made this big huge piece of corn and they all put their names on it and stuff. And then. Then each group was gonna make um, let's see, they were given, they were gonna cook, they had been given a recipe and they were going to cook something for a corn-based food from a particular culture or area and I think most of them did it with Indians. What'd you use: Mexico? Different places in the world that used corn. And so each group was gonna make a mural to show what the corn in their culture (inaudible). The mural might show how they grew the corn; it might show the marketplace, if they had it on the mural. And they all painted on the mural. And then each group had a poster report that gave the title, and had a map of the area and where the corn was grown and had an illustration of life in that region, how people in that region lived and stuff – and then their recipe was on the chart. And then they cooked the recipe the day before and we ate it. So it was *not* that complicated.

BH's effort was met with little enthusiasm. In fact, JL switched the topic by asking what dates were devoted to the cross-aged instruction. However, BH did not give up. She continued presenting her package and suggesting that the others should adopt a similar structure.

BH: And we could br, we could have like a, you could have a rice *metate* in your room, I got a corn *metate,* and you could have whatever, what are you doing? You're doing gathering. We could do our corn together. But we could have them actually grind the grains, and see what it looks like and then they could maybe take say I'm gonna make a plastic bag and put it on your poster. You could have the grain before it was ground and the grain after it was ground.

?: Mmhm.

At this point, JL confronted BH directly, in effect drawing a line between BH and the other teachers. She initiated a whole different line of discussion.

JL: Do y-? (to BH) I know that you've got this entire packet here of stuff to do with corn. (Turning away, to the others) Do you have, do you know what you're doing for your group yet? I don't know what I'm doing.

?: I have no idea.

JL: Yeah. And I'm wondering, maybe what we oughtta do is . . .

?: Oh, I . . .

JL: What I'm wondering is if, *maybe,* if we just meet together like I know this is probably not gonna be popular, like Thursday afternoon or something. Once we've had a chance to think about what we're going to do.

LL: Well, *I'm* really nervous about having that, I mean I don't wanna have, I need to know, before I plan this out, how many meetings we're gonna have.

JW: Right.

LL: That's what I need to know first.

JL: Right. But I, I mean going much beyond that, I, I can't talk about what I'm gonna do with the kids yet because I don't even know . . .

LL: Oh yeah.

JL: . . . *myself* what I'm gonna do. I have to go find something on *rice.*

Here JW stepped in and suggested that the team focus on the logistics of the actual cross-aged teaching. LL and JL joined her. So did eventually also BH.

LL: Well, if . . .

JW: But can we just like figure out the schedule so that we know how much we . . . ?

LL: Yeah. How much, I need to know if I need to plan two sessions or three sessions.

JW: I know. How big is it gonna be? How how long is it gonna take?

JL: Okay. Yeah.

JW: That's kinda . . .

LL: Oh I think that . . .

JW: I think if we can decide on that, that would be great for me too.

LL: I think if we have cooking be one session, and then two other sessions.

BH: And cooking has to be the day before two other sessions, right?

But the team did not yet turn BH away from her idea of selling her package to the others. BH took it up again, this time to be confronted quite directly by LL.

BH: Like if we had the same basic thing, we're gonna find out you know you're gonna work on get the grain in what region and have its region's usages because corn is used really differently in, there's a place in . . .
LL: But if you can do *corn* that way, but then I can't do use that same corn that way for hunting and gathering.
BH: But then you're not (inaudible) mix it up and do the corn in . . .
LL: I don't know. But I mean, what I mean is, is that *I* have to decide what hunting, what gathering does may be very different from what corn does.

BH tried one more time, now offering her model only for those doing the different grains. JL rejected the idea on grounds of not knowing enough of her theme yet. LL, too, rejected it again, pointing out that each teacher may do different things.

BH: Right. But if you're doing a grain . . . If you're doing a grain, couldn't all the grains just be the same format and make it easier? Then we could go to the library, and we could probably find stuff together. You were doing rice, rice in China and rice in ah Turkey. They might be different. And so they would step just look at that region to s- you know . . .
JL: I guess what I would like to do is I need to find out, I need to get the basic information myself, so I mean I can't, I'm drawing a total blank right now, because I don't have *any* information.
LL: Yeah, because it could it could be okay like if for example like if *her* group wants to prov . . . present a play that just happens to talk about rice in the play, and you do poster reports for yours, and I do . . .

BH now redirected her effort by asking about the logistics of the final presentations at the joint Harvest Festival of all the students. The ensuing discussion revealed that the model of the unit was indeed unclear on this point, especially for BH.

BH: How will you present this on the festival day?
JL: Well, that's what we need to others . . . (several people talking at once)
LL: All you need to do is to have to have some some ground rules. Like for example maybe each group has ten minutes in which to present their findings or their whatever.
BH: And we're gonna go through *how* many groups? I mean like if we have a hundred and fifty kids, that's five . . .
LL: There, there's *five* groups.

BH: I thought we were doing cooperative groups for each classroom!

JL: We are. But groups within have sub . . .

LL: Five groups. Five groups!

JL: So right.

BH: So they're only gonna present to the one class but it's five groups . . .

LL: No, no, no, no. See the whole . . .

BH: Five plus five is *twenty-five* . . .

LL: No it isn't, BH, it isn't, it isn't. My one group, your one group . . .

JL: You're gonna have subgroups within your one group.

LL: So like *corn* is gonna make a ten-minute presentation about corn.

BH: Uh huh. Uh huh.

LL: Wheat is gonna make a ten-minute presentation. Gathering will make a ten-minute presentation.

BH: Okay.

Interestingly enough, the early problem of grouping and figuring out the right numbers reappeared here. Symptomatically, BH claimed, "Five plus five is *twenty-five*," thus demonstrating again how problematic contingencies of an action may block a routine operation, leading to something that looks like an awkward error. What is even more interesting is that dialogue with the other teachers is not hampered by this error; they see right through it. When LL exclaims, "No it isn't, BH, it isn't, it isn't," she is not referring to the operational error but to the problematic action. The point is that not all the 25 cross-aged subgroups will have 10-minute presentations – only the five classrooms will each have 10 minutes, but each teacher can divide those 10 minutes between her subgroups as she finds appropriate. When this problem was collaboratively worked out, the object of the teachers' joint work emerged in a much clearer light.

Details of the schedule were then jointly worked out. For the team, this step resolved the problem posed by BH. There were now clear ground rules for the implementation of the cross-aged teaching sequence and the joint celebration as its culmination.

After a while, BH asked whether all the subgroups within a classroom were going to be doing the same things. This led her one final time to the idea of a package, this time referring to the limited amount of time the teachers would have left to prepare for diverse subactivities for the subgroups. The final rejection came from JL, who put the available preparation time in a different perspective by stating that Monday 1 week ahead is "light years away" for her.

The noncoordination between BH and the other members of the team involved direct confrontation and argumentation by means of negating what the other participant suggested. This argumentative talk seems to reflect a contradiction between principles — saving labor by means of prepackaged curriculum materials, on the one hand, versus creating unique and personally meaningful educational experiences "from below," on the other hand.

The lack of coordination was not just an unfortunate setback for the team. It forced them to focus on a serious gap in the model of the curriculum unit, namely, the missing or unclear "ground rules" for the final presentations on the harvest celebration day. In other words, elaboration and resolution of the noncoordination produced a more advanced model for the joint implementation effort facing the team.

Conclusion

In Figure 3.3 I presented the intended activity structure of the Global Education team. What do the empirical data and the analysis presented tell about that activity structure as it was actually practiced?

Three of the components of the teachers' activity system have been illuminated: the object, the instruments, and the rules. I have argued that behind the notion of Global Education, the emerging object of the teacher team's activity may be characterized as complexity and boundary crossing. These concepts themselves are only preliminary characterizations of the educational experience in the making.

We may now distinguish between two layers of instruments in the teacher team's activity: the "what" instruments (integrative curriculum units) and the "how" instruments, consisting of oral planning primarily with the help of conditional strings and spirallike circling back on issues (on "what" and "how" instruments, see Engeström 1990: 188–189). The "how" instruments were both a strength and a weakness. The use of conditionals and circling back was tremendously robust and powerful in keeping the doors of thought open to alternatives and in providing for complex "imagining together." On the other hand, the exclusive reliance on talk as a medium could make it difficult for the team to create and keep up a collective memory to be used as a resource when similar curriculum units are planned and implemented again in the coming years. The systematic rejection of prepackaged curriculum materials seemed to be a deeply held principle in the team. But without some form of documentation designed as an alternative to prepackaging, the team members could simply become overburdened. In the rules of the activity system, important additional rules were found. Along with the

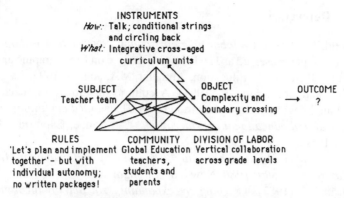

Figure 3.4. Contradictions in the activity system of the Global Education teacher team.

rule of doing things together, there existed the more traditional rule of individual autonomy as well as the rule of excluding written curriculum packages. These became manifest when BH's package was rejected. Packages could indeed easily become stultifying rules that standardize teachers' and students' work. On the other hand, insistence on individual autonomy may easily become an excuse for not engaging in substantive collaborative planning and teaching. Some of that flavor may be sensed in JL's insistence on her own timeline ("the way I function, that's still light years away") and her own individual gaining of mastery over the contents ("I know I'm gonna have stuff by the day we start this"). Such an approach could potentially lead to a form of teamwork in which collaboration is reduced to logistic and technical coordination.

In Figure 3.4, the two lightning-shaped arrows indicate contradictions in the activity system of the teacher team. The emergent object of global education as a new type of educational experience was pulling the teachers into a poorly charted zone of experimentation and innovation. Insistence on talk as the only "how" tool (arrow 1) and on the personal autonomy of the teacher as a rule (arrow 2) seemed to have the potential to restrict and disturb the team's journey through its collective zone of proximal development. At the same time, as shown in the resolution of the noncoordination between BH and the rest of the team, these very same tensions were also powerful sources of self-reflection and innovation. The picture will be significantly more complex when our analyses encompass the activity of the students in the classroom implementation of the curriculum unit.

References

Buchwald, C. (1993). *Collaborative curriculum construction without a textbook*. Paper presented at The Fifth Conference on Ethnographic and Qualitative Research in Education, Amherst, MA, June 5, 1993.

Cole, M., and Engeström, Y. (1993). A cultural–historical approach to distributed cognition. In G. Salomon (Ed.), *Distributed cognitions: Psychological and educational considerations*. Cambridge: Cambridge University Press.

Cuban, L. (1984). *How teachers taught: Constancy and change in American classrooms, 1890–1980*. New York: Longman.

Engeström, Y. (1987). *Learning by expanding: An activity-theoretical approach to developmental research*. Helsinki: Orienta-Konsultit.

Engeström, Y. (1990). *Learning, working and imagining: Twelve studies in activity theory*. Helsinki: Orienta-Konsultit.

Engeström, Y. (1991). *Non scolae sed vitae discimus:* Toward overcoming the encapsulation of school learning. *Learning and Instruction,* 1, 243–259.

Engeström, Y. (1994). Teachers as collaborative thinkers: Activity-theoretical study of an innovative teacher team. In I. Carlgren, G. Handal, and S. Vaage (Eds.), *Teachers' minds and actions: Research on teachers' thinking and practice*. London: The Falmer Press.

Engeström, Y., Engeström, R., and Kärkkäinen, M. (1995). Polycontextuality and boundary crossing in expert cognition: Learning and problem solving in complex work activities. *Learning and Instruction,* 5, 319–336.

Henry, J. (1963). *Culture against man*. New York: Vintage Books.

Høeg, P. (1994). *Borderliners*. New York: Farrar, Straus and Giroux.

Holt, J. (1964). *How children fail*. New York: Pitman.

Kobus, D. K. (1983). The developing field of global education: A review of the literature. *Educational Research Quarterly* 8, 21–28.

Lave, J. (1988). *Cognition in practice: Mind, mathematics and culture in everyday life*. Cambridge: Cambridge University Press.

Lektorsky, V. A. (1984). *Subject, object, cognition*. Moscow: Progress.

Leont'ev, A. N. (1978). *Activity, consciousness, and personality*. Englewood Cliffs, NJ: Prentice-Hall.

Leont'ev, A. N. (1981). *Problems of the development of the mind*. Moscow: Progress.

Nunes, T., Schliemann, A. D., and Carraher, D. W. (1993). *Street mathematics and school mathematics*. Cambridge: Cambridge University Press.

Paris, C. L. (1993). *Teacher agency and curriculum making in classrooms*. New York: Teachers College Press.

Perrow, C. (1986). *Complex organizations: A critical essay*. 3rd edition. New York: Random House.

Sarason, S. (1990). *The predictable failure of educational reform: Can we change course before it's too late?* San Francisco: Jossey-Bass.

Saxe, G. (1991). *Culture and cognitive development: Studies in mathematical understanding.* Hillsdale, NJ: Lawrence Erlbaum.

Snyder, B. R. (1971). *The hidden curriculum.* New York: Knopf.

Tharp, R. G., and Gallimore, R. (1988). *Rousing minds to life: Teaching, learning, and schooling in social context.* Cambridge: Cambridge University Press.

Tye, K. A. (Ed.) (1990). *Global education: From thought to action.* 1991 yearbook of the Association for Supervision and Curriculum Development. Alexandria: ASCD.

Tye, B. B., and Tye, K. A. (1992). *Global education: A study of school change.* Albany: State University of New York Press.

Vygotsky, L. S. (1978). *Mind in society: The development of higher psychological processes.* Cambridge: Harvard University Press.

Wagner, A. (1987). "Knots" in teachers' thinking. In J. Calderhead (Ed.), *Exploring teachers' thinking.* London: Cassell.

Wertsch, J. V. (Ed.) (1981). *The concept of activity in Soviet psychology.* Armonk: M. E. Sharpe.

4 The practice of teaching mathematics: Experimental conditions of change

LISA HEFENDEHL-HEBEKER

The following text is a report on an individual learning process concerning the author's own teaching practice. The experiences give rise to the hope that – within certain limits – mathematics teaching may be improved by increased awareness among teachers. So the underlying basic questions are

- How can I – in teaching mathematics – become aware of the initial richness of the subject, the viewpoint and the aims of my considerations, as well as the applied strategies and tools? and
- How can I reveal them to the learners, taking into account their viewpoints as well as possible?

Classical teacher education – burden or help?

My practical training as a mathematics teacher at a German gymnasium made claims that I even then felt to be hardly implemented. This training would lead the pupils "as smoothly and elegantly as possible . . . along systematically prepared paths to systematically predetermined outcomes" (Andelfinger 1987: 3). My instructor, whom I very much appreciated, signaled this norm by tending to assess the lessons according to their planning – which may or may not have been "right."

This manifested a view that the planning of lessons could be optimized with respect to a systematic implementation. That optimization was

- a neat mixture of anticipation of learning difficulties and control over the learning process by means of a methodically smooth and conclusive presentation of the subject matter as well as a questioning–developing teaching style
- location of the whole responsibility for the progress of the lesson in the hand of the teacher herself.

As in-depth analyses have shown (Andelfinger 1987; Andelfinger, Bekemeier and Jahnke 1983; Andelfinger and Voigt 1984; Fischer and Malle 1985; Maier and Voigt 1991; Voigt 1984), this teaching philosophy is based on the following three presuppositions:

1. *The subject matter:* Historically evolved, the subject-related approach shows strict, formal features. The subject matter is drawn up as a *formal and elegant long-term system.* Extensions of number systems are its main line at the secondary school level.

2. *Psychology of learning:* A view that construes learning as transfer, which "essentially consists in the fact that pupils are made aware, as well-prepared as possible, of topics that have been clarified in advance and that those topics are transferred to the pupils as trouble free as possible" (Fischer and Malle 1985: 78).

3. *Teaching the subject matter:* A *teacher-centered, questioning and developing style* has evolved as a main means of communication in the mathematics classroom. At its heart is a teacher who leads his or her pupils along a systematically prearranged chain of thought. The logical development of the subject matter somehow has to be elicited from the pupils. Suitable didactic reductions of topics, that is to say, simplifications, illustrations, descriptions in everyday language, etc., serve as content-related aids. They come together in so-called didactical models: conclusive, didactically established representations of self-contained topics, for instance, the operator model of fractions or the arrow model of negative numbers.

I found it burdensome to teach with such objectives and I experienced the underlying norms as being hardly realizable, all resulting in a didactic superego which I felt oppressive.

Setting free

Stefan's question

Some years ago I was to introduce a gymnasium class of 13-year-olds to the world of negative numbers (see Hefendehl-Hebeker 1988).

Without hesitation, though with some reservations (which I did not understand until later), I worked out the rules of addition and subtraction of rational numbers by means of illustrating them with geometric vectors proposed by the textbook. The following two rules are known to be fundamental to the method applied:

A. Vector addition ("putting the tail on the head"): Two vectors are added by putting the tail of the second vector on the head of the first. The sum vector points from the tail of the first vector to the head of the second.

B. Vector subtraction ("putting the head on the head"): Two vectors are subtracted by putting the head of the subtrahend on the head of the minuend. The difference vector points from the tail of the first vector to the tail of the second.

On this basis the textbook literature suggests, for example, the following teaching procedure by consulting the algebraic permanence principle:

1. The rules of arrow addition/subtraction will be reviewed, if necessary worked out again, treating them as a geometric correspondence to the currently familiar addition/subtraction of positive numbers.
2. The question of how to include negative numbers in graphic addition/subtraction leads to the decision to extend rule (A) and (B) to all number arrows.
3. Such observations, gained from the geometrical model, will be translated again into the language of arithmetic and condensed as rules about operating with rational numbers.

The philosophy underlying our arrow model of rational numbers appears to be very smooth. Nevertheless, I had reservations, if only because the formal rules of the model are not easy to deal with. Above all subtracting numbers with different signs requires considerable concentration on technique. It happened that a student teacher, who was to explain the model in a seminar, frankly confessed: "First of all I think about the result of a number operation and afterward check if my arrows have been properly placed." Conversely, the model is supposed to be good for pupils!

Notwithstanding, we went about drawing arrow pictures during the lesson. I led the class along a chain of tasks, well-designed – as I thought – with increasing degrees of difficulty. As expected the learning process reached a crucial stage when it came to subtraction. Presenting the calculation $(-3) - (+4) = -7$ required considerable attention, but its result appeared to be plausible at least. The task $(+2) - (-3) = +5$ did not differ from the previous task in terms of technique – and yet, should the result of a subtraction be greater than the minuend? And when it came to $(-5) - (-7) = +2$, my pupils were looking at me in a mixture of amazement and horror. Those three negative signs on the left-hand side must have given the impression that for something to be "more negative" than that was simply impossible! My insistence that what we had done was certainly formally correct failed to weaken the pupils' emotional reservations. At last it was Stefan – a boy clever enough to handle the formal rules that govern the arrow model – who put an end to the confused silence by saying, "I'm sure, Miss, there must be another explanation!"

By making this point Stefan did confirm the failure of my teaching efforts, but at the same time he set me free of the responsibility of forcing on the pupils an appreciation of negative numbers based solely on the arrow model. Stefan's objection became a challenge and impetus to me in many ways.

It entailed some rebuke from within the scientific community as well. When I gave a talk on my experience with Stefan at a conference, an older colleague reminded me, "After all, there are nomograms!" suggesting that they would provide a better model than the arrow one. And perhaps, if they had been used, Stefan's problem would not have occurred at all. But who knows? Objections of the type "if this had been done, that would not have happened" are often of little help in analyzing an actual teaching process. However, my questions and suspicions concerning the arrow model would not leave my mind. I realized that negative numbers could not be taken for granted at all, as it would appear to me in my routine actions.

The question revisited

Studies on epistemology and the history of mathematics (see Freudenthal 1983; Glaeser 1981; Hankel 1867) have brought to light that the evolution of mathematics required more than 300 years not only to handle negative numbers in a calculus-related way (partly with bad conscience) but to penetrate them philosophically.

At the very heart of the problem lay the relation between concrete interpretation and intellectual construction. In what ways should one interpret these new "fictional" numbers allowing such miraculous algorithms for solving equations? As long as mathematicians stuck to absolute magnitudes no really satisfying answer emerged. It was not until the past century that these obstacles could be overcome by a shift of view: the transition from the concrete to the formal. Hankel and his pioneers gave up a fruitless search for compelling, explanatory models, instead extending number fields regardless of content-related foundations like "number of" and "magnitude." This process took place within a framework of formal mathematics with the algebraic permanence principle as its central idea.

Eventually, C. F. Gauss raises an interpretation of negative numbers as relative numbers in which both the new formal-algebraic concept core and a concrete use are expressed: . . . the notion of setting quantities against each other with reference to a comparative mark. (Winter 1989: 142ff.)

From then on there was wide scope for interpretation and application. In particular the new numbers proved reliable within the realm of geometry, as Freudenthal often mentions.

Stefan, in his doubts, was therefore backed by the history of mathematics. He had to clear the same intellectual hurdles that manifested themselves during the lengthy discussion about the nature and interpretation of numbers. He therefore needed not only knowledge of the correct handling of these numbers but, more generally, enlightenment as to different points of view that can be taken toward numbers. Surely he also needed suitable help to overcome the didactic circle between conceptual core and means of representation (accomplished here through arrows).

Having had a critical look at Stefan's objection, I have at least gained an extended appreciation of negative numbers and the didactic problems involved – according to the principle "I, too, wish to study learning processes in order to gain a better understanding of mathematics" (Freudenthal 1974: 124). Seen in this way, I have learned a lot through Stefan's impulse. Stefan proved to be my colleague, and not the object of my endeavors.

Reappraising my teaching practice through interpretative studies

At the time I began thinking about the problems mentioned, the results of interpretative studies of classroom interactions (see Maier and Voigt 1991) made me increasingly sensitive to *Scheingespräch*[1] brought about by the routines inherent to our questioning–developing teaching style. A sequence of interpretative studies of my own lessons brought me both pain and relief – pain because it made me aware that I, too, clung to these routines; relief because I could get rid of the illusion that it would be possible to organize the learning of mathematics according to the linear structure that underlies concluding presentations of mathematically finished results.

Since then, I have been thinking about how to effect change, how to gain a new teaching culture in which authentic viewpoints are the basis of our dialogue and enlightenment takes the place of *Scheingespräch*. Such a changed culture is not achieved in one deliberate push; traditional behavior patterns are simply too firmly internalized.

A new, more released attempt

During my next teaching unit, I partly succeeded in ridding myself of "the method." I shall set out to explain in the following what I mean by this (see Hefendehl-Hebeker 1990).

1. I succeeded in maintaining my own right to exist – against my acquired didactic superego. This did not mean that I became careless. I kept on thoroughly preparing the overall program and each single lesson.
2. Having engaged in the subject matter and its historical evolution for a long time, I gained a vivid and close relationship with it. I could therefore speak about it in a spontaneous and authentic language.
3. Consequently, I made use of my right to be honest when holding classes. Now and then I dared to tell my pupils "what's what" instead of applying methodical tricks that did not convince me but led to the desired result through superficial working processes.
4. Accordingly, I granted my pupils the right to hold doubts and reservations as to the subject matter. I believed they were capable of being experts in their own right.
5. In doing so I represented an image of mathematics as a science evolving out of social interactions.

The following short dialogue may indicate this new style:

Teacher: Plus three plus minus three gives? What do you think?
Tim: . . . could it be that it's simply plus three minus three? I mean, minus three has got to be subtracted from three, so it comes to zero.
Teacher: This is *one* argument. Are there any others? – Nico!
Nico: That it comes to plus three! 'Cause if you do plus three plus zero, you get plus three as well, and minus three after all is much less than zero.
Teacher: Now, our task is in fact calculated as suggested by Tim. Why? That's what we've still got to understand. But Nico's point of view is justified too, because it says that it's impossible to add something less than zero. When negative numbers were discovered man was indeed thinking about such things.

Against this backdrop I want to discuss how we can get on the road to a changed culture in school and teacher education. These thoughts relate to how we deal with mathematics, pupils and student teachers.

Dealing with mathematics

There are myths about the relationship between proficiency in mathematics and teaching skills, myths that serve the purpose of obscuring rather than enlightening – as is the case in all fundamental aspects of life. The common assertion "Whoever has understood the way mathematics works is able to teach it well" is not differentiated enough to be helpful (though being evoked for condemning didactics). Certainly, there is no doubt that a good knowledge of mathematics is necessary for skillful teaching. But it is well known, on the other hand, that not everyone who is an expert in mathematics can give sensitive lessons. It may actually be worthwhile if we state the hypothesis in this way: "How are we to understand mathematics so as to teach it effectively?" in short, "What is a didactically sensitive appreciation of mathematics?" To answer this question I want to address some aspects of the following expositions, giving some instances which are drawn from my own practice of teaching. These aspects are interrelated and are divided only for the sake of clarity.

The abundant aspects of a mathematics theme

It was H. Freudenthal (1983) who revealed that mathematical ideas, concepts, and structures serve the purpose of describing phenomena of our physical, social, and intellectual world. An active appreciation of mathematics involves grasping these concepts, ideas, and structures in all their various relationships to the underlying phenomena. Freudenthal has given many interesting examples of the phenomenological variety of mathematics, which I want to call "wealth of aspects."

Feeling confident with the wealth of aspects in the realm of mathematics and moving flexibly between those aspects is not only a matter of thorough knowledge. Such ability relates to the psychology of learning and so to the practice of teaching. Two examples may illustrate this point:

Example 1: Dilatation and the rule of 3. A student teacher, giving a lesson to 15-year-olds, wants to apply homothetic transformation to a task that relates to land surveying (see Figure 4.1). The task is to determine the width of a lake. A pupil immediately suggests, "That's about the rule of three!" The teacher replies tentatively to this right intuition. Eventually, she herself gives the desired cue, "homothetic transformation," adding Figure 4.2 and describing it with the corresponding symbols of transformation geometry. When she reaches the final result $x = 40$ meters,

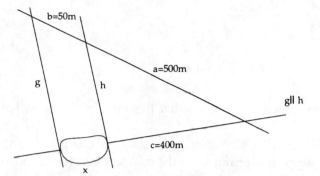

Figure 4.1.

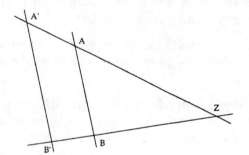

Figure 4.2.

$$A \xrightarrow{S(Z;m)} A' \qquad \overline{ZA'} = m\overline{ZA}$$
$$B \xrightarrow{S(Z;m)} B' \qquad \overline{ZB'} = m\overline{ZB}$$
$$\overline{A'B'} = m\overline{AB}$$

$$\boxed{\frac{\overline{ZA'}}{\overline{ZA}} = m = \frac{\overline{ZB'}}{\overline{ZB}} = \frac{\overline{A'B'}}{\overline{AB}}}$$

$$\frac{\overline{ZA} + \overline{AA'}}{\overline{ZA}} = \frac{\overline{ZB} + \overline{BB'}}{\overline{ZB}}$$

$$\frac{\overline{ZA}}{\overline{ZA}} + \frac{\overline{AA'}}{\overline{ZA}} = \frac{\overline{ZB}}{\overline{ZB}} + \frac{\overline{BB'}}{\overline{ZB}}$$

$$\boxed{\dfrac{\overline{AA'}}{\overline{ZA}} = \dfrac{\overline{BB'}}{\overline{ZB}}}$$

$$\overline{BB'} = \dfrac{\overline{AA'}}{\overline{ZA}} \cdot \overline{ZB} = \dfrac{50}{500} \cdot 400 = 40$$

the pupil concerned reacts somewhat frustratedly, "That's what I said before!"

In fact, the teacher's calculation is sluggish compared to the simplicity and suggestive character of the numbers involved. And after the pupil's remark she apparently could not see the wood for the trees. She failed to pick up the fundamental relationship between the subject of proportions and the subject of geometric similarity and to make use of it during the lesson. As a result she was unable to build on the pupil's viewpoint. An alternative procedure could have been:

1. The teacher picks up the pupil's intuitive suggestion and calculates x according to the "rule of 3" – presumably with a simple proportion:

   $x \div c = b \div a.$

   (After all, the teacher's approach amounts to precisely this.)
2. Following up on that, she could try to justify exactly her procedure, using, for example, elements of transformation geometry.

Example 2: The concept of vector. The concept of a geometric vector has many facets: vectors can be thought of as points in space, as triple coordinates, as pointers, or as classes of arrows (with equality of length and direction as an equivalence relation). One has to be able to mediate flexibly among these notions in order to grasp geometrical phenomena by means of vectors. This is shown by a task I like to deal with in exercises concerning the didactics of linear algebra.

1. Draw the following set of points into a coordinate system with three dimensions: $W = \{(x_1, x_2, x_3) \mid -1 \leq x_1, x_2, x_3 \leq 1\}$.
2. Assign the corresponding coordinates to each vertex of the cube W.
3. Determine the point–direction equation for its diagonals.

The solution (see Figure 4.3) shows how the different aspects of a vector mesh together.

- Fundamental to the edge model of the cube are vectors seen as arrows (edges) which connect points (vertices).
- The coordinates of the vertices (vector as an ordered triple of numbers) are most easily found via coordinate paths (chain of arrows).

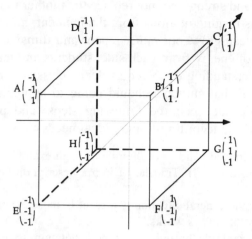

Figure 4.3.

- The point–direction equation involves the concepts "position vector" and "direction vector."

Steps in building a theory and the corresponding language

When holding a seminar for future primary school teachers, I was dealing with the equivalence class *modulo* 9. The common rule in elementary form reads, "Any positive integer has the same equivalence class *modulo* 9 as the sum of its digits." Exemplary proof was given with an abacus. Afterward I wrote the symbolic notation of the rule on the blackboard:

$$(*) \qquad \sum_{k=0}^{n} a_k 10^k \equiv \left(\sum_{k=0}^{n} a_k \right) \bmod 9 \quad (0 \leq a_k \leq 9)$$

and reminded my students that they would have heard about this notation during their lectures on elementary number theory. After this remark some students started laughing. They reacted to my puzzled inquiry with "We've never understood that this is the meaning of 'congruent modulo 9.' Somehow or other we've tried to make sense of theorems and proofs only in a formal way."

This is a typical example of a situation in which mathematics is done only formally and technically. As a result coherence between purpose and meaning no longer exists. As far as the language involved is concerned the principle that, according to M. Wagenschein, technical terminology should extend and strengthen, not replace the mother tongue, is violated. Otherwise, as is quite obvious from the incident, no real and rich mathematical life can grow, but rather a poor and flimsy one that demands a great deal of energy from each side: students and teachers. As far as a sound understanding of the equivalence class *modulo* 9 is concerned, it is helpful that students should come to appreciate the intellectual relationships between the following steps (and perhaps some other steps between, according to individual needs):

1. Giving reasons for the validity of the rule through examples on the abacus (concrete material). Translating into a description based on numbers.
2. Reasoning why the general situation is present in those examples of (1), that is, why "it always works."
3. Grasping the general rule through the symbolic notation, see equation (*), which is variable in respect both to the number of digits and how the digits are placed. Furthermore, using the summation sign shows mathematical economy and uniformity.

Such careful refinement of mathematical tools has to be done in the same way in geometry. It can be easily seen just by doing, for instance, that a parallelogram having the same base and altitude as a rectangle is also of the same area. Then you rebuild the parallelogram to get a rectangle by cutting off a triangle at one side and attaching it to the other so that "it fits." In the language of congruence geometry "cutting off" and "attaching to" are tantamount to "parallel shifting," "fitting" becomes "being congruent" and "rebuilding" means "cut into the same shape."

Shift of view

The evolution of mathematics took place in connection with various shifts of view and the fundamental shifts proved to be great advances. Take Galois theory as a case in point: This theory transforms the question of whether a polynomial equation can be solved by radicals into a question of what the corresponding Galois group is like. Thus, a problem previously viewed as being of an infinite nature (if we assume that the coefficients of the polynomial are elements of an infinite field) is now considered to be a finite one.

School mathematics requires such changes in perspective too. Even if their impact is not as far-reaching, they come along with intellectual hurdles that take some time to clear. As we have already seen, the introduction of negative numbers entails shifting the "absolute" point of view to a "relative" one.

Those who teach should be aware of this transformation. They should be able to talk about it. As steps of development of thought and shifts of view manifest themselves in the history of mathematics, studying them in terms of their didactic aspects is interesting, instructive, and enlightening.

Mathematical strategies

If we want to make mathematics more transparent, we need in particular the ability to talk at a metalevel about our strategies. The example I am about to give is a standard task for quadratic functions and it aims to show how many different general mathematical strategies and specific algebraic techniques are necessary to solve only this task. The task is:

Given a set of parabolas in the equation

$$y = p(b) = x^2 - (2b + 2)x + b^2,$$

(1) Change the form of this parabola function such that it clearly shows where its vertex lies.

(2) Determine the coordinates of the vertex S in terms of b.

(3) Calculate the coordinates of S for $b \in \{-3, -1, 1, 3\}$ and sketch the corresponding parabolas in a coordinate system.

(4) Determine the equation of the graph that lies on the set of all vertices.

(5) Which parabolas go through the point? $A = (2, 0)$?

A solution to this task involves the following strategies:

(1) We can obtain the equation $p(b) = [x - (b + 1)]^2 - 2b - 1$ by completing the square, thus *restructuring* the original equation.

(2) *Reading information* gives us the coordinates of the vertex: $S = (b + 1, -2b - 1)$.

(3) The coordinates of S for specific values of b can be obtained by calculating the values of the corresponding algebraic expressions (*specializing* through substitution) (see Figure 4.4).

(4) The graph that lies on the set of all vertices can be expressed as

$$T = \{(x, y) \mid x = b + 1 \wedge y = -2b - 1\}.$$

The equation $y = -2x + 1$ of the graph can be obtained by solving $x = b + 1$ for b (*restructuring*) and substituting the expression $x - 1$ for b in $y = -2b - 1$ (*substituting* for *elimination*).

(5) The last step begins with *specializing*: 2 is substituted for x and 0

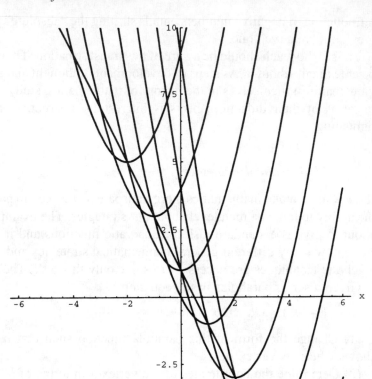

Figure 4.4. Plot of $x^2 - (2b + 2)x + b^2$ with parameter b ranging from -3 to 3 and increasing by 1.

is substituted for y, thus yielding the equation $b^2 - 4b = 0$, which is to be solved for $b(b = 0 \vee b = 4)$. This is accompanied by a *shift of view:* parameter b becomes the main variable. Finally, the specific values of b are being substituted again:

$$p(0) = x^2 - 2x$$
$$p(4) = x^2 - 10x + 16$$

The interface between mathematical theory and reality

Applying mathematics to real situations is a particularly important and comprehensive activity – usually done in school by working through tasks of practical arithmetic.

A class of 11-year-olds are posed the following problem:

"Felix wants to edge his patch with paving stones. The patch is 1.40 meters wide and 1.75 meters long."

The teacher has brought various tiles from home and is showing them to the class. Afterward she raises the question about which Felix has to think. The first reaction of the pupils is "Felix's gonna look which tile is the most beautiful one." The teacher rejects this suggestion kindly but determinedly: "Mathematics and colors, well, it's a bit of a problem. . . . Now, let's assume that the color isn't of such terrible importance." Her goal is the common divisor of 140 and 175, which gives her a suitable width for the tiles.

Pupils often learn only indirectly which aspects of reality can be described with mathematical means and which aspects should be ignored in doing so. However, it is important that they should consciously experience the power and limits of mathematics by comparing the subject to others.

In the same class a teacher student is dealing with another "garden task":

"A garden of 16.8 meters length and 9.6 meters width is supposed to be fenced in. The poles along the perimeter are supposed to be the same distance from one another. How long a distance are we allowed to choose to have as few poles as possible?"

The teacher's intended solution strategy amounts to the greatest common divisor (gcd) of 168 and 96. The pupils come up with the gcd by guessing: "We can't take the least common multiple 'cause we've only got the perimeter. It must be the gcd." The pupils' answer reflects how the mathematical techniques of the current topic and the given data of the task are squared by means of a plausible consideration. However, the tentative question of another pupil as to the width of the poles is dropped completely. But the width of the poles is likely to be essential for every pupil whose thinking is unspoiled. They need an experience of their own – be it drawing or practical acting – to decide that the width of the poles can be ignored in the very situation that is under consideration.

Treating the reality with mathematical tools amounts to idealization.

We should be able to talk about this process of idealizing and make pupils aware of it.

Our own relationship with mathematics

It is a platitude that you cannot excite anyone if you are not excited yourself. So it is difficult to excite pupils with a mathematical topic which does not seem to oneself to be worth talking about. During a teaching unit on quadratic functions a teacher student was dealing with the following practical task:

The Danish colors, the Danebrog, show a white cross against a red background. What is the width of the wide stripes on a 3 m times 2 m flag if the cross is supposed to cover one third of it? (See Figure 4.5.)

The teacher worked through the task quickly and uninterestedly, applying a preparated solution method. Afterward when our group was talking about his lesson he admitted that he himself had not seen anything exciting in the task. He lacked in particular the necessary key questions to integrate the task into the overall topic. For example:

- Why does this task emerge in connection with quadratic equations?
- What indicates that a variable width of stripes, expressed symbolically as x, will eventually amount to a quadratic equation?
- There are several approaches to express the area of the cross in terms of x (we worked out six during the accompanying seminar). All of these, however, yield the same quadratic equation $x^2 - 5x + 2 = 0$. Is that amazing?
- Why is only one of the solutions useful?

Teachers should be sensitive to these questions and encourage pupils to ask them. They must not just work through a topic but try to make pupils feel that what is being learned is worthwhile and is of interest for all the people involved.

An active learning of mathematics always goes hand in hand with individual interests, questions, and personal taste. I often get the impression that discussions related to the didactics of mathematics try to eliminate the question of personal mathematical taste. I once gave a talk which contained one of the many proofs of Pythagoras's theorem (see Figure 4.6).

I am fascinated by this proof in particular because the very pieces of area that are under consideration flow continuously together, thus rendering the proof a pure geometrical one without resorting to algebra. After the talk two colleagues beseeched me to acknowledge that there

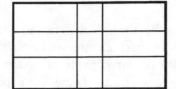

Figure 4.5.

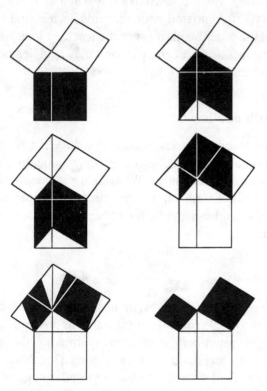

Figure 4.6.

were many "simpler" proofs of Pythagoras's theorem. But the proof applying the quantitatively least effort of reasoning does not necessarily contain the individually most impressive idea.

Thus, dealing authentically with mathematics involves the right to have a mathematical taste of one's own and the duty to respect the taste of others (peers, colleagues).

New burdens?

It has been shown in the first section that the hidden norms of traditional mathematics teaching impose on teachers an almost insoluble task: leading pupils along systematically prepared paths to systematically predetermined outcomes. It is time now to ask whether the expositions of this section amount to a new burden: Does the demand for an immensely differentiated knowledge of the subject matter take the place of the old burdens?

Certainly, this section was supposed to show that a knowledge of the subject matter, when differentiated enough, provides a sound basis for active teaching. On the other hand, it is not the teacher alone who has a responsibility for the learning process; part of it may be delegated to the pupils — as I shall discuss in the next section.

Dealing with pupils

My thoughts in the previous sections amounted to the view that teaching mathematics should deal with the subject matter authentically instead of working it out in ritualized procedures. What such a stance would imply when it comes to dealing with pupils has already become apparent. Once again I want to address explicitly some of the issues which I consider important.

Revealing the viewpoints involved in a talk

I have shown in the second section how, thanks to Stefan's question, my attempt to lead a class of 13-year-olds as smoothly as possible along the arrow model of negative numbers resulted in my thinking about change in teaching styles. The short dialogue there among Tim, Nico, and me showed the first signs of enlightening talk as opposed to suggestive talk.

Guiding principles of such enlightening talk might be the following:

- What point of view underlies our argumentation? What are our notions and our hidden presumptions?
- What is the objective? which aspects are neglected or ignored?
- How does one and the same situation look if our viewpoint changes or if it is described with different means of language?
- What are the advantages and disadvantages of a specific viewpoint?

You are invited to enrich the list. The didactics of mathematics can only gain from communication psychotherapy.

Regarding pupils as experts in their own viewpoints

With his question a self-confident Stefan took a view against negative numbers that was not intimidated by the elegance of the arrow model. He could advocate his natural viewpoint. Having had a good look at Stefan's objection, I learned about negative numbers what I deem to be well worth knowing.

If we take pupils seriously as experts in their own viewpoints, we gain valuable insights. Besides, we rid ourselves of the demands to anticipate all possible learning obstacles. We also come to appreciate the intellectual potential that lies in informal solutions.

Questions suitable for encouraging pupils to believe in their own expertise could be as follows:

- Why do you think your textbook gives this task at this point? What are you supposed to learn in the author's opinion?
- Consider the following list of tasks. Which are simple in your opinion? Which are difficult and why?
- In the following you will get a sequence of tasks together with their solutions. Find out whether the solutions are right. If you find errors, think about their possible cause. Modify the tasks so that the solution suggested becomes right.

Questions of this kind should be continued in the discussion.

Taking individual viewpoints and preferences seriously can have therapeutic effects. I once had a 17-year-old girl in the classroom who had to repeat the year because of poor marks in mathematics. Understandably, she had an aversion to the subject. Together with her I thought about what aspects of the topic (an introduction to analysis) might facilitate her approach.

As it turned out drawing was fun for her and the fact that it was possible to investigate the shape of a curve by analytical considerations astonished the girl. So we agreed to render the aspect of drawing a central idea for her during the course, emphasizing it whenever possible. Our agreement helped her to overcome her emotional barriers.

Dealing with errors in a constructive way

The attitude "everyone is an expert in his own view" requires us to deal with errors in a constructive way. A number of error studies in mathematics classes have shown that the cause of errors often lies in subjective but very plausible strategies of the pupils. Thus, we should refrain from viewing errors as irritating obstacles but regard them as vehicles enabling us to have a better look at the subject matter (see Fischer and Malle 1985). The list of questions in the preceding section also provides some ideas in this respect.

Sharing responsibility for the learning process

All the aspects mentioned so far are supposed to provide incentives for pupils to take on responsibility for the learning process. This sharing of responsibility must prove its worth particularly in phases when exercises are done. The traditional approach has been to grade sequences of exercises in small steps according to difficulty. It may be more fruitful to let pupils take part in the process of identifying difficulties and to think together with them about possible remedies. (See, for example, the fascinating suggestion for teaching the division algorithm in Winter 1989.)

Dealing with student teachers

If student teachers are to internalize the principles discussed here, they must experience them in their own education — it is not enough to acknowledge them only mentally. So the central ideas of both previous sections also apply to the students' professional training and the contact between them and lecturers.

I could give further instances from everyday courses at the university, making my list of examples even longer. But as my personal practice may already have revealed itself I shall skip that in favor of a concluding remark. In my last point I want to restrict myself to an experience I had last summer term. It has shown me how important it is that education in didactics apply to itself the principles it espouses.

I was lecturing on the didactics of geometry and in particular dealing with the subject of similarity. My approach was based on the following structure:

1. Subject-related fundamentals
2. Their treatment in the classroom

The introductory subject-related fundamentals met resistance from some students. Apparently they were not motivated to stock up with facts, and that is what they thought I demanded from them. The students smartly disguised their resistance by hinting to me that they were clever enough to learn those basic facts for themselves.

Though their performance throughout the accompanying exercise course did not always justify this claim, I nevertheless tried to overcome my anger at their lack of readiness to learn. I resorted to the example given in a previous section, where I described how a student got stuck when teaching dilatation, and analyzed how disastrous a lack of flexibility as to the subject matter could prove. In fact I succeeded in getting more attention from my audience after I had related my view to an everyday situation in class.

Education, in terms of both mathematics and its didactics, needs to ground its expositions and analyses in the reality it aims at. Active learning at all levels is necessary and possible.

Note

1. I am lacking a suitable English expression for the German *Scheingespräch*. Basically, it means a fake dialogue between the teacher and the pupil that is not really genuine, somehow superficial — the teacher puts on an act hiding his clear-cut agenda.

References

Andelfinger, B. (1987). Mathematik Sekundarstufe I: Umgebungen und Perspektiven für einen anderen Lehrplan. *Zentralblatt für Didaktik der Mathematik,* 1–14.

Andelfinger, B., Bekemeier, B., and Jahnke, N. (1983). *Zahlbereichserweiterungen als Kernlinie des Lehrplans – Probleme und Alternativen.* Occasional Paper No. 31, IDM Bielefeld.

Andelfinger, B., and Voigt, J. (1984). *Der Alltag des Mathematikunterrichts und die Ausbildung von Referendaren.* Occasional Paper No. 51, IDM Bielefeld.

Fischer, R., and Malle, G. (1985). *Mensch und Mathematik. Eine Einführung in didaktisches Denken und Handeln.* Unter Mitarbeit von H. Bürger. Mannheim: Bibliographisches Institut.

Freudenthal, H. (1974). Sinn und Bedeutung der Didaktik der Mathematik. *Zentralblatt für Didaktik der Mathematik,* 6, 122–124.

Freudenthal, H. (1983). *Didactical phenomenology of mathematical structures.* Dordrecht: Reidel.

Glaeser, G. (1981). Epistémologie des nombres relatifs. *Recherches en Didactique des Mathématiques,* 2, 303–346.

Hankel, H. (1867).*Vorlesungen über die Complexen Zahlen und ihre Funktionen.* Leipzig: Voss.

Hefendehl-Hebeker, L. (1988). ". . . das muß man doch auch noch anders erklären können!" Protokoll über einen didaktischen Lernprozeß. *Der Mathematikunterricht,* 34(2), 4–18.

Hefendehl-Hebeker, L. (1990). Wie "erholsam" darf Mathematikunterricht sein? In Landesinstitut für Schule und Weiterbildung (Ed.): *Die Zukunft des Mathematikunterrichts. Tagungsmaterialien, Beiträge, Diskussionssplitter und Nachgedanken eines Werkstattgesprächs.* Soest: LSW.

Maier, H., and Voigt, J. (Eds.) (1991). *Interpretative Unterrichtsforschung.* Köln: Aulis Verlag Deubner & Co.

Voigt, J. (1984). Der kurztaktige, fragend-entwickelnde Mathematikunterricht. Szenen und Analysen. *Mathematica didactica,* 7, 161–186.

Winter, H. (1989). *Entdeckendes Lernen im Mathematikunterricht.* Wiesbaden: Vieweg.

Part II

Classroom processes

5

Reasoning processes and the quality
of reasoning
ALBRECHT ABELE

Introduction

Language in mathematics instruction

Any study of processes of mathematical learning is confronted repeatedly with the interdependence of student and teacher behavior. Observed student errors will often be explained in terms of difficulties of understanding or of grasp. Teachers will describe this situation by saying, "The student does not understand the content (however often I explain)," whereas the student will complain, "The teacher is unable to explain." In positive terms, students tend to characterize a "good teacher" as one who "is good at explaining." Thus, how language is used as a means of communication is crucial for mathematics instruction, and, hence, for how mathematics is learned. This is why analyzing the communication processes in the mathematics classroom is such an important task for research on the didactics of mathematics. Differences in linguistic competence between teachers and students will lead to different language use, a fact described aptly by introducing the terms "teacher language" and "student language." Studies in this problem field will thus be concerned with the mutual dependencies in the relationship between teacher and student language and pursue, in particular, the question, "How can this student language, and, in particular, student reasoning, be described?"

The dual role of language in mathematics instruction becomes evident when we realize that mathematics is presented in this medium. However, the symbolic character of so-called mathematical language as its most salient feature often obscures its grammatical and semantic structure. Students often tend to understand elements of mathematical formulae as something extralinguistic and thus to treat them falsely.

Technical manipulations, which are frequently perceived as tricks, block activity or thinking oriented toward the *meaning* of signs. The resulting student errors are often not perceived in relation to their cause, and this leads to particular kinds of methodological problems in the handling of formulae in secondary school or all other kinds of schools as well when facts have to be described in the representational form of an equation. This situation results in a second focus of research concerned with verbalizations *within* mathematics, that is, with the question regarding the origins of formalized modes of writing or speech in elementary or secondary school mathematics instruction. An example is obtaining the shape of an equation in calculating $5 + 3 = 8$, or, in algebra, $5 + x = 8$.

When observing and recording student behavior and verbal habits while learning mathematics, it is advantageous to minimize teacher influence. This can be attained by asking the teacher to leave the class after a brief introduction to the problem, and only allowing him or her to return to monitor the closing discussion after the problem-solving process has ended. This method has made it possible to document both astounding as well as commonplace and expected things:

Reactions to camera and other studio equipment. As a rule, such diversions affect students' work attitudes only during the first few minutes.

Regression of social competence. Individual differences in the willingness of students to work in a group are affected by many interdependent factors. These include the class teacher, the classroom atmosphere, teaching methods, and even school level. As a result, a "regression of social competence" can be observed in many schools up to the tenth grade.

Striving for personal achievement. Students like to convince their peers of their own competence in the subject matter. The teacher usually intervenes to guide and help, stabilizing the ranking within the class by open or tacit approval. When the teacher is not present, students will try to obtain recognition in the group by themselves. They basically use two methods: imposing their own solution by convincing, reasoning, justifying, or explaining, that is, by means of *mathematical competence;* or insisting on their standing within the group or class, thus playing their social *competence* card.

This yields two focuses of research: investigating students' *reasoning behavior* in order to draw conclusions on their *ability* to reason and investigating the learning of mathematics as a *social process.*

There is an interplay between mathematical competence and social

Problem
Four students are traveling by bike.
Which of them is the fastest rider.

Uwe	30 km	2 hr
Heike	30 km	1 hr 30 min
Fred	40 km	2 hr
Bernd	40 km	1 hr 30 min

Figure 5.1. Problem (see text).

competence (cf. Abele 1992). An example may explain this interplay (see Figure 5.1). The students are quick to realize that Uwe is the slowest and Bernd the fastest. The debate then turns to who is second, Heike or Fred. The following correct approach proposed by Student S1 – although formulated only incompletely – is not accepted by her peers.

094. S1: Wait a minute . . . 30 km, . . . 30 km; . . . in 2 hours 40 km, in 2 hours. Then you would have needed 30 minutes for 10 km. That is, 30 minutes needed [for 10 km].
098. S4: No, 1 hour and 30[!]
099. S4: I think that this one [points to Fred] . . . no, that one [points to Heike] is third.

The controversy about who is second ends with the expressed consensus "Then Heike must be second." Student S1 has also temporarily abandoned her opposition and is obviously waiting for the teacher to reappear in order to gain support for her own view through the teacher's authority. As soon as the teacher enters the classroom, S1 immediately reopens the dialogue by exclaiming, "You know, actually, both should have been second!" a statement which disagrees markedly with the result tolerated just before. She counters her classmates' objections by reasoning: "But I find, with 1 hour 30 minutes (the time Heike took to ride 30 km), you would need 30 minutes for 10 km. Fred also needed more time for 10 km; half an hour more. Properly speaking, both should be second." The opposite position presented with a questioning look at the teacher by S2, "No, properly speaking . . . it is Heike," also seeks her confirmation. The teacher, however, refuses the usual role of arbiter assigned to her by the students by asking, "Who then?" This prompts another dialogue among the students determined by factual reasoning aimed at ensuring that the previous outcome of the discussion was correct. The teacher's decision not to abuse her own "authority of office" when students are tackling questions that they can answer on their own is bound to lead the students back to factual reasoning. Prejudices and opinions already voiced are checked. Ideas and notions are compared to the given data, various paths of thinking and solving are drawn upon for a justification, until finally even the last of the four students is convinced that the solution presented is correct and states, "I think two are second and nobody is fourth." This correct solution is now no longer conveyed exclusively by social dominance and accepted on this basis, but has been obtained by the students themselves in argument and counterargument within a contest of opinions on grounds of fact, and it becomes acceptable for this very reason.

A Pilot study on reasoning behavior

Vinçon's category system (1990) was used to analyze this study further by comparing speeds.

01 Facts
02 Conclusions
03 Assessments (opinions, norms)
05 Linkages in reasoning
06 Qualifications, relativizations
07 Conditions
08 Suppositions
09 Examples
10 Action proposals
11 Abstractions
12 Rejections
14 Challenges
15 Attacks
16 Defenses
17 Suggested compromises
18 Repetition, variations
19 Prompts, declarations of intentions
20 Questions

During an episode containing about 100 verbalizations, third-grade students created the following distribution of arguments.

The category observed most frequently was "Facts," followed by "Action proposals." This is also in line with cognitive development in 8- to 9-year-olds. Surprisingly, the next most frequent categories were Linkages in reasoning, followed by Conclusions. This distribution is situation-dependent, that is, problem-dependent, but it indicates the need to examine how reasoning behavior changes and develops from first to fourth grade, that is, between the ages of 6 and 10 years.

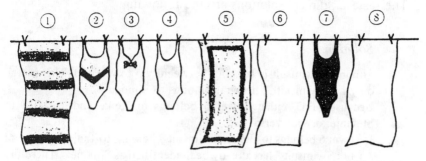

Figure 5.2. Which towels and swimsuits belong to which girls?

Comparative analysis of reasoning behavior in elementary school students

Analysis of facts: Which is the "most elegant" solution?

Problem: Five girls went for a swim in a lake. When they got back home, they hung their swimsuits and towels on the line to dry (see Figure 5.2).

The girls explain:

A. Tanja's swimsuit is next to Heike's towel.
B. Simone's swimsuit is hanging between two towels.
C. Ute's swimsuit is hanging to the left of Stefanie's.
D. Heike's towel is next to Stefanie's towel.

Conclusions have to be drawn according to the following rules:

1. Every swimsuit listed in A–D is assigned to its owner.
2. Each girl can be assigned only one swimsuit and only one towel.
3. We proceed from the simpler to the more difficult; that is, we first assign the proposition with the fewest alternatives.

For the problem-solving process, the four tasks (A, B, C, D) reveal various levels of difficulty. B is the easiest, because there is only *one* swimsuit hanging between two towels. Therefore, Swimsuit 7 belongs to Simone. This is followed by Propositions C and D with two possible allocations, and, finally, A with four. Thus, the students are not only free to choose one solution path among several, but are also faced with the strict *necessity* to decide in favor of one possibility or the other. It is this need to make a decision that is bound to lead to a discussion among the students even when they attempt to solve the problem as a group task without teacher guidance.

The fastest, "direct" solutions are the following:

Solution A

1. Simone's swimsuit is hanging between two towels (B). This is true only for Swimsuit 7; hence it belongs to Simone.
2. Because of D, either Towel 5 belongs to Heike and Towel 6 to Stefanie, or vice versa.
3. If Towel 6 belongs to Heike, Swimsuit 7 belongs to Tanja (because of A), but Swimsuit 7 has already been identified as Simone's. Therefore Towel 5 belongs to Heike and Towel 6 to Stefanie.
4. Then, because of A, Swimsuit 4 belongs to Tanja.
5. C states that Ute's swimsuit is hanging to the left of Stefanie's swimsuit. Swimsuits 2 and 3, which have not yet been allocated, show this constellation; hence Swimsuit 2 is Ute's and Swimsuit 3 is Stefanie's.
6. The students are finally free to choose how to distribute Towels 1 and 8 between Tanja and Ute.

Solution B

1. As in Solution A, Swimsuit 7 belongs to Simone.
2. Because of C, Swimsuit 3 belongs to Stefanie and 2 to Ute, or else 4 belongs to Stefanie and 3 to Ute.
3. If 4 belongs to Stefanie, Swimsuit 2 remains for Tanja. Then Towel 1 is Heike's because of A. In that case, however, D can no longer be fulfilled. Hence, Swimsuit 3 belongs to Stefanie, and 2 to Ute. And so forth

The further steps – as well as other solution paths – will not be given here. A different number of solution steps will lead to the solution given in A; the minimum number of steps needed has been attained in Solution A.

The initial question "Which is the most elegant solution?" is not intended to indicate a goal of learning; rather, it is addressed specifically to teachers willing to go into the problem in more depth and to use it to inquire into the didactic problem of the specific student behavior required. The "fastest" solution will not always be the "most elegant" one as well, and the "direct" path to the result (the teacher's representation) will not always be the one yielding the most extensive learning experience for the students who need to be encouraged and allowed to have their own experiences with mathematics.

The students in the test group were given two cardboard signs for each of the girls in the task. This enabled them to try out several possible solutions even when they did not know at first whether they were applying the right approach. If we want to analyze independent solution processes in students, they must be given the greatest possible latitude for making their own decisions. If the number of cards were limited to those required for the solution, the students would not have to check the last step in their solution and could allocate the remaining card to the remaining swimsuit or towel. With two cards for each of the five girls, they have to make mentally sure of their solution path right up to the last step.

How do students solve the problem?

General description of problem-solving processes. Let us consider possible descriptions of such processes:

Polya's (1949) focus is on the question "How is the solution sought?" He offers a framing plan for this that unfolds into four stages (Polya 1949, inside cover): (a) understanding the problem, (b) developing a plan, (c) executing the plan, (d) looking back.

Descartes (1960) establishes four rules that clearly bear the imprint of his biography and can also be considered as maxims of autonomous learning:

1. The first [rule] is never to accept a thing as true that I do not evidently recognize as true: that is, to carefully avoid haste and prejudices and not to judge about anything that is not so clearly present in my mind that I have no occasion to doubt it.
2. The second, to subdivide every problem I would examine into so many parts as are convenient and necessary to solve it more easily.
3. The third, to think in the appropriate order, that is, to begin with the simplest and most easily perceivable things in order to ascend by degrees, so-to-say in steps, up to recognition of the most composite ones, even to bring things into an order in which one follows upon the other most naturally.
4. The last, to establish lists and overviews that are so complete and so general that I were assured not to have forgotten anything. (Descartes 1960: 15–16, trans. from German trans.)

These four rules seem to be particularly appropriate for analyzing this problem. On the one hand, the concern is the truth content of propositions and conclusions that can be drawn from the given propositions. On the other hand, it is important for problem solving to decompose the complex problem into partial steps and to treat these by progressing from the simple to the difficult. In the present problem, the fourth rule would mean to recheck the solution found on the basis of Propositions A to D.

Biermann, Bussmann, and Niedworok (1977: 9–10) have compared the descriptions of stages in problem-solving processes given by several authors (e.g., Polya, Gagné, Pietrasinski, Duncker, Graumann, Merrifield, Johnson). They found that all approaches represent "problem-solving as a dynamic process" (p. 15). The order of the separate stages is roughly identical. However, they cannot be compared directly because each author separates the individual phases differently. Nonetheless, generally speaking, the problem-solving process can be broken down into the following stages:

1. Preparing for the problem.
2. Confronting the task or the problem.
3. Recognizing the problem.
4. Finding a solution by making a plan.
5. Executing the plan or solving the problem.
6. Checking the solution and correcting it if necessary.

Aspects of student behavior. The problem was presented to groups of four boys and girls in second, third, and fourth grade. All three student groups were able to solve it independently. However, substantial differences in problem-solving behavior were observed.

Analysis of the problem-solving behavior in the groups. Looking at the beginning of the problem-solving process, it is noticeable that the students seek a path of solution *individually.* This is not done jointly or according to rational principles. Although they are looking at the problem, it is too complex, and they do not know where to begin. Hence, each student looks for his or her own individual approach to the task and begins alone. Although the individual student tries to integrate the others by thinking aloud or by reading out propositions to them, the other students are so preoccupied with themselves or with their own thoughts that they do not adopt a common approach at first. An observer has the impression that the students are unwilling or even unable to respond because they do not understand their peers' ideas and approaches. This individualistic behavior changes into cooperation only after all students have established *their own* representation of the problem.

In their first step of problem solving, all three student groups assign Swimsuit 7 to Simone, an assignment in line with the optimal solution path. Obviously, all students recognize the definiteness of Proposition B and use it. Then, however, the solution paths of the individual groups

Table 5.1. *Problem-solving steps in Grade 2*

LS	1	2	3	4	5	6	7	8
1							Simone	
2		Ute					↓	
3		---	Ute				↓	
4		Stefanie	↓				↓	
5		Ute	Stefanie				↓	
6		↓	↓				↓	
7		Heike	Ute	Stefanie			↓	
8		↓	↓	↓	Heike		↓	
9		↓	↓	↓	↓	Stefanie	↓	
10	Ute	↓	↓	↓	↓	↓	↓	
11	Simone	Heike	Ute	Tanja	Heike	Stefanie	Simone	Ute
12	↓	↓	↓	↓	↓	↓	↓	↓
13	↓	↓	↓	↓	↓	Tanja	↓	↓
14	↓	Stefanie	↓	↓	↓	↓	↓	↓
15	↓	↓	↓	↓	↓	Ute	↓	Tanja
16	Simone	Stefanie	Ute	Tanja	Heike	Stefanie	Simone	Tanja
17	Simone	Ute	Stefanie	Tanja	Heike	Stefanie	Simone	Tanja

separate: Grades 2 and 3 require 16 further steps up to the correct solution, whereas grade 4 attains the correct overall solution after 4 more steps. An overview of the solution paths is given in Tables 5.1 to 5.3.

An initial, rather general description of student behavior reveals a situation that can be observed frequently in the classroom, particularly when solving text tasks. Polya's four stages in the problem-solving process (understanding the problem, developing a plan, executing the plan, retrospection) are not recognizable. The students' problem-solving process is structured differently:

Table 5.2. *Problem-solving steps in Grade 3*

LS	1	2	3	4	5	6	7	8
1							Simone	
2					Stefanie		↓	
3					↓	Heike	↓	
4	Heike	Tanja			↓	---	↓	
5	↓	↓		Ute	↓		↓	
6	↓	↓	Stefanie	↓	↓		↓	
7	↓	↓	↓	↓	↓	Heike	↓	
8	↓	↓	↓	↓	↓	---	↓	
9	Heike	Tanja	Stefanie	Ute	Stefanie	Tanja	Simone	Ute
10	Heike	Tanja	Ute	Stefanie	Stefanie	Tanja	Simone	Ute
11	↓	↓Ste-	↓	↓	↓	Heike	↓	↓
12	↓	fanie	↓	Tanja	Heike	Stefanie	↓	↓
13	Tanja	↓	↓	Heike	↓	↓	↓	↓
14	---	---	---	---	↓	↓	↓	↓
15				Tanja	↓	↓	↓	↓
16		Ute	Stefanie	↓	↓	↓	↓	↓
17	Simone	Ute	Stefanie	Tanja	Heike	Stefanie	Simone	Ute

First they try to find out everything that is possible or seems possible (*testing stage*). Suppositions (hypotheses), claims, questions, correct conclusions, which are partly justified and partly not, stand side by side without being sorted into any order. This is the work stage of free, unplanned search for solutions and partial solutions that is so essential for an independent process of student problem solving. It will yield – if the task formulation is open enough – a lot of good ideas and correct solution aspects that, nonetheless, stand side by side in an uncoordinated way.

A certain "perplexity" among the students then produces proposals as to "what should be done now," and this often brief planning stage

Table 5.3. *Problem-solving steps in Grade 4*

LS	1	2	3	4	5	6	7	8
1							Simone	
2			Stefanie				↓	
3		Ute	↓				↓	
4		↓	↓	Tanja	Heike		↓	
5		Ute	Stefanie	Tanja	Heike	Stefanie	Simone	

mostly shifts directly into the *final stage* of the problem-solving process. The partial solutions now known are related and linked to one another – often by a strategy of repeated and scrutinizing reflection – until a proposition accepted as "solution" has been obtained. This typical behavior can be observed particularly well in the third-grade group (see Table 5.2):

1. Unplanned search for a solution: The first proposal for an overall solution is presented to the teacher after the stage of "unplanned experimentation" following nine problem-solving steps (see Table 5.2). Simone's and Stefanie's swimsuits (7 and 3) have been determined correctly. As the teacher is not satisfied with this, the tenth step follows: It contains only one correct partial solution (Simone's swimsuit).

2. Planning stage: As the teacher does not accept this as a solution, but, on the other hand, is only sparing in giving the help the students expect, they try new approaches. Student 4 has the idea that every sign that occurs once in the solution must occur a second time as well. Student 2 thinks that one person's swimsuit and towel may have the same color. Although these ideas are rejected immediately by the other students, they show how the students try to find a structuring principle. Most of the ideas they introduce into the problem-solving processes are presented very convincingly. Qualifications like "I believe," "perhaps," or "maybe" are fairly rare in the third grade and appear mostly when students are unable to support their ideas by one of the Propositions A to D.

This transitory stage ends after the 12th step (see Table 5.2), when Student 3 formulates his plan: "Yes, let's do away with all this now. Now let us read . . . that, this, this we'll do it like this, and that. . . ."

Saying this, he removes the signs that had been placed with the clothing items (1), (2), (3), (4), leaving the names for (5), (6), and (7), because "this is correct, and then that is correct as well, because Simone's hangs between the two." His proposal to read Propositions A to D once more in correct order and thus to draw a tentative balance distinguishing between what has been established and what is still open releases new energies in all students, leading them to complete the overall solution.

The uncertainty about procedure within the groups does not occur only at the beginning; it is evident repeatedly in the ongoing problem-solving process. What is seen is not only a lack of orientation, but also the clear lack of a solution plan, and the two phenomena are interdependent. Polya (1949: 219) writes: "Success in problem-solving depends on selecting the proper aspect, on attacking the fortress from its most accessible side. In order to find out which aspect is the proper one, which side is most accessible, we try out several sides and aspects, we vary the task." This search for the best access has been mentioned as the rule "From the simple to the difficult." The students, however, do not abide by this rule. They do not even accede to the insight – at least this is not obvious to the observer – that the individual propositions given contain various degrees of difficulty if the intention is to derive a solution from them. The various degrees of difficulty arise from the fact that the respective proposition taken alone shows a different number of allocation possibilities. The fewer possible allocations a proposition has, the more easily it can be integrated it into the problem-solving process.

Proposition B is thus easiest to exploit because it has only one possible allocation. The second-grade students were quick to realize this, and they allocated swimsuit Number 7 to Simone in their first problem-solving step. After that, however, the other propositions seem equally confusing to them and they are at a loss regarding how to continue. The fact that there are, in part, several possible allocations is not verbalized by the students. Consequently, they were probably unaware of this. Only Student 2 at one time recognizes two possible allocations. Nonetheless, the students are aware that the further propositions can no longer be assigned unequivocally. This is where cognitive conflict and hence problem solving begins for the students. Lacking a solution plan, the students proceed according to a trial-and-error strategy. Two episodes illustrate this:

First episode (second solving step, cf. Table 5.1)

054.   S2: This is, I believe, Tanja's.

055. S4: Yes, I would say this is Tanja's.
056. S1: Wait a minute, Tanja, – Simone?
057. S2: Nope, Ute, this is Ute's!
058. S1: Simone! We only have . . .
059. S2: Which one is now on the left? That must be Ute's then. Left, yes, that one.
060. S4: Left.
061. S2: Left, nope, left here, this is Ute's, Ute's.
062. S4: Yes, this is Ute's.
063. S2: Ute.
064. S1: Ute, so.

The idea for allocating Tanja's swimsuit seems arbitrary and is not justified either. The same is true for the subsequent shift to Ute's swimsuit. Although Proposition C is also used to justify the second problem-solving step, the solution found is just *one* possibility that the students loosely test as well. At first, the students are not aware that there is a second possible allocation for Ute's swimsuit. As a result, they are initially fully convinced that the second step has been correct because they only see this one possibility.

Second episode: The students have already allocated Swimsuits 2, 3, and 7. Only Swimsuit 4 is still missing (seventh step, see Table 5.1).

107. S3: Hee, then this is where Heike's will be!
108. S4: Nope, Tanja's swimsuit is hanging next to . . .
109. S1: Now only one swimsuit is still missing, Heike, eh, that's easy, Heike
110. S3: Eh, that's what I said.
111. S1: Heike.
112. S2: Heike.

This allocation idea introduced by Student 3 and then taken up by Student 1 is based on none of the propositions and thus cannot be justified at all. It is difficult to ascertain how Student 3 got this idea, particularly because he had not offered any previous ideas and proposals. This makes it impossible to explain what he does through a familiar pattern of reasoning or a recurrent notion. With Student 1, in contrast, the emergence of this idea is easy to follow: She is aware that Stefanie, Ute, and Simone have already been allocated a swimsuit and that one is still free. Therefore, she checks through the remaining cards, and finds Heike's name, and as the latter does not have a swimsuit yet, she is allocated Number 4. In my opinion, however, Student 1 is unaware of the fact that this is *only* one possible allocation within the logic of her own approach, and that there is also another one. This is why she is so convinced that her own solution is correct.

Even though the "trial-and-error" strategy is very strong in the entire problem-solving process, the students do not realize this. When they have found a possible allocation and decide in its favor, they do not recognize that there are other possibilities as well, and that their own choice is only one among several possibilities. Verbalizations like "I believe," "I would say," "should properly," "I think," "let's try that for a change," "could be," or "perhaps" show clearly that the students are not always convinced by their own solution ideas, although they base these partly on the propositions.

Some solution ideas seem very arbitrary, but most of them are understandable to the observer because the students give their reasons for them. As a problem-solving strategy, the students try to combine trial and error with the ability to draw logical conclusions from the propositions. In doing this, they are by no means always uncertain when they offer their solution ideas to the group, but also take a very self-assured and convinced stance. They do this mostly when they have understood and penetrated a proposition to the extent that they are able to find a possible allocation for it which does not contradict the proposition. In this case, they can indeed justify their own idea with the proposition and thus convince their peers as well.

While the students still adhere to a proposition in the first example, their attempted solution in the second example has been taken entirely from thin air. Confirmation or error occurs only later in most cases. One reason for this is that students have to rely on themselves in the problem-solving process and are not pushed toward the solution by the teacher's suggestive questions – a very common practice in mathematics instruction. The students' approaches are thus strongly visual. As long as none of the cards has been allocated, they feel free to make suggestions for subsequent allocations. Then, however, the willingness to give up something found varies considerably among the students. Student 1 is extremely attached to the solutions found, whereas Student 2 is more ready and willing to check solutions and modify them if necessary.

A further salient point in problem solving is that while the students advance their own ideas, they are partly unable to justify them or say why this ought to be like that or how they got the idea. An observer gains the impression that choices are in part very arbitrary. Often, peers do not demand a precise explanation for how the problem-solving approach came into being, but rather assess things according to their own feeling whether the solution could be true and accept it on this basis.

This reveals very interesting modes of thought in the students. To give one example:

210.   S1: Well, let's write down Ute here.
211.   S2: But one of them does not fit here.
212.   S1: Eh, we've got it now, and here comes, eh, Simone.
213.   S4: Simone next to Simone?
214.   S1: Yes, could well be.
215.   S4: Could it be that . . . that you have confused them again?
216.   S1: Yes, precisely. Now we must . . .
217.   S2: Nope, well, ech.
218.   S4: Now it should fit.

The students are looking for a solution for Numbers 1 and 8, because these two towels have been impossible to allocate so far. However, as nothing can be stated about these towels at all, it is also impossible to justify the problem-solving ideas. It is interesting that Student 4 does not agree to the problem-solving idea advanced by Student 1 because Simone's towel would then hang beside her swimsuit. This obviously does not fit the image Student 4 has formed of the solution. This is also why he suggests switching Numbers 1 and 8, to which Student 1 agrees at once because she is in no position to justify her idea logically. Hence, she willingly responds to the discomfort shown by Student 4 and makes a switch between Numbers 1 and 8.

Now where does Student 4's discomfort arise from? This question cannot be solved here. What is clear, however, is that Student 4 has a representation of the task showing certain structures of order. Apparently, there is a rule that no cards bearing the same name may occur side by side, or, at least, that this is very improbable. It may be that Student 4 has derived this idea unconsciously from the solutions obtained so far because these also never show two cards side by side. Interpretation is made even more difficult by the fact that Student 4 says in the closing discussion that they have only "estimated" the two extreme towels. However, it could be that Student 4 became aware of this only at a later stage in the problem-solving process.

Another salient point in the students' problem-solving behavior is that they tackle only one proposition at a time and look for its solution. In doing so, they already have difficulties in recognizing that there are several solutions for this proposition; in most cases, they recognize just one possible allocation and are content with this. The ability to consider several things simultaneously or side by side has not yet evolved, or at least it does not appear when solving this task. In the entire problem-solving process, only one situation can be found in which a female student links two propositions:

164.   S2: Yes. Look here, Tanja's swimsuit is hanging next to Heike's towel, and Heike's towel is next to Stefanie's towel. Well, no . . . well if that's so, then Simone can, should be removed here. This is why Tanja is placed here.

This episode also shows that Student 2 has recognized that there are two possible allocations for Stefanie's and Heike's towels. She tries out these two mentally, additionally drawing on Proposition A. Her conclusion is that if Stefanie's towel is Number 5, and Heike's towel is Number 6, then Tanja's swimsuit must be Number 7. As this place, however, is already occupied by Simone's swimsuit, and as this allocation is fixed by the clarity of Proposition B, Stefanie's towel must be Number 6, and Heike's towel must be Number 5, and, hence, Number 4 must be Tanja's swimsuit.

It is interesting to see the context that prompted this insight. Prior to that, Student 2 has tried to make the other students see that Tanja's swimsuit is still unaccounted for but belongs with the others because of Proposition A. However, as she has been unable to convince all her peers of this fact, she continues to think about how she could make her view clear to the others and justify it, arriving at this conclusion.

Analysis of the problem-solving process raises the question of how far the students master some of the stages of problem-solving named. Let us draw on Biermann's raw subdivision into stages for this. According to this, the students have already completed the first two stages in the teacher's presence: the first stage (preparing for the task) through the story, and the second stage (confronting the problem) through the picture and Propositions A to D. Matters already become more difficult for Stage 3 (recognizing the problem). The problem with the task is that there are various possible allocations for the various propositions. However, the students do not seem to be aware of this. In trying to translate the propositions into problem-solving steps, they take the first possibility they recognize as the only possible one and immediately fix it. They do not see, or do not want to see, that there may be other possibilities as well, as this would make the task even more difficult for them. If the students had recognized the problem more clearly, the observer would have been able to see this in their statements. Only Student 2 suggests twice that she recognizes several possible allocations and thus the ambiguity of the propositions. As the students, however, never take up this point during the entire problem-solving process — at least not verbally — the conclusion is that either it was so evident to them that they did not care to mention it, or that they did not recognize the problem contained in the task after all. Probably the latter is true for the second-grade students. If we take into account that "recognizing" happens within a progressive process while the students grapple with problem solving, we cannot very well say that the second-grade students did not "recognize" the problem at all. Nonetheless, their process of cognition clearly seems to be in the initial stage. Although they experience the various possible allocations, because they are forced to revise their own, this does not guide them to the insight that the ambiguousness of Propositions A, C, and D creates a fundamental problem. Thus, one hasty allocation is replaced by the next hasty allocation, as soon as the error has been realized. The ability to consider the total number of possibilities simultaneously and select the correct one among these (by means of additional propositions) does not become visible in the problem-solving process.

When examining the fourth stage (finding a solution by establishing a plan), the question arises what can be understood as establishing a plan in the context of this task. As the students cannot fall back on familiar, similar tasks to find a solution (as suggested by Polya 1949), they must find a solution path themselves. Therefore, in this case, establishing a plan means deducing a problem-solving path. At first, this seems easy: applying the propositions and thus obtaining a solution. However, to achieve this directly and without detours, it would be necessary to proceed as follows:

1. First use the proposition with the fewest possible allocations for the solution.
2. Reduce the possible allocations for the remaining propositions by fixing the allocations of the objects already recognized unambiguously.

Even if this approach is not established by the students in advance as a plan, it could become visible during the course of the problem-solving process as intuitive action.

As has been stated the students do not consciously realize the ambiguity of several propositions. This is why they fail to find a direct solution path that represents the execution of a plan (previously established or formed during the problem-solving process). Instead, they stumble from step to step by trial and error, making many detours and not being able to direct the process from one stage to the next.

The fifth stage (executing the plan) is therefore not recognizable as an independent stage. Rather, establishing and executing the plan are reflected in the students' iterative solving process by trial and error, although we cannot speak of a plan in the proper sense here.

The sixth stage (testing the solution and correcting it if necessary) is realized to different degrees by the students. Whereas Student 1 is quickly satisfied with a result, Student 2 is very keen on checking the result again and correcting it. This partly gives rise to conflict in the group, and so the important Stage 6 is interrupted or abandoned, as can be seen particularly well from the following episode:

Third episode

221. S1: Well, there is no Tanja.
222. S2: Now, let's read, let's read, well, let's read all the parts, let's read everything.
223. S1: We'll call the teacher now, but only after I have tied my shoelace.
224. S2: Nope, first again Tanja's swimsuit
225. S1: No, no! Now Tanja's missing.
226. S2: Yes, Tanja's swimsuit is next to Heike's towel. Then that must be . . .
227. S1: Tina, open the door, will you, and tell her she can come in.
228. S2: Oh, but look, if it says here that Tanja's swimsuit is next to Heike's towel . . .
229. S1: Tiiiiiiinaaaaa – hurry!
230. S2: You hurry! (short pause). Well, if Tanja's is already written there.

If Student 1 had not objected so strongly to checking the solution, the teacher's intervention at this moment would probably have been superfluous, because the students would have recognized their error on their own.

Comparing the problem-solving processes

We shall compare the problem-solving processes in the three age groups in terms of (a) time required, (b) problem-solving paths, (c) reasoning behavior, and (d) social competence.

Comparison of the time required

Temporal comparisons of the problem-solving processes reveal interesting aspects. The times required for the entire solving process are given in Table 5.4. The time measure was the duration of independent student work (without teacher presence) until the correct result was obtained.

If we neglect the sidetracks and the teacher's recurrent checks, taking

Table 5.4. *Duration of problem-solving processes*

Grade	Time (min)
2	19:40
3	13:23
4	4:20

Table 5.5. *Duration of problem-solving processes (without sidetracks)*

Grade	Revised time (min)	Sidetracks
2	12:05	7:35
3	6:55	6:28
4	3:20	1:00

into account merely the actual problem-solving stages until the correct result was obtained, the students required the "revised times" given in Table 5.5. Grade 2 thus needed almost twice as much time as grade 3, whereas grade 3, in turn, took twice as long as grade 4.

If we consider the time given as "sidetracks" in Table 5.5, it is evident that it increases from grade 4 to 2. Grade 3 uses almost as much time for sidetracks and teacher checks as for the problem-solving stages proper. In grade 2, the problem-solving stages amount to almost two-thirds, whereas a good third of the time was used for sidetracks and checks. Relatively speaking, grade 3 has "squandered" the most time, a fact which is certainly due to this working group's being very unruly and sometimes lacking in concentration.

Comparison of the problem-solving paths

As a first step, all three groups allocated Simone's name to Swimsuit 7. They are immediately aware of the unambiguity of Proposition B. After this, paths separate in the different groups. The reason for this remains uncertain, because the children do not give reasons for their mode of proceeding. Freytag-Loeringhoven (1980: 61) thinks that the problem-solving path depends mainly on the contact partners' behavior. The second-grade students require 16 further steps until they attain the correct solution, grade 3 the same, whereas grade 4 attain the correct solution after only 4 further steps. Tables 5.1 to 5.3 provide an overview of the individual problem-solving steps. In grades 2 and 3, it is interesting to go through the columns for the diverse swimsuits and towels from top to bottom, thus following the solution path for the individual objects. It is notable that the students frequently found the same allocation in earlier solving steps as the final solution in the table's last line. These correct allocation options, however, were changed again during the course of the problem-solving process. This confirms the description given: The students recognize correct partial aspects again and again, but are still unable to unify these into a conclusive overall solution. The allocation remains unchanged until the solution in the final line only in columns 5 and 7 for grade 2 (see Table 5.1) and in columns 7 and 8 for grade 3 (see Table 5.2).

Although the third-grade students required less time for their problem-solving process than those in grade 2, they still needed the same number of problem-solving steps up to the correct solution. The tables show very clearly that the problem-solving paths are very different. There is one spot where they cross, however: The twelfth solving step is nearly identical for grades 2 and 3. But as the two groups have attained this step via different paths, they also proceed differently from there.

The path taken by grade 4 is relatively short. This is certainly due to the fact that the second step *just happens* to be correct, and that the students are able to base their further path on this step. This is naturally why they achieve the goal so rapidly. In the third grade, in contrast, allocations are false from the second up to the fifth step. This means that the students have allocated nearly all the objects in the picture incorrectly, making further proceedings considerably more difficult: First they must recognize that "something is wrong," and then they must find new possibilities of allocation. It can be seen clearly that the students had difficulties in removing signs that had already been assigned to objects even after they had recognized that "something was amiss." This explains why the students required a relatively long time and many solving steps until attaining the correct result after this bad start. Most significant is the fourteenth step, in which all the questionable cards (1) to (5) are removed, a step which proves to be the key to solution! From this point on, the students attain the correct solution rather quickly. This also shows how blinded the students were by the previous false allocations.

In second grade, the allocation of the second problem-solving step is correct, but it is changed again immediately in the third step. In the fifth step, the students swap the allocations for numbers 2 and 3, making all allocations correct again. They do not build on this correct solution consistently, however, but rather make another swap between the allocations in their seventh step, making them false again. The students do not show a clear conception in their approach. Whereas they were momentarily certain about an allocation and able to justify it, they again change the situation in their next step without any thorough reflection. It is therefore no surprise that they require so much time for the correct result. In comparison, the third-grade students, while also beginning with false allocations, show a clear conception in their own approach. If we follow the problem-solving process from the sixth step (see Table 5.1), we can see how long it takes before the second-grade students revise the false allocation in the second column (Heike's swimsuit). The students are very confined in their further proceedings by this error; for a long time, they do not succeed in uncovering this error and in removing the sign "Heike." However, as soon as they have done this, they require only four additional steps to attain the correct solution.

In conclusion, all three student groups show the following three main stages in their problem-solving processes: (a) unstructured trials: a solution plan does not yet exist; (b) planned testing; (c) composition of an overall solution from the partial solutions. Age-dependent developmental stages are reflected in typical details found when comparing the problem-solving processes on the basis of these stages.

The problem-solving processes were prompted by a relatively general problem formulation containing several paths of access to the solution as a typical feature of "openness." The decision-making processes required from the students are bound to induce them to structure their solving process themselves.

Comparison of student reasoning

Fourth-grade students behave differently from second graders in many respects because of their difference in age. To plan and implement *age-appropriate* instruction require the teacher to be informed about specific, age-dependent changes in children's learning behavior. Justifying, arguing, and concluding are important basic cognitive qualifications

Table 5.6. *Categories*

	Category	Grade 2	Grade 3	Grade 4
01	Facts			
02	Conclusions	46 (100%)	35 (100%)	13 (100%)
	correct	14 (30.4%)	12 (34.3%)	8 (61.5%)
	false	21 (45.7%)	16 (45.7%)	4 (30.8%)
	unjustifiable	11 (23.9%)	7 (20.0%)	1 (7.7%)
03	Opinions			
04	Linkages			
05	Explanations	2	3	5
06	Qualifications			
07	Conditions			
08	Assumptions	21	15	4
09	Examples			
10	Action proposals			
11	Abstractions			
12	Agreements			
13	Rejections			
14	Questioning			
15	Challenging			
16	Insisting			
17	Compromise			
18	Repetitions			
19	Prompts			
20	Questions			

that take a key role in mathematics instruction – but of course not only there. Conversely, however, mathematics also provides an excellent field in which to develop and practice this ability.

In the study of the age-dependent differences in reasoning behavior, the 20 categories established by Vinçon (1990) provide an effective aid to describe and evaluate these differences. Table 5.6 shows the frequencies of the aspects of "conclusions," "explanations," and "presumptions" in the three student groups; the first column gives the absolute frequency; and the second, the percentage based on the total number of conclusions.

The categories are limited to the three given in Table 5.6 for two reasons: First, we are interested primarily in how students handle conclusions; and, second, statements that can be assigned to other categories (i.e., linking statements, formulating claims, etc.) are, in part, so rare that the only thing that can be established is that the student groups observed make linkages only rarely or not at all. No development in these abilities was observable from grade 2 to grade 4. It must be stressed, however, that all these statements apply only to this situation and this task. We shall now look at the three categories individually.

1. Conclusions: The absolute number of conclusions decreases continuously from grade 2 (46), across grade 3 (35), to grade 4 (13). This is explained by the simultaneous decrease in time required from grade 2 (12.5 min), across grade 3 (6.55 min), to grade 4 (3.20 min), but also by the "small" share of correct conclusions in grade 2 (30.4%) and grade 3 (34.3%) compared with the "high" share of correct solutions in grade 4 (61.5%). This signifies that the more or less unplanned search strategy pursued by "trial and error" in grade 2 and marked by many false trials is replaced increasingly across grade 3 to grade 4 by a clearly structured, goal-oriented search strategy involving conscious feedback from the givens of the task and from connections already found. This conclusion is supported by the number of quotes or direct references to Propositions A to D (see Table 5.7).

If we consider the difference in time taken to solve the problem, this table shows impressively the strong "task orientation" in the fourth grade's work (showing three times the number of references per minute). With 4.5 references per minute, grade 3 takes an intermediate position.

A content analysis of the conclusions recorded shows that in grade 2 a relatively large number (11; 23.9%) of these "conclusions" cannot be justified on the basis of the given propositions. These unjustifiable conclusions were counted separately for each grade. They decrease from 11 (23.9%) for grade 2, across 7 (20%) for grade 3, to 1 (7.7%) for grade 4. Justifiable conclusions reveal a similar relationship to false conclusions: Their number remains initially rather constant from grade 2 (21; 45.7%) to grade 3 (16; 45.7%), and then drops significantly in grade 4 (4; 30.8%).

2. Assumptions: The number of assumptions also decreases from grade 2 to 4: Whereas grade 2 requires a total of 21 assumptions, and

Table 5.7. *Number of direct references to propositions*

	Grade 2	Grade 3	Grade 4
Number of propositions	29	31	24
Number of propositions per minute	2.4	4.5	7.2

Table 5.8. *Number of assumptions*

	Grade 2	Grade 3	Grade 4
Assumptions	21	15	4

Table 5.9. *Number of explanations*

	Grade 2	Grade 3	Grade 4
Explanations	2	3	5

grade 3 still 15, grade 4 needs only 4 until the conclusion is attained. This also shows that the second- and third-grade students are still uncertain in their reasoning behavior; their overview of the task is more restricted, and they align their own responses more strongly to agreement among their peers. Often, they already formulate their problem-solving ideas in a way likely to challenge their fellow students to take a position – and ideally to agree (see Table 5.8).

3. Explanations: In the reasoning used for their own problem-solving ideas, fourth-grade students use *explanations* most frequently (5) compared with third- (3) and second-graders (2). The problem-solving processes reveal that the fourth-grade students do not shirk questions, as observed frequently in grades 2 and 3. The attempt to answer questions by *explanations* provides the fourth-grade students with more clarity on the individual problem-solving steps. This means that they are likely to make fewer false allocations because they tend to reflect on their individual steps more intensely (see Table 5.9).

Comparison of social behavior

Comparison of the problem-solving processes in the three grades shows that the fourth-grade students cooperate most closely in their approach. When one of the students introduces a problem-solving idea, the others listen and then discuss the proposal until each student agrees; they do not deviate and interrupt one another with new ideas.

When they are compared with grade 3, it is seen that these students already have more difficulties in staying with a problem-solving proposal and checking it thoroughly. There are more frequently situations in which students have different ideas on to how to proceed and solve the problem, and each student tries to make his or her own ideas prevail – sometimes they even do this simultaneously. A further difference from grade 4 is that solution attempts are sometimes abandoned because a problem has arisen for which they do not see an answer; they show less willingness to reflect on this and seek a solution in joint discussion.

When the comparison is shifted to grade 2, this behavior is still worse. The students in this age group most frequently encounter the problem of having different problem-solving ideas and conceptions about how to proceed that they are unable to coordinate. Compared with that of the other two groups, their own mode of work is the least cooperative. It is also observable in part that they do not listen to one another, and then it is no wonder that they do not understand the problem-solving proposals introduced. Communication and cooperation within the groups are thus different: When it is less qualified, students need more time to attain consensus on a solution.

The students are often uncertain when advancing their own ideas and seek confirmation from their peers. This is shown by formulations like "or?" and "perhaps." Grade 3 uses these terms most frequently (9), followed by grade 2 (6) and grade 4 (2). How dependent the students are on their peers' confirmations is revealed even more clearly by observing their behavior in the video recordings: When a student draws a conclusion from Propositions A to D, he or she presents it to the group. If it does not receive the confirmation expected, but rather objections and rejection, often the student withdraws the proposal instead of proving the idea from the propositions and thus making a stand in front of the others. This behavior is particularly observable in grade 2, but also in grade 3. In grade 4, this student behavior changes: The students hold their ground against the others in presenting their own problem-solving ideas; they stick to them and try to explain them to the others. Obviously, the fourth-grade students are better able to cope with situations of conflict and to defend their own point of view by justifying it.

For some approaches to solutions in grades 2 and 3, what the class says about them seems to be more important than whether they are connected to Propositions A to D. In these cases, the problem-solving idea is not tested against the propositions, but is subject to the socially motivated evaluation by the group members. This behavior is more prominent in grade 2 than in grade 3; it is not observable in grade 4. Hörner (1991) thinks that everybody has a desire to expose his or her own views to the opinion of others and to communicate about them with them. In doing so, "interpretations and explanations must eventually be tested and recognized intersubjectively if one wants to take one's place in the environment" (p. 85, translated). The students are more concerned with holding their own ground and attaining recognition by the group than in pursuing their own problem-solving idea and justifying it. Generally, a group's objective is to arrive at a more or less common point of view (Battegay 1991: 18). In this process, the individual often abandons his or her own view and adapts to the group.

When observing the groups, it is conspicuous that there is one student in each group who, in some ways, takes over the role of leader. This is revealed in the frequency of speaking up and trying to exert power over the group through this as well as through behavior (cf. McLeish, Matheson, and Park 1975: 126–131). The students need not be aware of this behavior but can take over this role unconsciously.

In grade 2, Student 2 speaks up most frequently (139 times), closely followed by Student 1 (110 times). When doing so, Student 2 introduces her own statements mainly with regard to the solving process in order to further it. The contributions of Student 1, in contrast, are not very constructive for the problem-solving process. However, she attempts to make her influence prevail over her peers' and to decide about them or the problem-solving process. This rivalry over leadership in grade 2 naturally consumes a lot of the group's time and energy and has a negative influence on the problem-solving process. In the third grade, Student 3 speaks up most often (130 times). He is also the one who most significantly determines the group's procedure and gives the strongest judgments about the others' problem-solving ideas. In the fourth grade, Student 3 speaks up most often (52 times). Some of his fellow students' contributions are even addressed directly to him, a tendency that further confirms his leadership. He generally has a good overview of the problem, and, at the end, volunteers for the task of explaining everything again to one of the other students.

Hence, it is not observable in any group that the students attempt to solve their task in equal responsibility as a team, but rather structures always tend to emerge that can be observed in any kind of group work. The role assumed by the individual student within a group is dependent mostly on his or her personality.

Summary

As this case study is concerned with only three examples, the results may not be generalized without further ado. However, they may help to recognize general tendencies from an individual case. Some trends can be shown, and connections to similar studies can be established. The students, particularly those in grades 2 and 3, showed a very strong visual attachment to the signs in their approach. They were in no position to abstract. For these students in their stage of concrete–operative reasoning, it is important that mathematics instruction presents tasks to them in such a way that they are enabled to connect them with simultaneous or immediately prior experience. By extending the horizons of their experience, students learn to focus their own reasoning performance increasingly toward the process of abstraction and generalization. In selecting task materials, it is worth keeping these as open as possible in order to enable the students to integrate their own representations and modes of thought.

In this connection, it seems to be important to evaluate correctly the significance of interaction within the group for the solving process. Hörner (1991) thinks that the individual student will be prompted to advance his or her ideas only by the other members of the group. This verbalization or reshaping of own thoughts makes him or her more conscious of them. "The group not only makes it possible and legitimate to speak, but it also furthers the clearness of reasoning" (Hörner, 1991: 102, translated). Krummheuer (1992) demands an innovative problem solving because of the changed childhood conditions: "Developing a culture of reasoning marked by formats of interaction in school not only creates favorable conditions for cognitive processes of learning but also establishes positive evaluations of cooperation with peers. Social learning and learning subject matter could be integrated in this way" (p. 224, translated). Fuhrmann (1986) considers that one of the advantages of cooperative group work is that students mutually enrich one another "by everyone expecting and experiencing information, prompting, support, recognition, opportunities of self-confirmation, and so forth from the others and are personally willing . . . to help attain certain goals" (p. xx, translated). This assertion is clearly supported by the present episodes. Many a problem-solving idea would not have been explained again or justified and thus reflected as well if no questions or objections had been raised by the other group members. Besides, the continuous confrontation of every student with the views, proposals, and objections offered by the other group members as well as with the necessity of taking a position against them creates an extremely intense communication on the level of everyday student language that will attain only intermittently the level of mathematical language. This latter level, however, also cannot be the most important goal of learning mathematics in elementary school. As has been stated elsewhere (see Abele 1981), the three case studies examined once more confirm how strongly students will relate in cooperative problem solving, how positively the group situation influences the children's willingness to speak, and how intensely communication between the students is required and enhanced by this.

The communication observed permits conclusions on the motivation provided by problem tasks. Besuden (1985: 77, translated) advances the thesis that "problem-rich methods of instruction are those best-suited and pedagogically most promising to motivate students." He points out that this provides not only short-term but also long-term student motivation: "Mathematics itself no longer presents itself to them as a paved exercise field, but rather as an experimental playground in which flowers can be found as well" (p. 81, translated).

The video recordings of the problem-solving processes show large differences in student participation – as can be observed in any classroom. It could therefore be presumed that the students differed in their motivation: the "better" student more, the "weaker" student less. Hence, the "weaker" student would not receive any considerable encouragement from cooperative problem-solving processes, because this means that other students would do his or her work. This time-honored prejudice is clearly refuted by the present case study because all students integrate themselves into the process of learning on the basis of their own potential (see, also, Bruder 1992; Kiesswetter 1983; Leneke 1992; Sommer and Viet 1987).

The errors occurring within the problem-solving process mainly do not result from a lack of attention or concentration of the students, but are rather errors of reasoning that may clarify the thinking processes within students (see Hasemann 1986). Where the difficulties in the child's grasp and reasoning lie can be observed only in problem-oriented approaches. Kothe and Gnirk (1978) have already pointed out that a new basis of learning may be attained by analyzing student errors, a fact again confirmed by this case study. To pursue this problem further in the field of textual tasks and to evaluate it in an age group comparison would be a promising continuation of earlier studies (see, also, Lorenz 1982; Radatz 1978).

Bauersfeld (1978), Krummheuer (1992), and Voigt (1984) provide an explanatory model for the partial student disorientation observed during the problem-solving process. The teacher's response expectation, his or her interference with the students' thinking processes by personally eliciting, confirming, and challenging student utterances, strongly curtails the free opportunities of students to act and argue and guides them in a way clearly perceptible to the students into a certain direction desired by the teacher. This was not true for this case study, so that these students who were unexperienced in independent work have been required to show an uncommon measure of independence.

References

Abele, A. (1981). Gruppenunterrichtliche Ansätze im Mathematikunterricht der Grundschule. In W. Klafki, E. Meyer and A. Weber (Eds.), *Gruppenarbeit* im *Grundschulunterricht.* Paderborn.

Abele, A. (1992). Schülersprache – Lehrersprache. *Mathematische Unterrichtspraxis,* 13(1), 18. Donauwörth.

Battegay, R. (1991). Autonomie in der Gruppe und durch die Gruppe. In E. Meyer and R. Winkler (Eds.), *Unser Konzept: Lernen in Gruppen. Grundlagen der Schulpädagogik.* (Bd.2). Hohengehren.

Bauersfeld, H. (1978). Analysen zur Kommunikation im Mathematikunterricht. In H. Bauersfeld (Ed.), *Analysen zum Unterrichtshandeln (IDM* Bd. 5). Köln.

Bauersfeld, H. (1982). Analysen zur Kommunikation im Mathematikunterricht. In H. Bauersfeld (Ed.), *Analysen zum Unterrichtshandeln.* Köln.

Besuden, H. (1985). Motivierung im Mathematikunterricht durch problemhaltige Unterrichtsgestaltung. *Der Mathematikunterricht* 3, 75–81.

Biermann, N., Bussmann, H., and Niedworok, H.-W. (1977). *Schöpferisches Problemlösen im Mathematikunterricht.* München, Wien, Baltimore.

Bruder, R. (1992). Problemlösen lernen – aber wie? *mathematik lehren,* 52, 6–12.

Descartes, R. (1960). Von der Methode des richtigen Vernunftgebrauchs und der Wissenschaftlichen Forschung (L. Gäbe, Ed.). Meiner.

Freytag-Loeringhoven, W.-D. v. (1980). Über Taktiken der Lösungshilfe im unterrichtlichen Lernprozeß. In K. Boeckmann (Ed.), *Analyse von Unterricht in Beispielen.* Stuttgart.

Fuhrmann, E. (1986). *Problemlösen im Unterricht.* Berlin.

Hasemann, K. (1986). *Mathematische Lernprozesse.* Braunschweig.

Henning, H. (1992). Kooperatives Arbeiten beim Lösen von Aufgaben – zwei Unterrichtsbeispiele. *Mathematik in der Schule,* 4, 193–199.

Henning, H., and Schuster, E. (1992). Mathematisches Experimentieren beim Problemlösen. *Mathematik in der Schule,* 7/8, 385–393.

Hörner, H. (1991). Verstehensprozesse in Gruppen. In E. Meyer and R. Winkel (Eds.), *Unser Konzept: Lernen in Gruppen. Grundlagen der Schulpädagogik* (Bd.2). Hohengehren.

Kiesswetter, K. (Ed.) (1983). Materialien zum Problernlösen. *Der Mathematikunterricht* 3.

Kothe, S., and Gnirk, H.-J. (1978). Breidenbachs Forderungen: Sachgerecht und kindgemäß was heißt das heute? In H. Bauersfeld (Ed.), *Fallstudien und Analysen zum Mathematikunterricht.* Hannover.

Krummheuer, G. (1991). Argumentationsformate im Mathematikunterricht. In H. Maier and J. Voigt (Eds.), *Interpretative Unterrichtsforschung* (pp. 57–78). Köln.

Krummheuer, G. (1992). *Lernen mit "Format."* Weinheim.

Leneke, B. (1992). Kommunikation beim Aufgabenlösen – nicht nur erlaubt, sondern notwendig! *Mathematik in der Schule,* 6, 340–344.

Lorenz, J.H. (1982). Lernschwierigkeiten im Mathematikunterricht der Grundschule und Orientierungsstufe. In H. Bauersfeld (Ed.), *Analysen zum Unterrichtshandeln (IDM* Bd.5). Köln.

Maier, H., and Voigt, J. (Eds.) (1991). *Interpretative Unferrichtsforschung.* Köln.

Maier, H., and Voigt, J. (1992). Teaching styles in mathematics education. *Zentralblatt für Didaktik der Mathematik,* 24(7), 149–253

McLeish, J., Matheson, W., and Park, J. (1975). *Lernprozesse in Gruppen.* Ulm.

McNeal, M. G. (1991). *The social context of mathematical development.* Doctoral dissertation. West Lafayette, Ind.

Polya, G. (1949). *Schule des Denkens.* Bern.

Radatz, H. (1978). Einige methodologische Probleme bei der Analyse von Schülerfehlern. In H. Bauersfeld (Ed.), *Fallstudien und Analysen zum Mathematikunterricht.* Hannover.

Sommer, N., and Viet, U. (1987). Psychologische Probleme bei der Förderung mathematikschwacher Schüler. *Der Mathematikunterricht,* 1, 51–61.

Steinbring, H. (1991). Mathematics in teaching process: The disparity between teacher and student knowledge. *Recherches en Didactique des Mathématiques,* 11(1), 65–108.

Steinbring, H. (1994). The relation between social and conceptual conventions in everyday mathematics teaching. In L. Bazzini and H.-G. Steiner (Eds.), *Proceedings of the Second Italian-German Bilateral Symposium on Didactics of Mathematics.* IDM – Materialien und Studien, Vol. 39 (pp. 369–383). Bielefeld.

Vinçon, I. (1990). Argumentationen zwischen Schülern und Lehrer im Deutschunterricht. In A. Abele (Ed.), *Neuere Entwicklungen in Lehre und Lehrerbildung* (pp. 128–137). Weinheim.

Voigt, J. (1984). *Interaktionsmuster und Routinen im Mathematikunterricht.* Weinheim.

Voigt, J. (1991). Interaktionsanalysen in der Lehrerfortbildung. *Zentralblatt für Didaktik der Mathematik, 23,* 161–168.

Wittmann E. (1981). *Grundfragen des Mathematikunterrichts* (6. Aufl.). Braunschweig.

6

A constructivist perspective on the
      culture of the mathematics classroom
      PAUL COBB AND ERNA YACKEL

Several European theorists have argued that the social dimension is intrinsic to mathematical development (Balacheff 1986; Bauersfeld 1980; Bishop 1985; Brousseau 1984; Bartolini Bussi 1991). However, sociological perspectives have only come to the fore in mathematics education in the United States in the last few years. As recently as 1988, Eisenhart could write with considerable justification that mathematics educators

are accustomed to assuming that the development of cognitive skills is central to human development, [and] that these skills appear in a regular sequence regardless of context or content. (1988: 101)

The growing trend to go beyond purely psychological analyses is indicated by an increasing number of texts that question an exclusive focus on the individual learner (Brown, Collins, and Duguid 1989; Greeno 1991; Lave 1988a; Newman, Griffin, and Cole 1989; Nunes, Schliemann, and Carraher 1993; Saxe 1991). One of the central issues that emerge from this shift toward sociological perspectives is that of analyzing the culture of the mathematics classroom. In this regard, a range of closely related theoretical notions have been introduced, including the classroom discursive practice (Walkerdine 1988), the tradition of classroom practice (Solomon 1989), and the classroom microculture (Bauersfeld, Krummheuer, and Voigt 1988). We, for our part, have spoken of the classroom mathematics tradition (Cobb et al. 1992). This term was chosen to emphasize that the culture established in a mathematics classroom is in many ways analogous to a scientific research tradition. For example, both are created by a community and both profoundly influence the individual's construction of scientific or mathematical ways of knowing by both supporting and constraining what can count as a problem, a solution, an explanation, and a justification (Barnes 1982; Knorr-Cetina 1982; Lampert 1990; Tymoczko 1986).

In the first of the two major sections of this chapter, we outline a constructivist approach to the culture of the mathematics classroom by elaborating this notion of the classroom mathematics tradition. In the second section, we compare and contrast this approach with those developed in the Vygotskian and activity theory traditions. It will become clear in the course of the discussion that the version of constructivism we describe is both a theoretical position and a practical viewpoint. As a theoretical position, constructivism provides an orientation when addressing issues that arise while working with teachers and their students in the classroom. This classroom-based research in turn leads to modifications in theoretical suppositions and assumptions. The relationship between theory and practice is therefore reflexive in that theory guides practice, which feeds back to inform theory. The major theoretical changes we have made while collaborating with teachers involve a shift away from a purely individualistic position and toward an interpretive stance that coordinates both psychological and sociological perspectives. It is to this latter position that we refer in the remainder of this chapter when we use the term "constructivism."

A constructivist view

Basic assumptions

The theoretical position that underlies our view of the classroom culture coordinates a psychological constructivist perspective on individual students' mathematical activity with an interactionist perspective that draws on symbolic interactionism (Blumer 1969) and ethnomethodology (Mehan and Wood 1975). The theoretical basis for the psychological perspective has been developed by von Glasersfeld (1989a) and incorporates both the Piagetian notions of assimilation and accommodation and the cybernetic concept of viability. Von Glasersfeld uses the term "knowledge" in "Piaget's *adaptational* sense to refer to those sensory—motor and conceptual operations that have proved viable in the knower's experience" (1992: 380). Further, he dispenses with traditional correspondence theories of truth and instead proposes an account that relates truth to the effective or viable organization of activity: "Truths are replaced by viable models – and viability is always relative to a chosen goal" (1992: 384). In this model, perturbations that the cognizing subject generates relative to a purpose or goal are posited as the driving force of development. As a consequence, learning is characterized as a process of self-organization in which the subject reorganizes his or her activity in order to eliminate perturbations (von Glasersfeld 1989b). As von Glasersfeld notes, his instrumentalist approach to knowledge is generally consistent with the views of contemporary neo-pragmatist philosophers such as Bernstein (1983), Putnam (1987), and Rorty (1978).

Although von Glasersfeld defines learning as self-organization, he acknowledges that this constructive activity occurs as the cognizing individual interacts with other members of a community. Thus, he elaborates that "knowledge" refers to

conceptual structures that epistemic agents, given the range of present experience within their tradition of thought and language, consider *viable*. (1992: 381)

Further, he contends that "the most frequent source of perturbations for the developing cognitive subject is interaction with others" (1989b: 136). The interactionist perspective developed by Bauersfeld and his colleagues (Bauersfeld 1980; Bauersfeld, Krummheuer, and Voigt 1988) complements von Glasersfeld's cognitive focus by viewing communication as a process of mutual adaptation wherein individuals negotiate meanings by continually modifying their interpretations. However, whereas von Glasersfeld tends to focus on individuals' construction of their ways of knowing, Bauersfeld emphasizes that

the descriptive means and the models used in these subjective constructions are not arbitrary or retrievable from unlimited sources, as demonstrated through the unifying bonds of culture and language, through the intersubjectivity of socially shared knowledge among the members of social groups, and through the regulations of their related interactions. (1988: 39)

Further, he contends that "learning is characterized by the subjective reconstruction of societal means and models through negotiation of meaning in social interaction" (1988: 39). In accounting for this process of subjective reconstruction, Bauersfeld focuses on the teacher's and students' interactive constitution of the classroom microculture. Thus, he argues that

participating in the processes of a mathematics classroom is participating in a culture of using mathematics. The many skills, which an observer can identify and will take as the main performance of the culture, form the procedural surface only. These are the bricks of the building, but the design of the house of mathematizing is processed on another level. As it is with culture, the core of

what is learned through participation is *when* to do what and *how* to do it . . . .
The core part of school mathematics enculturation comes into effect on the
meta-level and is "learned" indirectly. (1993: 24)

Bauersfeld's discussion of indirect learning clarifies that the occurrence
of perturbations is not limited to those occasions when participants in an
interaction believe that communication has broken down and explicitly
negotiate meanings. Instead, for Bauersfeld, communication is a process of often implicit negotiations in which subtle shifts and slides of
meaning frequently occur outside the participants' awareness. Newman, Griffin, and Cole (1989), speaking within the Vygotskian and
activity theory tradition, make a similar point when they say that in an
exchange between a teacher and a student,

the interactive process of change depends on . . . the fact that there are two
different interpretations of the context and the fact that the utterances themselves serve to change the interpretations. (1989: 13)

However, it should be noted that Newman, Griffin, and Cole use
Leont'ev's (1981) sociohistorical metaphor of appropriation to define
negotiation as a process of mutual appropriation in which the teacher
and students continually coopt or use each others' contributions. In
contrast, Bauersfeld uses an interactional metaphor when he characterizes negotiation as a process of mutual adaptation in the course of which
the participants interactively constitute obligations for their activity
(Voigt 1985). It can also be noted that in the account of Newman,
Griffin, and Cole, the teacher is said to appropriate students' actions into
the wider system of mathematical practices that he or she understands.
Bauersfeld, however, takes the local classroom microculture rather than
the mathematical practices institutionalized by wider society as his
primary point of reference. This focus reflects his concern with the
process by which the teacher and students constitute the classroom
microculture in the course of their interactions. Further, whereas
Vygotskian theorists give priority to social and cultural processes, analyses compatible with Bauersfeld's perspective propose that individual
students' mathematical activity and the classroom microculture are reflexively related (Cobb 1989; Voigt 1992). In this view, individual
students are seen as actively contributing to the development of the
classroom microculture that both allows and constrains their individual
mathematical activities. This reflexive relation implies that neither an
individual student's mathematical activity nor the classroom microculture can be adequately accounted for without considering the other.

School mathematics and inquiry mathematics

We can best clarify why we have come to the position outlined by describing the issues we have attempted to address in our work. For the past 9 years, we have been involved in a classroom-based research and development project in elementary school mathematics. Our overall goal has been to investigate the feasibility of changing what it might mean to know and do mathematics in school. The motivation for conducting this project stemmed from the depressing consequences of traditional mathematics instruction documented by our own and others' prior research. In contemporary terms, it appears that many students develop what Sfard and Linchevski (1994) call semantically debased or pseudostructural conceptions in which they associate a sequence of symbol manipulations with various notational configurations (Thompson 1994). This conclusion is consistent with Walkerdine's (1988) claim that the purpose of doing mathematics in school is to produce formal statements that do not signify anything. Similarly, Nunes, Schliemann, and Carraher (1993) observe that

school mathematics represents a syntactic approach, according to which a set of rules for operating on numbers is applied during problem solving. Meaning is set aside for the sake of generality. (1993: 103)

Analyses of the social interactions in traditional American classrooms indicate that students' construction of semantically debased rules enables them to be effective and thus to be judged as mathematically competent (Gregg 1993; McNeal 1992; Schoenfeld 1987; Yang 1993). We have followed Richards (1991) in calling the mathematics tradition established in such classrooms the school mathematics tradition. The taken-as-shared mathematical practices jointly established by the teacher and students appear to have the quality of one of the five types of classroom norms identified by Much and Shweder (1978), that of instructions. Support for this claim is derived from analyses which document that the consequence of transgressing an established classroom mathematical practice is ineffectiveness, rather than merely error per se (Cobb, Yackel, and Wood 1992). We have called these practices *procedural instructions* to emphasize that public discourse is such that symbol-manipulation acts do not appear to carry the significance of acting mentally on taken-as-shared mathematical objects. As a consequence, mathematics as it is constituted in these classrooms appears to be a depersonalized, self-contained activity that is divorced from other aspects of students' lives (Confrey 1986).

Against the background of this characterization of traditional mathematics instruction, our pragmatic goal when collaborating with teachers has been to help them initiate and guide the development of different microcultures in their classrooms. To this end, we have supported their attempts to renegotiate classroom social norms so that they and their students together constitute communities of validators. Further, our personal beliefs about what it means to know and do mathematics are such that we have attempted to ensure that the manipulation of symbols comes to carry the significance of acting on taken-as-shared mathematical objects. In such a classroom, an explanation that does not go beyond the symbols to describe signified actions on mathematical objects typically transgresses the standards of argumentation jointly established by the teacher and students. Thus, the ideal to which we have worked is that of the teacher and students together acting in and elaborating a taken-as-shared mathematical reality in the course of their ongoing negotiations of mathematical meanings. We, as constructivists, therefore want the teacher and students to act as Platonists who are communicating about a mathematical reality that they experience as objective (Hawkins 1985). This, of course, is not to say that Platonism offers an adequate account of how students come to experience such a reality. Instead, we claim that what is colloquially called meaningful mathematical activity involves the Platonist experience of acting with others in a mathematical reality.

We have again followed Richards (1991) in calling the microculture established in such a classroom the inquiry mathematics tradition. Interactional analyses conducted in project classrooms where a microculture of this type has been established indicate that the classroom mathematical practices are, in Much and Shweder's (1978) terms, truths rather than instructions. At first glance, this view that truths are normative might seem questionable given that truths appear to tell us how the world is, not how it ought to be. However, as Much and Shweder and numerous others have noted, individuals are typically challenged by other members of their community when their actions transgress a currently accepted truth. Further, if their actions continue to conflict with a truth and they cannot justify their conduct in ways that satisfy the standards of argumentation of their community, they eventually cease to be members of that community. Thus, in this view, members of a community such as the teacher and students in a classroom interactively constitute the truths that tell them how the world is or ought to be, and these truths constrain their individual activities.

The conclusion that arguments in an inquiry mathematics classroom establish *mathematical truths* is, of course, consistent with the view that the teacher's and students' negotiations are about the nature of an emerging mathematical reality. In addition, interactional analyses indicate that the consequence of transgressing an established classroom mathematical practice is error per se rather than ineffectiveness, because students in such a classroom continue to be effective when they engage in mathematical argumentation (Cobb, Yackel, and Wood 1992). In general, the distinction that we have made between inquiry mathematics and school mathematics, and between mathematical truths and procedural instructions, is succinctly captured by Davis and Hersh's (1981) observation that

mathematicians know they are studying an objective reality. To an outsider, they seem to be engaged in an esoteric communication with themselves and a small group of friends. (1981: 43–44)

The public discourse of traditional school mathematics in which symbols need not signify anything beyond themselves might well seem to be an esoteric communication to many students. In contrast, the students in an inquiry mathematics classroom appear to know that they are studying an objective mathematical reality.

Psychological and interactional analyses

One of our primary research goals has been to account for students' mathematical learning as they interact with the teacher and their peers in inquiry mathematics classrooms. In addressing this issue, we have continued to find psychological analyses extremely valuable and yet have also found it essential to coordinate them with a strong sociological perspective. This contention concerning the value of psychological analyses clashes with Solomon's (1989) and Walkerdine's (1988) claims that psychological models are largely irrelevant to analyses of students' classroom learning. In developing her argument, Solomon characterizes mathematics as an activity in which students learn to follow situated mathematical rules that, in Much and Shweder's (1978) terms, appear to have the quality of instructions. As has already been noted, Walkerdine, for her part, proposes that the purpose of doing mathematics in school is to produce formal statements that do not signify anything beyond themselves. In these accounts, there is no place for what we take to be the key feature of meaningful mathematical activity – the mental manipulation of abstract yet experientially real mathematical objects. Thus, in our view, Solomon's and Walkerdine's arguments appear to be tied to the school mathematics tradition. Within the confines of this classroom microculture, their claim that psychological analyses are irrelevant has considerable merit. In our experience, it is frequently impossible to infer the quality of any individual student's mathematical thinking when analyzing video recordings and transcripts of traditional school mathematics lessons. However, Solomon's and Walkerdine's arguments can be questioned when inquiry mathematics is considered. As we and others have attempted to illustrate elsewhere, psychological models can make an important contribution to analyses of the learning–teaching process as it occurs in inquiry mathematics classrooms (Cobb 1993; Cobb, Yackel, and Wood 1992). It is in fact primarily for this reason that approaches which subordinate the individual to social and cultural processes are not entirely relevant given our interests and purposes. We find it more useful to propose that individual students' mathematical interpretations of tasks and of others' actions are allowed and constrained by and yet contribute to the development of the classroom mathematical microculture.

We would readily acknowledge that the claims we have made concerning the relevance of psychological analyses reflect our beliefs about what it ought to mean to know and do mathematics in school (cf. Voigt 1992). As a point of clarification, we should also stress that the type of psychology to which we refer falls outside the mainstream American information-processing tradition. In this regard, we accept Lave's (1988a) critique of the mainstream position and concur that it assumes

that children can be taught general cognitive skills (e.g., reading, writing, mathematics, logic, critical thinking) *if* these "skills" are disembedded from the routine contexts of their use. Extraction of knowledge from the particulars of experience, of activity from its context, is the condition for making knowledge available for *general* use. (1988a: 8, italic in the original)

Thus, in the mainstream view, there is a "binary opposition between 'abstract, decontextualized' knowledge and immediate, 'concrete, intuitive' experience" (1988a: 41). In contrast, the psychology that we attempt to practice is interpretivist and aims to account for individuals' inferred experiences rather than their cognitive behaviors. This focus on experience is crucial for our purposes given that learning in an inquiry mathematics classroom involves constructing experientially real mathematical objects. Such an approach explicitly rejects the representational view of mind and instead treats mathematics as a human activity (Cobb, Yackel and Wood 1992). The focus is therefore on the development of

Table 6.1. *A framework for analyzing individual and collective activity in the classroom*

Social perspective	Psychological perspective
Classroom social norms	Beliefs about our own role, others' roles, and the general nature of mathematical activity
Sociomathematical norms	Specifically mathematical beliefs and values
Classroom mathematical practices	Mathematical conceptions and activity

mathematical ways of knowing, rather than on mathematical knowledge as the static product of development. The specific models we use were developed by Steffe and his colleagues (Steffe, Cobb, and von Glasersfeld 1988; Steffe et al. 1983) to explain how, in particular situations, students might come to experience numbers and composite units of 10 as arithmetical objects that exist independently of their activity. In place of a binary opposition between abstract, decontextualized knowledge and concrete, intuitive experience, the approach we are proposing assumes that conceptual activity is necessarily situated and that what a student experiences as concrete and intuitive becomes increasingly abstract in the course of mathematical development.

Thus far, we have discussed why, for our purposes, it is useful to take a psychological perspective. We should also stress that this perspective is, by itself, inadequate even when the immediate goal is to account for an individual student's mathematical activity. As will become apparent, we concur with the view that both the learning process and its products, mathematical ways of knowing, are social through and through. Further, we contend that there is no inherent conflict between this view and the adoption of a psychological perspective. Later we discuss three aspects of the culture of inquiry mathematics classrooms that have been the focus of our research and, in each case, illustrate the coordination of psychological and sociological perspectives. These various aspects are summarized in Table 6.1.

Classroom social norms

In the course of our research, we have conducted a series of year-long classroom teaching experiments in collaboration with both second- and third-grade teachers of 7- and 8-year-old students. The social arrangements in these classrooms typically involved small-group collaborative activity followed by whole-class discussions of students' interpretations and solutions. At the beginning of the first of these experiments, it soon became apparent that the teacher's expectation that students would verbalize how they had interpreted and attempted to solve tasks ran counter to their prior experiences of class discussions in school. The students had been in traditional classrooms in the previous school year, and there they had been steered toward officially sanctioned solution methods during discussions (Cobb, Wood, and Yackel 1991). As a consequence, the students took it for granted that they were to infer what the teacher had in mind rather than to articulate their own understandings. The teacher with whom we collaborated coped with this conflict between her own and the students' expectations by initiating the renegotiation of classroom social norms. We have documented this process in some detail elsewhere (Cobb, Yackel, and Wood 1989). For our current purposes, it suffices to note that the social norms for whole-class discussions that became explicit topics of conversation included explaining and justifying solutions, attempting to make sense of explanations given by others, indicating agreement and disagreement, and questioning alternatives in situations in which a conflict in interpretations or solutions has become apparent. We should also stress that the teacher did not have an explicit list of issues for renegotiation and, indeed, no such list could ever be complete (Mehan and Wood 1975). Instead, these issues emerged as she and the students did and talked about mathematics and were therefore interactional accomplishments. From our perspective as observers, the manner in which the teacher capitalized on unanticipated events by framing them as paradigmatic situations in which to discuss the obligations she expected the students to fulfill constituted a crucial aspect of her expertise. Significantly, these interventions did not appear to be and were not reported by the teacher as conscious decisions (Cobb, Wood, and Yackel 1993). Instead, they seemed to evidence her situated knowing-in-action (Lave 1988a; Schön 1983) – her resonance with the interactionally constituted situation in which she was a participant. It therefore appears that renegotiations of social norms were not consciously induced by thought but instead originated in joint activity as the teacher and students did and talked about mathematics.

Thus far, we have discussed classroom norms primarily from a social perspective and have said little about what we take to be their psychological correlates, the teacher's and students' individual beliefs about their own role, others' roles, and the general nature of mathematical activity in school (Cobb, Yackel, and Wood 1989). An analysis of social norms focuses on regularities in classroom social interactions that, from the observer's perspective, constitute the grammar of classroom life (cf. Voigt 1985). Such an analysis treats these regularities as manifestations of shared knowledge (Gergen 1985). However, when the interactions are viewed from a cognitive perspective that focuses on the teacher's and students' interpretations of their own and others' activity, it becomes apparent that there are differences in their individual beliefs. From this perspective, the most that can be said when interactions proceed smoothly is that the teacher's and students' beliefs fit in that each acts in accord with the other's expectations (Bauersfeld 1988; von Glasersfeld 1984). In cognitive terms, renegotiations of social norms occurred in project classrooms when there was a lack of fit – when either the teacher's or a student's expectations were not fulfilled. In sociological terms, these renegotiations occurred when there was a perceived breach of a social norm (Much and Shweder 1978). It was by capitalizing on such breaches that the teacher and, to an increasing extent, the students initiated the renegotiation of social norms and, in the process, influenced others' beliefs. These beliefs in turn found expression in their individual interpretations of situations that arose in the course of social interactions. This relationship between social norms and beliefs can be summarized by saying that individual interpretations that fit together constitute the social norms that both allow and constrain the individual interpretations that generate them. This, for us, is an instantiation of the reflexive relationship between the communal culture and individual experience of, and action in, the lived-in world (cf. Lave 1988a).

Sociomathematical norms

In considering the second aspect of the classroom culture that has been of interest to us, it can be noted that the classroom norms discussed thus far are not specific to mathematics, but instead apply to almost any subject matter area. For example, one would hope that students will challenge each other's thinking and justify their own interpretations in science and literature lessons as well as in mathematics lessons. As a consequence, we have also analyzed normative aspects of whole-class discussions that are specific to students' mathematical activity (Yackel and Cobb 1996). To clarify this distinction, we speak of sociomathematical norms rather than mathematical norms. Examples include what counts as a different solution, a sophisticated solution, an efficient solution, and an acceptable explanation.

As part of the process of guiding the development of an inquiry mathematics tradition in their classrooms, the teachers with whom we have worked regularly ask the students whether anyone has solved a problem a different way and then explicitly reject contributions that they do not consider different. It was in fact while analyzing classroom interactions in these situations that our interest in sociomathematical norms first arose. The analysis indicated that, on the one hand, the students did not know what would constitute a mathematical difference until the teacher accepted some of their contributions but not others. Consequently, in responding to the teacher's request for a different solution, the students were simultaneously learning what counts as a mathematical difference and helping to constitute interactively what counts as a mathematical difference in their classroom. On the other hand, the teachers in these classrooms were themselves attempting to develop an inquiry form of practice and had not, in their prior years of teaching, asked students to explain their thinking. Consequently, the experiential basis from which they attempted to anticipate students' contributions was extremely limited. They therefore had to respond to the students' contributions even though they had not consciously decided what constituted a mathematical difference. Thus, as was the case with general classroom social norms, the negotiation of this and of other sociomathematical norms was not consciously induced by thought, but instead emerged in the course of joint activity (Yackel and Cobb 1996).

In the classrooms in which we have worked, the process of negotiating what counts as a sophisticated or an efficient solution was typically more subtle and less explicit than was the case for different solutions. For example, the teachers rarely asked whether anyone had a more sophisticated way or more efficient way to solve a problem. However, their reactions to students' solutions frequently functioned as implicit evaluations that enabled students to infer which aspects of their mathematical activity were particularly valued. As Voigt (1993) notes, these implicit judgments made it possible for students to become aware of more developmentally advanced forms of mathematical activity while leaving it to the students to decide whether to take up the intellectual challenge. Students could therefore develop a sense of the teacher's expectations for their mathematical learning without feeling obliged to imitate solutions that might be beyond their current conceptual possibilities. From our perspective as observers, we can in fact view this as one of the ways in which the teachers attempt to cope with a tension inherent in teaching, that between mathematical learning viewed as a process of active individual construction, and mathematical learning viewed as a process of enculturation into the mathematical practices of a wider society.

The sociomathematical norms we have discussed thus far all involve the development of a taken-as-shared sense of when it is appropriate to contribute to a discussion. In contrast, the norm of what counts as an acceptable explanation deals with the actual process by which students go about making a contribution. Given that the teachers with whom we have collaborated were attempting to establish inquiry mathematics traditions in their classrooms, acceptable explanations had to involve symbolized actions on mathematical objects rather than following of procedural instructions. However, it was not sufficient for a student merely to describe personally real mathematical actions. Crucially, for it to be acceptable, other students had to be able to interpret the explanation in terms of actions on mathematical objects that were experientially real to them. Thus, the currently taken-as-shared basis for mathematical communication constituted the background against which students explained their thinking. Conversely, it was by means of argumentation that this constraining background reality itself evolved. This implies that the process of argumentation and the taken-as-shared basis for mathematical communication are reflexively related.

As a final point, we note that the analysis of sociomathematical norms clarifies the process by which teachers foster the development of intellectual autonomy in their classrooms. This issue has special significance for us because the development of student autonomy was an explicitly stated goal of the project at the outset. We originally characterized intellectual autonomy in terms of students' awareness of and willingness to draw on their own intellectual capabilities when making mathematical decisions and judgments. This view of intellectual autonomy was contrasted with intellectual heteronomy, wherein students rely on the pronouncements of an authority to know how to act appropriately (Piaget 1973; Kamii 1985). The link between the growth of intellectual autonomy and the development of an inquiry mathematics tradition becomes apparent when we note that, in such a classroom, the teacher guides the development of a community of validators and thus encourages the devolution of responsibility. However, students can only take over the traditional teacher's responsibilities to the extent that they have developed personal ways of judging that enable them to know-in-action both when it is appropriate to make a mathematical contribution and what constitutes an acceptable contribution. This requires, among other abilities, that students can judge what counts as a different solution, an insightful solution, an efficient solution, and an acceptable explanation. But these are the types of judgments that the teacher and students negotiate when establishing sociomathematical norms. Thus, the development of an inquiry mathematics tradition necessarily involves the negotiation of sociomathematical norms. It is while students participate in this process of negotiation that they construct specifically mathematical beliefs and values that enable them to participate as increasingly autonomous members of an inquiry mathematics community (Yackel and Cobb 1996). We therefore take these beliefs and values to be the psychological correlates of the sociomathematical norms and consider the two to be reflexively related (see Table 6.1).

It is apparent from the account we have given that we revised our conception of intellectual autonomy in the course of the analysis. At the outset, we defined autonomy in psychological terms as a characteristic of individual activity. However, by the time we had completed the analysis, we came to view autonomy as a characteristic of an individual's participation in a community. Thus, although the development of autonomy continues to be a central pragmatic goal for us, we have redefined our view of what it means to be autonomous by going beyond our original purely psychological position.

Classroom mathematical practices

The third aspect of our research on the classroom microculture has focused on the taken-as-shared mathematical practices established by the classroom community. For example, various solution methods that involve counting by 1s are established mathematical practices in second-grade project classrooms at the beginning of the school year. Further, some of these students are also able to develop solutions that involve the conceptual creation of units of 10 and of 1s. However, when they do so, they are obliged to explain and justify their interpretations of number words and numerals. Later in the school year, solutions based on such interpretations are taken as self-evident by the classroom community. The activity of interpreting number words and numerals in this way is an institutionalized mathematical practice that is beyond justification. From the students' point of view, numbers simply are composed of 10s and 1s — it is a mathematical truth. Significantly, despite this apparent unanimity, both classroom observations and individual interviews indicate that there are qualitative differences in individual students' numerical interpretations. Thus, from the observer's perspective, their numerical interpretations are not shared, but instead fit, or are compatible, for the purposes at hand in that they are able to communicate effectively without these differences becoming apparent. It is in fact for this reason that we speak of a taken-as-shared rather than a shared basis for mathematical communication. Each student assumes that his or her interpretation is shared by the others, and nothing occurs in the course of ongoing interactions that leads the student to question this assumption.

In our view, the relationship between individual students' mathematical activity and classroom mathematical practices is reflexive. On the one hand, students actively contribute to the development of classroom mathematical practices. On the other hand, their participation in these practices both allows and constrains individual activity and learning. In this account, neither individual students' activity nor the classroom mathematical practices are paramount. Consequently, the teacher's role is characterized as that of initiating and guiding the development of both individual students' mathematical ways of knowing and the classroom community's taken-as-shared mathematical practices so that they become increasingly compatible with those of the wider society. Thus, the teachers with whom we collaborate are necessarily authorities in their classrooms who attempt to fulfill obligations to the school and to the wider society. The central issue concerns the ways in which they express that authority in action. In this regard, we will touch on two issues. The first concerns the manner in which project teachers capitalize on students' contributions to discussions, and the second concerns the instructional tasks they pose. We noted when discussing sociomathematical norms that the project teachers' implicit evaluations of students' contributions enabled the students to become aware of more sophisticated forms of mathematical activity. A second way in which the teachers expressed their authority in action was repeatedly to redescribe students' contributions in terms that the students would not have used, but that nonetheless might have made sense to them (Cobb, Perlwitz, and Underwood 1994). It was, of course, critical that the teachers initiate the negotiation of mathematical meanings in this way in that they were the only members of their classroom communities who could judge which aspects of their students' activity might be significant with respect to their enculturation into the mathematical practices of wider society. As these redescriptions often involved the use of conventional written mathematical symbols, it could be argued that these cultural tools served as carriers of meaning. However, at a more detailed level of analysis, students could be seen to give meaning to the teacher's use of conventional symbols within the context of their ongoing activity (Cobb, Wood, and Yackel 1993). Further, some of the students usually reconceptualize their own prior activity in the process of interpreting their teacher's redescriptions. In doing so, they actively contribute to the classroom community's development of taken-as-shared ways of using and interpreting the symbols. Thus, in our view, the teachers' use of conventional symbols to redescribe their students' mathematical activity was one of the ways in which they guided both the students' mathematization of their initially informal activity and the classroom community's establishment of increasingly sophisticated mathematical practices.

The types of instructional tasks the teachers pose are crucial if they are to achieve their pedagogical agendas by intervening in the ways described. Several investigations indicate that well-intentioned teachers who genuinely want to build on their students' thinking find it necessary to steer or funnel them to the predetermined responses they have in mind all along (Bauersfeld 1988; Voigt 1985). However, it appears that interactions of this type frequently degenerate into social guessing games that mitigate against the students' development of conceptual understanding (Brousseau 1984; Steinbring 1989). Teachers can prevent funneling only to the extent that students' contributions constitute a resource that they can use to achieve their agendas. This in turn requires that instructional developers analyze the various ways in which individual students solve specific tasks in instructional settings where an inquiry mathematics tradition has been established. Only then is it possible to anticipate the range of contributions that students may typically make, and thus the ways in which teachers may be able to use their activity as a resource. In such an approach to instructional development, psychological analyses of the individual students' mathematical thinking can make an important contribution to the development of an inquiry mathematics microculture in which the teacher guides the emergence of individual and collective systems of meaning.

Comparisons and contrasts with sociocultural approaches

Thus far, we have attempted to illustrate a constructivist approach to the culture of the mathematics classroom by focusing on general social norms, sociomathematical norms, and mathematical practices. In each case, we emphasized that a constructivist analysis involves the coordination of psychological and sociological perspectives. We now compare this constructivist approach with those developed in the Vygotskian and activity theory tradition, and then consider situations in which the various approaches may be relevant. Although our intent is to be even-handed in making this comparison, we would readily acknowledge that we have not discovered a transtheoretical metalanguage. The influence of our theoretical perspective is most apparent in the selection of the issues that we discuss. Most deal with reform at the classroom level, a concern that is central to our research program. We should therefore acknowledge that classroom processes are embedded in encompassing systems of activity at the school and societal levels. We will argue that one of the strengths of sociocultural approaches is the manner in which they take account of these broader systems of activity. These approaches therefore come to the fore when addressing a range of issues including diversity and reform that involve the restructuring of the school.

Basic assumptions

Sociocultural theorists who work in the Vygotskian and activity theory tradition typically assume that cognitive processes are subsumed by social and cultural processes. Empirical findings cited in support of this position are from paradigmatic studies such as those of Carraher, Carraher, and Schliemann (1985), Lave (1988a), and Scribner (1984). These investigations demonstrate that an individual's arithmetical activity is profoundly influenced by his or her participation in encompassing cultural practices such as completing worksheets in school, shopping in a supermarket, selling candy on the street, and packing crates in a dairy. It can be noted in passing that these findings are consistent with and add credence to the claim that mathematics as it is realized in the school mathematics and inquiry mathematics microcultures constitutes two different forms of activity. Further, the findings are compatible with the view that mathematical practices are negotiated and institutionalized by members of communities.

In making the assumption that priority should be given to social and cultural processes, theorists working in this tradition adhere to Vygotsky's (1979) contention that "the social dimension of consciousness is primary in fact and time. The individual dimension of consciousness is derivative and secondary" (p. 30). From this, it follows that "thought (cognition) must not be reduced to a subjectively psychological process" (Davydov 1988: 16). Instead, thought should be viewed as

something essentially "on the surface," as something located . . . on the borderline between the organism and the outside world. For thought . . . has a life only in an environment of socially constituted meanings. (Bakhurst 1988: 38)

Consequently, the individual-in-social-action is taken as the basic unit of analysis (Minick 1989). The primary issue addressed in this tradition is then that of accounting for psychological development in terms of participation in social interactions and culturally organized activities.

This issue has been formulated in a variety of different ways. For example, Vygotsky (1978) emphasized both social interaction with more knowledgeable others in the zone of proximal development and the use of culturally developed sign systems as psychological tools for thinking. In contrast, Leont'ev (1981) argued that thought develops from practical, object-oriented activity or labor. Several American theorists have elaborated constructs developed by Vygotsky and his students and speak of cognitive apprenticeship (Brown, Collins, and Duguid 1989; Rogoff 1990), legitimate peripheral participation (Lave and Wenger 1991; Forman 1992), or the negotiation of meaning in the construction zone (Newman, Griffin, and Cole 1989). In each of these contemporary accounts, learning is located in coparticipation in cultural practices. As a consequence, educational recommendations usually focus on the types of social engagements that increasingly enable students to participate in the activities of the expert rather than on the psychological processes and conceptual structures involved (Hanks 1991).

It is apparent from this brief account that there are several points of contrast between the sociocultural and constructivist positions. For example, sociocultural theorists give priority to social and cultural processes, whereas constructivists typically contend that these and psychological processes are reflexively related, with neither having precedence over the other. Further, sociocultural theorists locate the mind in participation in social action, whereas constructivists, when they take a psychological perspective, locate it in individual activity. Later we further clarify these differences by considering how the two groups of theorists may analyze a classroom episode.

An illustrative example

The sample episode that we will present occurred in a third-grade project classroom and involved instructional tasks developed by Streefland (1991) in which children were asked to divide pizzas fairly. The teacher's intent was that the equivalence of fractional partitionings would become both a focus of individual children's activity and a topic of conversation. In one task, the children worked in pairs to share two pizzas among four people. The task statement showed a picture of two circles. The children's explanations during the subsequent whole-class discussion indicated that some pairs had divided each pizza in half whereas others had divided each into quarters. The teacher recorded these solutions by writing "1/2" and "1/4 + 1/4" to symbolize the portion that each person would receive. One child then commented on the solutions as follows:

Richard: Yeah, but instead, in this one [the drawing of two pizzas divided into fourths], you'd get two pieces, or you'd get a big half.
Teacher: Well, do they still get the same amount [of pizza to eat]?
Richard and Dawn: Yeah.
Richard: Yeah, they still get the same amount. Both of those equal a half.
Teacher: What could you find out here? Do you know?
Richard: They're both the same, but just done differently.
Teacher: So two-fourths is the same, or equal to, one-half, right? (writes "1/4 + 1/4 = 1/2")

Here, the question of whether a person would receive the same amount of pizza in the two cases appeared to be both meaningful and relevant to the children, and, as a consequence, the issue of the equivalence of different partitionings became a topic of conversation. It can also be noted that the teacher redescribed Richard's response by talking of two-fourths as being equal to one-half, and by using conventional written symbols to record the solutions. It is, of course, not clear at this point whether terms such as two-fourths meant numerical quantities or amounts of pizza for the children.

In a subsequent task, the children shared four pizzas among eight people. Their various solutions involved dividing pizzas into halves, quarters, and eighths. Referring to these solutions, the teacher asked, "Do you get more pizza one way than another?"

Jenna: It wouldn't make a difference.

John: Well, in a way they're the same, and in a way they're different.
Teacher: How are they the same?
John: They're the same because if you put four-eighths together, it equals a half.

We speculate that in saying that "in a way they're the same," John was referring to the context of doing mathematics in school, and that when he said "in a way they're different" he was referring to the out-of-school context of a pizza restaurant. He subsequently elaborated his claim that four-eighths together make a half as follows:

John: I know a way you can tell. Four plus four equals eight, and one plus one equals two.
Teacher: Four plus four equals eight. Now how does that help us?
John: It tells us that it's just half.
Teacher: So, in other words, you're saying that if we add four-eighths and four-eighths (writes "4/8 + 4/8 = 8/8"), that would equal eight-eighths, would be the whole thing.
Jenna: What could we do with the [solution that involved dividing pizzas into] fourths?

Jenna's final question indicates that the task for her now was not simply partitioning pizzas fairly, but also involved demonstrating the equivalence of different partitionings. As the episode continued, it became apparent that this shift in the nature of the task was quite widespread in the classroom. It can also be observed that, in the preceding exchange, the drawings of pizzas now seemed to signify fractional quantities of some type for the children. However, although neither John nor Jenna referred to pieces of pizza, we would speculate that the emergence of these quantities was supported by situation-specific imagery.

Despite this developing consensus, there were differences in the ways in which children explained their solutions. For example, during a subsequent discussion, one child explained that 3/6 was equal to 1/2 by dividing a circle in half and then partitioning one of the halves into three equal pieces called "sixths." However, another child explained, "I do it backwards."

Six-sixths. . . . Half of six-sixths is three-sixths and so that would be. . . . Since six-sixths is a whole, then three-sixths is one-half.

As a consequence of these differences, the process of establishing the equivalence of different partitionings itself became an explicit topic of conversation.

In analyzing this sample episode, a sociocultural theorist might both locate it within a broader activity system that takes account of the function of schooling as a social institution and focus on immediate interactions between the teacher and students. This dual focus is explicit in Lave and Wenger's (1991) claim that their

concept of legitimate peripheral participation provides a framework for bringing together theories of situated activity and theories about the production and reproduction of the social order. (1991: 47)

Consequently, for Lave and Wenger, an "analysis of school learning as situated requires a multi-layered view of how knowing and learning are part of social practice" (1991: 40). One of the issues that such an analysis might address is that of identifying the community of practice that is in the process of reproduction when students engage in mathematical activity in school. For Lave and Wenger, this is the community of mathematically schooled adults rather than the community of mathematicians (1991: 99–100).

A sociocultural analysis might also focus on the practice of doing mathematics in school. Given Walkerdine's (1988) contention that this practice aims at the production of formal, symbolic statements, it could be argued that the simulated activity of sharing pizzas was transformed into that of establishing the equivalence of symbolic expressions. It might also be noted that there are significant differences between actual sharing of pizzas in a restaurant and the simulated sharing activity in the classroom (Lave 1988b; Walkerdine 1988). John's comments that "in a way they're the same" and "in a way they're different" would then take on particular significance. Further, it might be observed that none of the students explicitly questioned the convention that each person eats the same amount of pizza even though, in a restaurant, they might not expect to receive the same portion as an adult. In addition, the observation that the students seemed to view both the emergence of the issue of equivalence and the shift away from talk of pizzas as natural developments in the classroom might be taken as further indications of what it means to do mathematics in school.

When attention turns to the specific classroom interactions, a sociocultural theorist might investigate how the teacher and students participated in the practices of the classroom, and how, as a consequence, students become able to participate in the practices of mathematically schooled adults to varying degrees (Lave and Wenger 1991). Alternatively, the sample episode could be analyzed from a perspective which emphasizes that individual thinking is "a function of social activity in which the individual internalizes the ways of thinking and acting that have developed in sociocultural history" (Rogoff 1990: 36).

Here, the analysis would reflect the view that children's participation in mathematical activities "with more skilled partners allows children to internalize tools for thinking and for taking more mature approaches to problem solving" (Rogoff 1990: 14). Newman, Griffin, and Cole (1989) might also focus on "the *transformations* between the interpsychological and the intrapsychological" (1989: 62), but by considering the teacher's and students' mutual appropriation of each other's contributions. Thus, it might be noted that the teacher in the sample episode consistently appropriated the students' explanations into a wider system of mathematical practices when she drew circles to record their solutions. In general, sociocultural theorists would tend to emphasize both the teacher's introduction of cultural tools such as circles and written fraction symbols and the way in which they were subsequently used in the classroom. These tools are particularly significant because, in Lave and Wenger's words, "the artifacts used within a cultural practice carry a substantial portion of that practice's heritage" (1991: 101). This view, which appears to be shared by Newman, Griffin, and Cole (1989) and Rogoff (1990), is consistent with the fundamental claim of activity theory that "mediating objects carry in themselves reified socio-historical experiences of practical and cognitive activity" (Lektorsky 1984: 137).

Sociocultural theorists offer a variety of explanations for the process by which students gain access to the heritage of mathematical practices carried in conventional mathematical symbols. For example, Rogoff (1990) views it as a process of internalization, and Newman, Griffin, and Cole (1989) characterize it as a process of appropriation wherein the symbols become psychological tools for thinking. However, Lave and Wenger (1991) contrast this conception of learning as internalization with their view of learning as increasing participation in communities of practice (1991: 49). In analyzing the sample episode, Lave and Wenger might therefore focus on the students' increasing participation in the ways of perceiving and manipulating fraction symbols that characterize the practices of mathematically schooled adults (1991: 102). In doing so, they might treat the symbols as mediating technologies and delineate their role in "allowing [a] focus on, and thus supporting [the] visibility of, the subject matter" (1991: 103).

In comparing these possible sociocultural accounts, it can be noted that Rogoff discusses psychological processes when she speaks of internal, psychological tools for thinking whereas Lave and Wenger appear to avoid referring to this psychological domain to the greatest extent possible. Newman, Griffin, and Cole take an intermediate position by using Leont'ev's notion of appropriation "without denying the importance of individual representations (or the difficulties intrinsic to determining their form)" (1989: 62). Despite these differences between the various theorists, the view of the teacher that emerges in each case is that of an enactor of established mathematical practices whose role is to guide students' learning by initiating opportunities for increasing participation in these practices.

Given sociocultural theorists' assumption that social and cultural processes subsume individual cognitive processes, one of their primary goals is to account for the social and cultural basis of individual experience. In contrast, constructivists are typically concerned with the quality of individual students' interpretive activity and with the teacher's and students' interactive constitution of classroom norms and practices. With regard to the sample episode, an analysis conducted from the psychological perspective might focus on the meaning that the fraction symbols and circles introduced by the teacher had for individual children (Pirie and Kieren 1994; Steffe and Wiegel 1994; Streefland 1991). Here, differences in the ways that individual students established equivalences would be stressed. For example, the way in which one child explained that 3/6 was equal to 1/2 by partitioning a circle might be contrasted with another's reasoning that "since six-sixths is a whole, then three-sixths is one-half." In particular, it might be argued that the second child had internalized partitioning activity to a greater extent and could conceptually compose and decompose fractional units of some type.[1] The goals of a psychological analysis of this type would be to infer the quality of individual children's mathematical experiences and to posit conceptual operations that account for those ways of experiencing (Confrey 1990; Thompson 1992, 1994). In the course of such an analysis, the researcher seeks to identify the perturbations that might arise as students attempt to achieve their goals and thus to develop models of the students' self-organization in interaction.

When the focus shifts to classroom social interactions, a constructivist analysis might attempt to clarify aspects of the classroom microculture established by the teacher and students. In the sample episode, it could be noted that the students were obliged to explain and justify their thinking. Further, it appeared to be legitimate for the students to raise issues and questions. Thus, Richard interjected, "Yeah, but instead, in this one, you'd get two pieces, or you'd get a big half," and Jenna asked, "What could we do with the fourths?" In addition, an analysis of this type might focus on standards of mathematical argumentation by attempting to tease out what counts as a problem, a solution, an explanation, and a justification. Although it is usually impossible to develop a single convincing explanation of a brief episode, the students' frequent use of terms such as "put" and "make" suggests the metaphor of acting in physical reality. The tentative conjecture that an acceptable explanation had to carry the significance of acting on experientially real mathematical objects might therefore be tested when analyzing further episodes from this classroom.

An analysis conducted from the sociological perspective might also be concerned with the evolution of classroom mathematical practices. For example, in the sample episode, there was a shift from sharing pizzas, to establishing the equivalence of partitionings, to comparing and contrasting different ways of establishing equivalences. Such an analysis would stress the children's active contributions to the interactive constitution of these practices. For example, it has already been noted that Jenna raised the issue of "What could we do with the fourths?" In addition, it was John who first explicitly introduced the notion of sameness when he responded to the question of whether there would be "more pizza one way than another" by saying, "Well, in a way they're the same, and in a way they're different." Further, it would be stressed that the teacher's and students' activities and the classroom mathematical practices are reflexively related in that one does not exist without the other. This interest in the mathematical practices established in particular classrooms can be contrasted with sociocultural theorists' concern for students' participation in culturally organized practices.

Thus far, this discussion of a constructivist analysis has focused almost exclusively on the students. An account of this type would also acknowledge that the teacher is necessarily an authority in the classroom. For example, the teacher's role in initiating and guiding the emergence of certain issues and not others as topics of conversation would be stressed. Further, the analysis would emphasize both the teacher's routine of redescribing the students' explanations and her introduction of conventional written fraction symbols. However, in contrast to the sociocultural view that cultural tools are carriers of established meanings or of a cultural practice's heritage, it would be argued that the students actively participated in the negotiation of the ways of symbolizing established by the classroom community. Such an analysis would therefore reflect the assumption that "everything in social interaction can be loaded with meaning and thus develop into a socially taken-as-shared 'mediator'" (Bauersfeld 1990: 5). This characterization of the teacher's role as that of initiating and guiding the emergence of individual and collective systems of meaning can be contrasted with the sociocultural view of the teacher as one who guides students' initiation into established mathematical practices.

Individual differences and different communities

The discussion of the sample episode indicates that constructivist and sociocultural theorists typically address different problems and issues. Although most sociocultural theorists acknowledge that learning is a constructive process, their overall goal is to account for the ways in which participation in culturally organized activities and face-to-face interactions largely determine psychological development. Qualitative differences in individual activity are typically accounted for in terms of the individuals' participation in different practices. For example, in the course of their discussion of four different examples of apprenticeship, Lave and Wenger (1991) compare and contrast the ways in which the four groups of apprentices participated in their communities of practice. An analysis of qualitative differences in the activities of apprentices within the same community is simply not relevant to the issue they have chosen to address.

Newman, Griffin, and Cole (1989) come closer to the issue of qualitative differences in individual thinking than do most contemporary sociocultural theorists when they consider the learning outcomes of high-group and low-group children on a social studies unit. The thrust of their argument is that the high-group children, from the outset, made categorical–taxonomic interpretations that are characteristic of both Western technological societies and the school curriculum and discourse, whereas the low-group children initially made functional–relational–thematic interpretations that are characteristic of non-Western societies. Newman, Griffin, and Cole therefore contend that the low-group children start out doing a different task and, as a consequence, "had to do an extra 'hidden' piece of intellectual work in the course of the lessons" (1989: 114). Here, in keeping with the basic tenets of the sociocultural perspective, qualitative differences in individual children's activity in school are accounted for in terms of their participation as members of different out-of-school communities. Thus, as is the case with Lave and Wenger's analysis, it is the individual's participation in particular culturally organized practices that carries the explanatory burden when accounting for psychological development. A basic assumption underpinning this work is that it is inappropriate to isolate qualitative differences in students' interpretations of school tasks because these differences reflect qualitative differences in the communities in which the students participate (Bredo and McDermott 1992).

The contrast between sociocultural and constructivist approaches is nowhere sharper than on this issue of individual differences. A primary goal for constructivists is to account for qualitative differences in both individual students' mathematical activity and classroom microcultures. The burden of explanation in such accounts falls on models of individual students' psychological self-organization and on analyses of the process by which they and the teacher, as actively cognizing individuals, constitute the local social situation of their development. Thus, whereas a sociocultural theorist might view classroom interactions as an instantiation of the culturally organized practices of schooling, a constructivist would see an evolving microculture that does not exist apart from the teacher's and students' attempts to coordinate their individual activities. In addition, whereas a sociocultural theorist might see a student appropriating the teacher's contributions, a constructivist would see a student adapting to the actions of others in the course of ongoing negotiations. In making these differing interpretations, sociocultural theorists would tend to invoke sociohistorical metaphors such as appropriation whereas constructivists would typically employ interactional metaphors such as accommodation and mutual adaptation. Further, whereas sociocultural theorists typically stress the homogeneity of members of established communities and eschew analyses that isolate individual differences, constructivists tend to stress heterogeneity and to eschew analyses that isolate pregiven social and cultural practices. From one perspective, the focus is on the social and cultural basis of personal experience. From the other perspective, it is on the constitution of social and cultural processes by actively interpreting individuals.

Beyond essentialism

The contrasts that we have drawn between sociocultural and constructivist approaches to the culture of the mathematics classroom reflect differences in underlying theoretical assumptions. We have seen, for example, that sociocultural theorists locate mind in the individual-in-social-action, whereas constructivists stress the importance of subjective experience and locate mind in individual activity. At times, these two approaches appear to be in direct opposition, with proponents of each claiming hegemony for their characterization of mathematical activity. Further, the arguments that adherents to each approach advance in support of their position are frequently essentialist in nature. In effect, adherents on each side claim that they have the individual and the community right – this is what they are, always were, and always will be, independent of history and culture. In his critique of essentialist claims of this type, Rorty (1983) observes that

the idea that only a certain vocabulary is suited to human beings or human societies, that only that vocabulary permits them to be "understood," is the seventeenth-century myth of "nature's own vocabulary" all over again. (1983: 163)

He instead proposes a more pragmatic approach which acknowledges that the vocabularies we use, or the perspectives we take, are instruments for coping with things rather than ways of representing their intrinsic nature. In making these arguments, Rorty follows Dewey's and Kuhn's contention that we should

give up the notion of science traveling towards an end called "corresponding with reality" and instead say merely that a given vocabulary works better than another for a given purpose. (1983: 157)

Thus, "to say something is better understood in one vocabulary than another is always an ellipsis for the claim that a description in the preferred vocabulary is most useful for a certain purpose" (p. 162).

This pragmatic approach implies that we should consider what sociocultural and constructivist approaches might have to offer relative to the problems or issues at hand. In this regard, we suggest that the sociocultural theories are concerned with the conditions for the possibility of learning (Krummheuer, personal communication), whereas constructivist theories focus on both what students learn and the processes by which they do so. For example, Lave and Wenger (1991), who take a relatively radical position by attempting to avoid any reference to mind in the head, say that "a learning curriculum unfolds in *opportunities for engagement* in practice" (1991: 93, emphasis added). Consistent with this formulation, they note that their analysis of various examples of apprenticeship in terms of legitimate peripheral participation accounts for the occurrence of learning or failure to learn (1991: 63). In contrast, a constructivist analysis would typically focus on the process by which students learn as they participate in a learning curriculum, and on the processes by which this curriculum is interactively constituted in the local situation of development. In our view, both these approaches are of value in the current era of educational reform that stresses both students' meaningful mathematical learning and the restructuring of the school while taking issues of diversity seriously. Constructivists might argue that sociocultural theories do not adequately account for the process of learning, and sociocultural theorists might retort that constructivist theories fail to account for the production and reproduction of the practices of schooling and the social order. In our view, the challenge of relating individual students, the local microculture, and the established practices of the broader community requires that adherents to each perspective move beyond such negative gainsaying and acknowledge the potential positive contributions of the other perspective. In following this approach, constructivists would accept the relevance of work that addresses the broader sociopolitical setting of reform. Conversely, sociocultural theorists would acknowledge the tension that teachers such as Lampert (1985) and Ball (1993) experience between attending to individual students' understandings and initiating them into the mathematical practices of wider society.

Acknowledgments

The research reported in this paper was supported by the National Science Foundation under grant No. RED 9353587. The opinions expressed do not necessarily reflect the views of the Foundation.

Several notions central to this paper were elaborated in the course of discussions with Heinrich Bauersfeld, Götz Krummheuer, and Jörg Voigt at the University of Bielefeld, Germany.

Notes

1. Clearly, it is not possible to make a definitive psychological interpretation of a single utterance such as this. In practice, psychological analyses are developed by accounting for a student's activity across a number of episodes. Alternative explanations of a particular utterance can then be ruled out by referring to regularities in the child's activity across episodes. In general, the goal of a psychological analysis is to develop a wholistic account of a child's activity in a number of situations.

References

Bakhurst, D. (1988). Activity, consciousness, and communication. *Quarterly Newsletter of the Laboratory of Comparative Human Cognition,* 10, 31–39.

Balacheff, N. (1986). Cognitive versus situational analysis of problem-solving behavior. *For the Learning of Mathematics,* 6 (3), 10–12.

Ball, D. L. (1993). With an eye on the mathematical horizon: Dilemmas of teaching elementary school mathematics. *Elementary School Journal, 93,* 373–397.

Barnes, B. (1982). *T. S. Kuhn and social science.* New York: Columbia University Press.

Bartolini Bussi, M. B. (1991). Social interaction and mathematical learning. In F Furinghetti (Ed.), *Proceedings of the Fifteenth Conference of the International Group for the Psychology of Mathematics Education* (pp. 1–16). Genoa, Italy: Program Committee of the 15th PME Conference.

Bauersfeld, H. (1980). Hidden dimensions in the so-called reality of a mathematics classroom. *Educational Studies in Mathematics, 11,* 23–41.

Bauersfeld, H. (1988). Interaction, construction, and knowledge: Alternative perspectives for mathematics education. In T. Cooney and D. Grouws (Eds.), *Effective mathematics teaching* (pp. 27–46). Reston, VA: National Council of Teachers of Mathematics and Erlbaum Associates.

Bauersfeld, H. (1990). *Activity theory and radical constructivism: What do they have in common and how do they differ?* Bielefeld, Germany: University of Bielefeld, Institut für Didaktik der Mathematik, Occasional Paper 121.

Bauersfeld, H. (1993). *"Language games" in the mathematics classroom: Their function and the education of teachers.* Unpublished manuscript, University of Bielefeld, Germany, Institut für Didaktik der Mathematik.

Bauersfeld, H., Krummheuer, G., and Voigt, J. (1988). Interactional theory of learning and teaching mathematics and related microethnographical studies. In H. G. Steiner and A. Vermandel (Eds.), *Foundations and methodology of the discipline of mathematics education* (pp. 174–188). Antwerp: Proceedings of the TME Conference.

Bernstein, R. J. (1983). *Beyond objectivism and relativism: Science, hermeneutics, and praxis.* Philadelphia: University of Pennsylvania Press.

Bishop, A. (1985). The social construction of meaning – a significant development for mathematics education? *For the Learning of Mathematics, 5* (1), 24–28.

Blumer, H.(1969). *Symbolic interactionism: Perspectives and methods.* Englewood Cliffs, NJ: Prentice Hall.

Bredo, E., and McDermott, R. P. (1992). Teaching, relating, and learning. *Educational Researcher, 21*(5), 31–35.

Brousseau, G. (1984). The crucial role of the didactical contract in the analysis and construction of situations in teaching and learning mathematics. In H. G. Steiner (Ed.), *Theory of mathematics education* (pp. 110–119). *Occasional paper 54.* Bielefeld: IDM.

Brown, J. S., Collins, A., and Duguid, P. (1989). Situated cognition and the culture of learning. *Educational Researcher, 18*(1), 32–42.

Bussi, M. B. (1991). Social interaction and mathematical knowledge. In F. Furinghetti (Ed.), *Proceedings of the Fifteenth Conference of the Interna-*

tional Group for the Psychology of Mathematics Education (pp. 1–16). Genoa, Italy: Program Committee of the 15th PME Conference.

Carraher, T.N., Carraher, D.W., and Schliemann, A.D. (1985). Mathematics in streets and in schools. *British Journal of Developmental Psychology, 3,* 21–29.

Cobb, P. (1989). Experiential, cognitive, and anthropological perspectives in mathematics education. *For the Learning of Mathematics,* 9(2), 32–42.

Cobb, P. (1993, April). *Cultural tools and mathematical learning: A case study.* Paper presented at the annual meeting of the American Educational Research Association, Atlanta.

Cobb, P., Perlwitz, M., and Underwood, D. (1994). Construction individuelle, acculturation mathématique et communauté scolaire. *Revue des Sciences de l'Éducation,* 20, 41–62.

Cobb, P., Wood, T., and Yackel, E. (1991). Learning through problem solving: A constructivist approach to second grade mathematics. In E. von Glasersfeld (Ed.), *Constructivism in mathematics education* (pp. 157–176). Dordrecht: Kluwer.

Cobb, P., Wood, T., and Yackel, E. (1993). Discourse, mathematical thinking and classroom practice. In E. Forman, and A. Stone (Eds.), *Contexts for learning: Sociocultural dynamics in children's development* (pp. 91–119). Oxford: Oxford University Press.

Cobb, P., Wood, T., Yackel, E., and McNeal, G. (1992). Characteristics of classroom mathematics traditions: An interactional analysis. *American Educational Research Journal,* 29, 573–602.

Cobb, P., Yackel, E., and Wood, T. (1989). Young children's emotional acts while doing mathematical problem solving. In D. B. McCleod and V. M. Adams (Eds.), *Affect and mathematical problem solving: A new perspective* (pp. 117–148). New York: Springer-Verlag.

Cobb, P., Yackel, E., and Wood, T. (1992). A constructivist alternative to the representational view of mind in mathematics education. *Journal for Research in Mathematics Education,* 23, 2–33.

Confrey, J. (1986). A critique of teacher effectiveness research in mathematics education. *Journal for Research in Mathematics Education,* 17, 347–360.

Confrey, J. (1990). A review of the research on student conceptions in mathematics, science, and programming. In C. B. Cazden (Ed.), *Review of Research in Education* (Vol. 16, pp. 3–55). Washington, DC: American Educational Research Association.

Davis, P. J., and Hersh, R. (1981). *The mathematical experience.* Boston: Houghton Mifflin.

Davydov, V. V. (1988). Problems of developmental teaching (part I). *Soviet Education,* 30(8), 6–97.

Eisenhart, M. A. (1988). The ethnographic research tradition and mathematics education. *Journal for Research in Mathematics Education,* 19, 9–114.

Forman, E. (1992, August). *Forms of participation in classroom practice.* Paper presented at the International Congress on Mathematical Education, Québec City.

Gergen, K. J. (1985). The social constructionist movement in modern psychology. *American Psychologist, 40,* 266–275.

Greeno, J. G. (1991). Number sense as situated knowing in a conceptual domain. *Journal for Research in Mathematics Education, 22,* 170–218.

Gregg, J. (1993, April). *The interactive constitution of competence in the school mathematics tradition.* Paper presented at the annual meeting of the American Educational Research Association, Atlanta.

Hanks, W. F. (1991). Foreword. In J. Lave and E. Wenger, *Situated learning: Legitimate peripheral participation* (pp. 13–26). Cambridge: Cambridge University Press.

Hawkins, D. (1985). The edge of Platonism. *For the Learning of Mathematics, 5*(2), 2–6.

Kamii, C. (1985). *Young children reinvent arithmetic: Implications of Piaget's theory.* New York: Teachers College Press.

Knorr-Cetina, K. D. (1982). *The manufacture of scientific knowledge.* Oxford: Pergamon Press.

Lampert, M. L. (1985). How do teachers manage to teach? Perspectives on the problems of practice. *Harvard Educational Review, 55,* 178–194.

Lampert, M. (1990). When the problem is not the question and the solution is not the answer: Mathematical knowing and teaching. *American Educational Research Journal, 27,* 29–63.

Lave, J. (1988a). *Cognition in practice: Mind, mathematics and culture in everyday life.* Cambridge: Cambridge University Press.

Lave, J. (1988b, April). *Word problems: A microcosm of theories of learning.* Paper presented at the annual meeting of the American Educational Research Association, New Orleans.

Lave, J., and Wenger, E. (1991). *Situated learning: Legitimate peripheral participation.* Cambridge: Cambridge University Press.

Lektorsky, V. A. (1984). *Subject object cognition.* Moscow: Progress Publishers.

Leont'ev, A. N. (1981). The problem of activity in psychology. In J. V. Wertsch (Ed.), *The concept of activity in Soviet psychology.* Armonk, NY: Sharpe.

McNeal, B. (1992, August). *Mathematical learning in a textbook-based classroom.* Paper presented at the International Congress on Mathematical Education, Québec City.

Mehan, H., and Wood, H. (1975). *The reality of ethnomethodology.* New York: John Wiley.

Minick, N. (1989). *L. S. Vygotsky and Soviet activity theory: Perspectives on the relationship between mind and society.* Literacies Institute, Special Monograph Series No. 1. Newton, MA: Educational Development Center, Inc.

Much, N. C., and Shweder, R. A. (1978). Speaking of rules: The analysis of culture in breach. *New Directions for Child Development*, 2, 19–39.

Newman, D., Griffin, P., and Cole, M. (1989). *The construction zone: Working for cognitive change in school*. Cambridge: Cambridge University Press.

Nunes, T., Schliemann, A. D., and Carraher, D. W. (1993). *Street mathematics and school mathematics*. New York: Cambridge University Press.

Piaget, J. (1973). *To understand is to invent*. New York: Grossman.

Pirie, S., and Kieren, T. (1994). Growth in mathematical understanding: How can we characterise it and how can we represent it? *Educational Studies in Mathematics*, 26, 61–86.

Putnam, H. (1987). *The many faces of realism*. LaSalle, IL: Open Court.

Richards, J. (1991). Mathematical discussions. In E. von Glasersfeld (Ed.), *Radical constructivism in mathematics education* (pp. 13–52). Dordrecht, Netherlands: Kluwer.

Rogoff, B. (1990). *Apprenticeship in thinking: Cognitive development in social context*. Oxford: Oxford University Press.

Rorty, R. (1978). *Philosophy and the mirror of nature*. Princeton, NJ: Princeton University Press.

Rorty, R. (1983). Method and morality. In N. Haan, R. N. Bellah, and P. Robinson (Eds.), *Social science as moral inquiry* (pp. 155–176). New York: Columbia University Press.

Saxe, G. B. (1991). *Culture and cognitive development: Studies in mathematical understanding*. Hillsdale, NJ: Lawrence Erlbaum Associates.

Schoenfeld, A.H. (1987). What's all the fuss about metacognition? In A.H. Schoenfeld (Ed.), *Cognitive science and mathematics education* (pp. 189–216). Hillsdale, NJ: Lawrence Erlbaum Associates.

Schon, D. A. (1983). *The reflective practitioner*. New York: Basic Books.

Scribner, S. (1984). Studying working intelligence. In B. Rogoff and J. Lave (Eds.), *Everyday cognition: Its development in social context* (pp. 9–40). Cambridge: Harvard University Press.

Sfard, A., and Linchevski, L. (1994). The gains and the pitfalls of reification: The case of algebra. *Educational Studies in Mathematics*, 26, 87–124.

Solomon, Y. (1989). *The practice of mathematics*. London: Routledge.

Steffe, L. P., Cobb, P., and von Glasersfeld, E. (1988). *Construction of arithmetical meanings and strategies*. New York: Springer-Verlag.

Steffe, L. P., von Glasersfeld, E., Richards, J., and Cobb, P. (1983). *Children's counting types: Philosophy, theory, and applications*. New York: Praeger Scientific.

Steffe, L. P., and Wiegel, H. G. (1994). Cognitive play and mathematical learning in computer microworlds. *Educational Studies in Mathematics*, 26, 111–134.

Steinbring, H. (1989). Routine and meaning in the mathematics classroom. *For the Learning of Mathematics*, 9(1), 24–33.

Streefland, L. (1991). *Fractions in realistic mathematics education: A paradigm of developmental research.* Dordrecht, Netherlands: Kluwer.

Thompson, P. W. (1992). Notations, principles, and constraints: Contributions to the effective use of concrete manipulatives in elementary mathematics. *Journal for Research in Mathematics Education, 23,* 123–147.

Thompson, P. W. (1994). Images of rate and operational understanding of the fundamental theorem of calculus. *Educational Studies in Mathematics, 26,* 229–274.

Tymoczko, T. (1986). Introduction. In T. Tymoczko (Ed.), *New directions in the philosophy of mathematics* (pp. xii–xcii). Boston: Birkhauser.

Voigt, J. (1985). Patterns and routines in classroom interaction. *Recherches en Didactique des Mathematiques, 6,* 69–118.

Voigt, J. (1992, August). *Negotiation of mathematical meaning in classroom processes.* Paper presented at the International Congress on Mathematics Education, Québec City.

Voigt, J. (1993). *Thematic patterns of interaction and mathematical norms.* Unpublished manuscript, University of Bielefeld, Germany, Institut für Didaktik der Mathematik.

von Glasersfeld, E. (1984). An introduction to radical constructivism. In P. Watzlawick (Ed.), *The invented reality.* New York: Norton.

von Glasersfeld, E. (1989a). Constructivism. In T. Husen and T. N. Postlethwaite (Eds.), *The International Encyclopedia of Education* (1st ed., supplement, Vol. 1, pp. 162–163). Oxford: Pergamon.

von Glasersfeld, E. (1989b). Cognition, construction of knowledge, and teaching. *Synthese, 80,* 121–140.

von Glasersfeld, E. (1992). Constructivism reconstructed: A reply to Suchting. *Science and Education, 1,* 379–384.

Vygotsky, L. S. (1978). *Mind and society: The development of higher psychological processes.* Cambridge, MA: Harvard University Press.

Vygotsky, L. S. (1979). Consciousness as a problem in the psychology of behavior. *Soviet Psychology, 17* (4), 3–35.

Walkerdine, V. (1988). *The mastery of reason: Cognitive development and the production of rationality.* London: Routledge.

Yackel, E., and Cobb, P. (1996). Sociomathematical norms, argumentation, and autonomy in mathematics. *Journal for Research in Mathematics Education, 27,* 458–477.

Yang, M. T-L. (1993, April). *A cross-cultural investigation into the development of place value concepts in Taiwan and the United States.* Paper presented at the annual meeting of the American Educational Research Association, Atlanta.

7 The culture of the mathematics classroom: Negotiating the mathematical meaning of empirical phenomena

JÖRG VOIGT

One of the reasons why we study the culture of the mathematics class-room is that the classroom appears to live a life of its own. Although mathematics educators and other agents try to improve it and provide for well-substantiated orientations, proposals, and conditions, changing everyday classroom practice appears to be toilsome and accompanied by unintended side effects. Everyday the participants in the classroom develop unreflected customs and stable habits that enable them to cope with the complexity of classroom life while functioning as a resistance to educational reform.

If mathematics educators and teachers collaborate in changing the culture of experimental classes, the change is time-consuming, and the hope of a swift transfer to other classrooms keeps us in suspense. We know from the history of classroom teaching that the fundamental change of teaching styles in regular classrooms has occurred over a period of several generations of teachers (Maier and Voigt 1989).

One could be tempted to blame the resistance on teachers' reluctance. However, school and university teachers often do not notice their habits. If they are given the opportunity to examine videotapes of their own classrooms "under the microscope," they generally express their surprise about what happens in the classroom microculture. The tradition of "Socratic catechism" still seems to be alive in their classrooms (Maier and Voigt 1992). This involves the teacher's setting a sequence of brief questions that elicit bits of knowledge. Presumably, when acting in this way, teachers reproduce routines that have been developed unintentionally during their schooldays.

The stability of life in the mathematics classroom has been analyzed in several empirical studies that reconstruct interactional regularities of the classroom discourse. When solving mathematical problems, the teacher and the students are entangled in traditional patterns of social interaction (Bauersfeld 1982; Jungwirth 1991; McNeal 1991; Steinbring 1989; Voigt 1985) that are established by the participants' routines. The stability of these patterns would be astonishing if one expected that classroom discourses are generally structured by rational argumentation. However, mathematical discourses are also socially structured; even mathematical arguments that should lead to the joint understanding of a mathematical topic often seem to be replaced by social adjustments, for example, by teachers' suggestive hints and by students' tactical behavior.

The customs of the classroom microculture are not only so dominant because they are effects of the conditions of the macroculture, of the environment of the classroom situation, but because they are hidden. As professionals of classroom life, the teacher and the students act routinely without being conscious of their routines. Those routines intermesh in smooth interactions. In the first grades, of course, there must be conflict situations, because these students have not had many opportunities to develop routines specific to the mathematics classroom. Therefore, episodes from a first- and a second-grade classroom are prominent examples in this study.

The hidden regularities have to be understood in order to improve the mathematics classroom. The present study contributes to this understanding; it complements studies that present new ways of teaching or that call for a change of the environment of classroom processes.

The empirical topic of this study is the interactional regularities that occurred when the classroom members took real-world problems as the starting points for mathematical consideration. According to one of the main claims of mathematics education, students should experience mathematics as being related to the "real world." The mathematical concepts and operations that the teacher wants to introduce should be embedded in real-world situations (Griesel 1975; Oehl 1962). The present study explores how the classroom members ascribe mathematical meanings to the empirical phenomena of real-world situations. Interactional regularities as well as spontaneous improvisations can be reconstructed. The study particularly focuses on how classroom members negotiate mathematical meanings when they ascribe different meanings to phenomena. Several classroom episodes are used as illustrations. These episodes were observed in different case studies that used microethnographic methods (Voigt 1990). With one exception, all are taken from German classrooms.

Analyses are based on an interactional theory of teaching and learning mathematics. These theoretical concepts are outlined. Their use enables the observer to describe some general characteristics of the processes through which the classroom microculture evolves.

This interest in the interaction between teacher and students suggests that their relationships should be discussed. Does the teacher act as a dogmatic expert who views the students as ignorant novices, or does the teacher tend toward a more symmetrical relationship? Does the teacher suppress those students' contributions that imply other interpretations of the real-world phenomena than those expected by the teacher? Or does the teacher accept alternative interpretations and strive explicitly to negotiate mathematical meaning with dissenting persons?

The mathematics teacher is educated as an expert in mathematics. However, when interpreting real-world situations, the expert also has to make use of her or his commonsense knowledge. Confronted with an unexpected mathematical interpretation, the expert in mathematics may be tempted to overrely on her or his position as an expert and to act as an authoritarian. The following case provides the first example.

The problem of mathematical modeling

The layman's innumeracy or the expert's restrictedness?

Paulos (1988) presents a paradigm case in order to complain about innumeracy.

Innumeracy, an inability to deal comfortably with the fundamental notions of number and chance, plagues far too many otherwise knowledgeable citizens. The same people who cringe when words such as "imply" or "infer" are confused react without a trace of embarrassment to even the most egregious of numerical solecisms. I remember once listening to someone at a party drone on about the difference between "continually" and "continuously." Later that evening we were watching the news, and the TV weathercaster announced that there was a 50 percent chance of rain for Saturday and a 50 percent chance for Sunday, and concluded that there was therefore a 100 percent chance of rain that weekend. The remark went right by the self-styled grammarian, and even after I explained the mistake to him, he wasn't nearly as indignant as he would have been had the weathercaster left a dangling participle. In fact, unlike other failings which are hidden, mathematical illiteracy is often flaunted: "I can't even balance my checkbook." "I'm a people person, not a numbers person." Or "I always hated math." (Paulos 1988: 3–4)

Paulos seems to be very confident about passing a negative judgment on both the weatherman and the person to whom he was talking. He speaks of "a perverse pride in mathematical ignorance" (p. 4). When I first read this text, I was tempted to agree with Paulos. The probability of obtaining two heads in two flips of a coin is $1/2 \times 1/2 = 1/4$. Consequently, the probability of at least one tail in two flips is $1 - 1/4 = 3/4$. Therefore, the chance of rain on the weekend should be 75 percent, not 100 percent.

But, is the supposition valid that on 2 successive days, the weather situations are independent, that is, the outcome of one event has no influence on the outcome of the other? Often, meteorologists can predict a short period of bad weather for the weekend without knowing how quickly the front of rain will arrive. In this case, the weatherman can be sure that it will rain that weekend, and the probability that the area of bad weather will move through may be the same on Saturday and on Sunday. The corresponding model would be the unique flip of a coin: The tail corresponds to rain on Saturday, and the head to rain on Sunday. On the basis of this alternative supposition, the chance of rain on the weekend would indeed be 100 percent.

We do not know which mathematical model the weatherman or Paulos's partner had in mind. In cases of mathematical modeling, we should be cautious about viewing ourselves as experts and assessing the statements of so-called laymen or novices. Possibly, the school custom of using stereotyped text problems for mathematical topics narrows the mathematics teacher's and educator's perspective on empirical phenomena. Instead of the people's innumeracy, our limited intellectual interest can cause the problem of mathematical modeling called mathematizing.

In meteorology, complex mathematical models are developed in order to describe and to predict the weather. In mathematics education, it is legitimate to reduce complex models in order to explicate basic ideas on how mathematical concepts and operations can be related to real-world problems. Nevertheless, several classroom episodes presented in the next sections confirm that the choice of the suppositions of a mathematical model is a critical point in the classroom, too. If we were to ensure that classroom discourse was intellectually honest, mathematical modeling would be an important topic, at least in cases of conflicting mathematical interpretations.

In classroom practice, the problem of mathematical modeling usually does not become explicit, except when it is thematized as a part of applied mathematics at secondary level (e.g., Steiner 1969). Because teachers often anticipate a real-world situation as a starting point for the introduction of a mathematical concept or as an occasion for the teaching of mathematical operations, they determine which suppositions of the mathematical model are relevant. When they take specific suppositions for granted, they may not recognize alternative suppositions implied in the students' divergent mathematical statements. In this case, students can experience their teacher as a dogmatic authority. In their experience, the mathematical reasons involved in a model are mixed with nonmathematical reasons for the choice of a model.

As future citizens, students will have to cope with many real-world problems that seem to be mathematically intransparent, or that they would spontaneously mathematize differently from the expert (e.g., life insurance, levels of pollution in the environment, taxes). Is the citizen competent to distinguish between the necessary mathematical inferences and the suppositions of modeling that depend on interests? It could be hoped that paying more attention to the quality of the negotiation of mathematical meaning in the classroom could improve the education of the "competent layman."

Different mathematical interpretations of empirical situations

In this study, mathematical meaning is taken as a product of social processes, in particular, as a product of social interactions. From this point of view, mathematical meanings are primarily studied as emerging between individuals. This view is based on microsociology, with particular emphasis on symbolic interactionism (Blumer 1969; Goffman 1974) and on ethnomethodology (Garfinkel 1967; Mehan 1979). However, the sociological concepts have been modified in order to deal with teaching and learning mathematics (Bauersfeld, Krummheuer, and Voigt 1988).

In order to apply the interactional theory to the problem of mathematical modeling, the discussion has to be prepared by several lines of argumentation. First, the problem is discussed with regard to the ambiguity of empirical phenomena. Second, the differences between the teacher's and the students' perspectives are taken as a background of this ambiguity. Third, an analytical epistemological consideration offers an explanation for why, in school, the distinction between the freedom of modeling and the obligations of mathematical inferences is difficult as well as necessary.

Ambiguity

In order to view mathematical meanings as a matter of negotiation, it is helpful to take into account the ambiguity of objects in the mathematics classroom. From a naive point of view, tasks, questions, and so forth, have definite, clear-cut meanings. If one looks at microprocesses in the classroom carefully, they seem to be ambiguous and call for interpretation. The following picture (Figure 7.1) presented in a textbook for first graders (*Mathemax 2* 1984: 3) provides an initial example. According to the teacher's manual, the picture is a representation of subtraction. Because pictures in regular textbooks are highly stereotyped, presumably, most teachers would interpret the picture as $9 - 3 = 6$, that is, three persons are leaving the swimming pool. German textbooks usually display subtraction as a movement of objects away from a group of other objects; and addition, as a movement toward the group.

Many authors of textbooks as well as mathematics teachers seem to assume that such pictures have unambiguous mathematical meanings and represent tasks that have definite solutions. Usually, only one mathematical statement should fit a picture that is presented. "Pictures . . . that reproduce real-world situations . . . are almost self-explanatory" (Schipper and Hülshoff 1984: 56, translated). Teachers often believe that students can and should discover "the" mathematical task in the empirical phenomena, as if the students were incapable of discovering many others. Especially at the elementary level, there is a strong tendency to minimize the ambiguity of empirical issues. There is a hope that students could learn mathematics easily if mathematical meanings were bound unambiguously to specific dealings with concrete things.

One assumption made by interactionists is that every object or event in human interaction is plurisemantic ("indexical," see Leiter 1980: 106–138). In order to make sense of an object or event, subjects use their background knowledge to form a meaningful context for interpreting the object. For example, a first grader can interpret the first picture as the invitation to share a personal experience at a swimming pool, or the teacher can take the picture as an opportunity for motivating students to apply subtraction. The individual subject, however, does not necessarily experience the ambiguous object as plurisemantic. Rather, if the background understanding is taken for granted, the subject experiences that object as unambiguous.

In an experiment, several pairs of second graders were confronted with such pictures, and they were asked to find the numerical problem.

Figure 7.1. Textbook picture intended as a representation of subtraction.

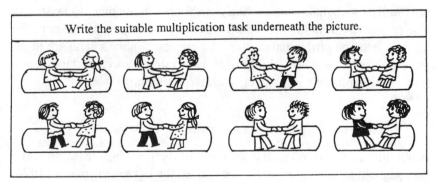

Figure 7.2. Textbook picture intended as a representation of 8 times 2 or 2 times 8.

When confronted with Figure 7.1, Steffi and Jenni gave different answers:

Jenni

6 + 3   *There are 6 persons in the swimming pool, and 3 outside.*

6 − 3   *There are 3 more persons in the pool than outside.*

Steffi

5 + 4   *There are 5 persons playing with a ball. And there are 4 more persons.*

5 − 4   *There is one more player than nonplayers.*

In the interview, the children were asked to explain why they had indicated addition. Jenni looked irritated and immediately reacted by suggesting 6 × 3. Then, immediately, Steffi added 5 × 4. The children were not able to explain these terms empirically. Nevertheless, they said that they believed that they could also have been expected by the author of the textbook and that they had to discover his expectation.

The next picture (Figure 7.2) was viewed as the representation of 2 × 8 or 8 × 2 by the author of the textbook (*Rechnen–Schritt für Schritt 2– Arbeitsheft* 1985). Jenni and Steffi suggested 8 + 8 (persons) and 4 + 4 (pairs), and, in addition, they offered 8 × 8 and 4 × 4 as provisional tries.

On the one hand, the previous emphasis on ambiguity can be viewed as an intellectual game. On the other hand, this ambiguity can give rise to conflicts among the classroom members. One of the findings of a research project (Neth and Voigt 1991; Voigt 1993) was that, in principle, such pictures, text problems, games, and stories have multiple meanings when the children interpreting these objects and events are not familiar with their specific functions for the mathematics classroom. The processes of mathematization taken for granted by the expert, the teacher, become problematic when the empirical phenomena are interpreted by subjects whose thinking is not so disciplined by the conventions and regulations of a specific classroom culture.

Different individual perspectives

Ambiguity is an essential characteristic of the teaching–learning situation when individual perspectives differ. Referring to Goffman's (1974) frame analysis, Krummheuer (1983) reconstructed different "framings" between the teacher and the students of an algebra class in classroom observations over several months. Comparing the interactions during collaborative learning with the interactions during frontal class teaching, he demonstrated that "misunderstandings" between teacher and students were quite typical. Steinbring's epistemological study (1991) explained why the "disparity" between the teacher's and the students' background knowledge has to exist when the teacher is introducing new mathematical concepts that require a change in the students' background understanding of the nature of mathematical concepts. These findings reveal the need to explore different perspectives between the teacher and the students. Such a difference becomes explicit in the next episode.

The teacher had introduced the multiplication of numbers by repeated addition; the first factor of the product was identified as the multiplier, and the second as the multiplicand. This introduction is in line with the textbook for second grade, and it represents common practice in German schools today. During the course of this introduction, the teacher held up three packs each containing 10 pens. On the basis of the common practice, it can be assumed that, from the teacher's perspective, these materials represent "3 times 10."

Teacher: Can you find a multiplication task for these packs?
Natalie: 10 times 3.
The students become noisy; the teacher calls them to order.

Teacher: Yes, but, is that really correct?
Student A: No!
Teacher: Do you all agree?
Student B: No.
Student C: That's just like a swap task.

(The students had experienced the swap task in the context of addition, e.g., 3 + 4 = 4 + 3. It had not been applied yet in the context of multiplication.)

An interpretation: At first, Natalie did not meet the teacher's expectation of mathematization; the teacher expected that the first factor represented the amount of packs. She seemed to produce a negative evaluation of Natalie's answer. With regard to the "weaker" students, the teacher may have wanted to eliminate the risk of ambiguity. Two students opposed Natalie's statement. However, student C hinted at the mathematical identity of the different forms.

In this episode, some participants assumed that each number of a product was represented unambiguously by specific materials. But Natalie offered a divergent statement, and one student justified its correctness as well as the equality of the different terms by drawing an analogy to a mathematical argument established previously in the context of addition.

In the case of the picture of a swimming pool, Jenni's, Steffi's, and the textbook author's interpretations of the picture were based on the quantification of different categories. As in Paulos's case, the persons could not come to an agreement by pure mathematical reasoning. If a classroom member mathematically models an empirical situation in a way that is incompatible with the others' perspectives, an intellectually honest discussion necessitates the negotiation of the mathematical meaning of the situation.

In the case of the packs of pens, the participants' interpretations differed merely in terms of the order of the same quantifications. A pure mathematical argument offered the chance of an agreement. This mathematical argument changed the perspective; attention was now focused on the inner-mathematical relations, not on the relation to the empirical phenomena.

Mixing empirical and theoretical aspects

An epistemological consideration provides an additional explanation of why ambiguity and the negotiation of meaning are essential features of mathematics in school. The epistemological bases of school mathematics differ from those of university mathematics. In his epistemological study, Struve (1990) elaborated the nature of mathematics peculiar to school and compared it with the characteristics of modern mathematics. A pure mathematician who tries to discover truths about a mathematical reality can understand meanings of mathematical objects as unambiguous, because of the formal interrelatedness among the objects in the context of a mathematical theory. In order to understand the meaning of a mathematical object, the mathematician can content herself or himself with its formal definition and its theoretical use. The mathematician's understanding of a mathematical object does not have to depend on empirical, physical existence of the object (see, e.g., Hilbert's understanding of geometry). However, especially in elementary school, the meaning of a mathematical object is related to empirical issues: The number concept is related to materials, geometrical concepts are related to physical space, and so forth.

Also, if the participants agreed on a mathematical statement, a negotiation of meaning that refers to the difference between theoretical and empirical aspects might be necessary. This is because the validity of a mathematical statement can be justified either empirically or theoretically. The earlier picture of the swimming pool serves as an example of this. Let us assume that the participants agreed on $9 - 3 = 6$. Why are 9 persons minus 3 persons 6 persons? On the one hand, it is true, because 6 persons can be counted in the pool. On the other hand, it is true, because $10 - 3$ equals 7, with the consequence that $9 - 3$ has to be 1 less than 7. In the first case, an empirical argument is given, and, in the second case, a theoretical argument relates mathematical objects without being based on empirical evidence.

Although empirical plausibility can represent inner-mathematical reasons, students should eventually become familiar with the mathematical rationality. Therefore, the teacher must ensure that students do not restrict their thinking to empirical evidence (Steinbring 1994a). Through processes of negotiation of what counts as a reason, the teacher can stimulate the students to develop a sense of theoretical reasoning, even if empirical reasons are convincing and seem to be sufficient.

The next episode illustrates the mixture of empirical and theoretical arguments. It is taken from a classroom in which first graders had to learn to add numbers by mathematizing shopping stories. The teacher set a story problem: "Peter wants to buy a magic ball. The magic ball costs 30 deutschmarks (DM). Peter has saved 24 DM." At the beginning, the students had stated that the price of a magic ball was higher, and they had discussed how Peter got the missing money. Then the teacher led the children to solve the intended numerical problem 30 DM = 24 DM + ?. But one student expressed a problem:

Student: But I haven't got a coin of 6 DM.
Teacher: That's another problem. You can think about that later.

The reason for the problem may be that, 3 days before, the participants had divided amounts of money by handling coins. In that context, the teacher had not accepted the written equation 80 cents = (50 cents) (30 cents). The numbers had been circled, in order to represent a 50-cent coin and a 30-cent coin. The teacher had objected that there was no such thing as a 30-cent coin. However, in the present lesson, the equation 30 DM = 24 DM + 6 DM became the official solution, and it was written on the blackboard.

The teacher had used empirical facts, coins, as a means of introducing a mathematical topic. But the relation to empirical facts, taken as a helpful means, caused a problem. The student was confused by the clash between theoretical reasons and empirical conditions.

The negotiation of mathematical meaning

Implicit and explicit negotiation

From the perspective of symbolic interactionism, interaction is more than a sequence of actions and reactions. A participant in an interaction monitors her or his actions in accordance with what she or he assumes to be the other participants' background understandings and expectations.

At the same time, the other participants make sense of this action by adopting what they believe to be the actor's background understandings and intentions. The subsequent actions of the other participants are interpreted by the former actor with regard to her or his expectations; this can prompt a reconsideration, and so forth. For example, a student can orient her or his mathematical thinking by taking the teacher's supposed claims into account and by expecting a specific reaction from the teacher. But the teacher's factual reaction may irritate the student. Then, the student can take the teacher's action as an occasion to change her or his perspective.

The participants' background understandings, their expectations, and their mutual interpretations can be hidden. From the interactionist point of view, mathematical meanings are negotiated, even if the participants do not explicitly argue from different perspectives. Nevertheless, we can study negotiation directly, if we focus our attention on obvious conflicts between the teacher and the students. In cases of conflict, the accomplishment of intersubjective meanings, taken as mathematical meanings, becomes problematic.

In a previous episode, the student Natalie offered the mathematical term 10 times 3 that other participants had not expected. Then, student C used an argument that, from the observer's theoretical point of view, implied the commutative law. This situation was tricky, because the attempt to make the commutative law empirically plausible would demand the establishment of a close relationship between the order of mathematical symbols of the term and the concrete materials. The teacher reacted as follows:

Teacher:
That's a swap task. If I empty the packs, and line up the pens in threes, then that is also 30. In lines of three! But for these packs [*she points to the three original packs*] you would really have to write 3 times 10. [*The original German phrase is "müßtet ihr eigentlich schreiben."*]And then, you have the same amount of pens again.

The teacher mentions that the difference between the orders of pens does not affect the total amount. Implicitly, the teacher points out that the different terms have something in common on the level of concrete objects. The obligation that the number of packs has to represent the first factor seems to be weakened. This obligation is not definite. For example, in her instruction, the teacher used the subjunctive mood, *irrealis*. That is, the German phrase *müßtet ihr eigentlich schreiben* is sometimes used in order to express an expectation on which one no longer insists. The students could take the phrase as a hint that Natalie's interpretation was not wrong, but merely unexpected.

Mathematical meanings taken-to-be-shared

The understanding of the negotiation of meaning as the accomplishment of intersubjective meanings does not imply that the teacher and the students gain a "shared knowledge." From the symbolic interactionist and the radical constructivist points of view, only mathematical meanings taken-to-be-shared can be produced through the negotiation of meaning. Goffman (cited in Krummheuer 1983) and von Glasersfeld (cited in Cobb 1990) have used the terms "working interim" and "consensual domain" to describe how the participants interact *as if* they interpreted the mathematical topic of their discourse in the same way. The problem is that a person cannot be actually certain that her or his understanding is consistent with those of the other participants, even if the participants collaborate without conflict.

What is meant by knowledge taken-to-be-shared? It emerges during processes of negotiation. From the observer's point of view, meanings taken-to-be-shared do not indicate a partial match of individuals' constructions. A meaning taken-to-be-shared is not a cognitive element; it exists at the level of interaction.

Symbolic interactionism views meaning . . . as arising in the process of interaction between people. The meaning of a thing grows out of the ways in which other persons act toward the person with regard to the thing. . . . Symbolic interactionism sees meanings as social products. (Blumer 1969: 5)

This theoretical understanding of classroom processes emphasizes the dynamics and the openness of classroom situations, in contrast to an understanding of classroom situations determined by the environment, by the pregiven culture, including the stock of mathematical knowledge and social structures.

In the preceding episode, the meaning of a swap task emerged in the context of multiplication. It is not clear whether the students understood the argument presented by Student C. However, the subsequent interactions confirm the assumption that the participants arrived at a taken-to-be-shared understanding of the swap task, the commutative law of multiplication; they acted as if they used it implicitly, without provoking a conflict. Later in the same lesson, even the teacher transcended her previous rules, when she interpreted "15 times 2" as 2 packs, each containing 15 pens.

The mathematical theme

In the course of negotiation, the teacher and the students create a network of mathematical meanings taken-to-be-shared. From the observer's point of view, this network of meanings can be called a mathematical "theme." In the episode, the theme is the relationship of materials to mathematical terms.

Because "a teacher is not safeguarded against the students' creativity" (Bauersfeld, personal communication), the students can make original contributions to the theme; the theme may not merely be a representation of the mathematical content that the teacher intended to establish. If the teacher tries to establish mathematical knowledge, she or he is dependent on the students' indications of understanding. Likewise, the students are dependent on the teacher's understanding and acceptance of their contributions. So the theme is not a fixed body of knowledge, but as the topic of discourse, it is constituted interactively; it changes through the negotiation of meaning. From this theoretical point of view, the development of classroom knowledge is not understood as the addition of new knowledge to the stock of knowledge, but as a dynamic interactive process in which knowledge is fragile and open to change.

The two classroom episodes presented next illustrate the interactive constitution of themes accomplished by participants with different perspectives. In the first episode, the teacher achieves her claims smoothly, whereas, in the second episode, the students insist on their divergent perspectives.

The first episode is taken from a second-grade teaching experiment run by Cobb, Wood, and Yackel (1991) in the United States; the author participated in its analysis. Prior to the episode, the participants had discussed fractions as parts of the area of circles. Now, the teacher is using pieces of apples to represent fractions.

Teacher: If we have 20 kids and 25 apples. How are we going to split it up so everyone gets the same amount . . . equal pieces?
Student: One.
Teacher: We can give everybody one and what are we going to do with the other 5 apples?
Student: Throw them away.
Teacher: We're going to throw them away. Well we can throw them away, but that's kind of wasteful.
Student: Split them in half.

Teacher: *(Speaking simultaneously)* What are we going to do? Split them in
    half.
Students: Split them in fourths, split them in fourths.
Matt: Split them into 20 more pieces
Teacher: Right. Split them into 20 more pieces. So sometimes a fraction is not
    only a whole thing or a group it has . . . it has extra pieces.
Alan: You split the apples into 5 pieces because . . .
Bke: No into fourths.
Teacher: Wait a minute. Sh *(to rest of the class)* OK.
Bke: Five apples. 5 times 4 is 20.
Teacher: 5 times 4.
Bke: It would be fourths. Split the apples into fourths.
Teacher: We would split the apples into fourths. And so how much would
    everybody get?
Bke: One and a fourth.
Teacher: One and a fourth apples. Or we would get 5/4ths. *(She circles 5/4 on
    the overhead projector.)* Wouldn't we?

On the one hand, the story could be understood as a pragmatic problem;
from this perspective, the distribution of only 20 apples is reasonable.
At first, the students suggest that every child gets 1 apple, and that 5
apples are left over. In real life, it is not necessary to distribute the
remaining apples immediately. On the other hand, the story could be
taken as a representation of how fractions are handled. Intending to
thematize fractions, the teacher has to cope with the pragmatic solution
that is reasonable from the first perspective. The teacher speaks of
wastefulness. The student who thought of the problem as true to life has
to change her or his perspective in order to understand the intellectual
problem intended by the teacher. Through the subsequent negotiation of
meaning, the participants arrive at a provisional agreement. During the
process of interaction, two different perspectives become adjusted, so
that a joint theme is achieved.

Of course, the accomplishment of a theme does not necessarily imply
that the participants share the particular mathematical meanings. In the
episode, the teacher and the students talk about fourths. They act as if
they are thinking of the same concept, but it is possible that the participants make divergent interpretations and change their understandings.
Thus, a fourth can be interpreted as a part of an apple, as a part of five
apples, or as a number.

If students originally contribute to the classroom discussion, and if
the teacher does not direct and evaluate students' actions in a rigid step-

Figure 7.3. Pictures of apples and worms as a representation of numerical statements.

by-step way, then the theme can be described as a river that produces its own bed. The outcome of the discussion is not clear from the outset. The next example illustrates this.

Beforehand, the teacher had written two equations on the blackboard, $4 = 3 + 1$ and $4 = 2 + 2$. The students had placed stickers on the blackboard in line with the equations. Following the first equation, 3 airplanes with a red tail and 1 airplane with a silvery tail have been attached to the board. Following the second equation, 2 mushrooms with a collar and 2 mushrooms without a collar have been chosen. During these activities, the participants do not encounter any major conflicts. But then a problem arises.

The teacher writes $4 = 1 + 3$ on the board and then immediately attaches 5 apple stickers: 1 apple with two worms, 3 apples, each with one worm, and 1 apple without a worm; all worms are smiling cheerfully (see Figure 7.3).

The teacher expects the students to realize that 1 apple, the last one, has to be removed. But the students begin a heated debate. The sequence presented in the following is only a small part of the discussion that arose:

Patrick: There are four apples with one worm, and one does not have one.
Teacher: There, kids have just put their hands up. Matthias!
Matthias: There are two in the apple. And, in the other three, there are only one. And, there is one without. There, we have to write a zero, because there is no worm inside. Zero worm.
Student D: There, one has to write one.
Students: *(simultaneously shouting)* No, one, no, one, no zero.
Student E: A zero is, if there is nothing at all.
Matthias: Hey, hey, but there is no worm on it.

Then, the teacher tries unsuccessfully to emphasize the counting of the apples. She gives suggestive hints, and she confirms those students' statements that are in line with her expectations. For example, the student Katrin, who often tries to assist the teacher, provides her with an opportunity to do so.

Katrin: It is not about worms; it is about apples.
Teacher: And, what do you think, if it is about apples?
Katrin: Then, one has to write a four and there a one.
Matthias: But there is not a single worm in it.

Finally, the participants determine that different mathematical statements fit the empirical phenomena. The additional statement 5 = 2 + 1 + 1 + 1 + 0 is written on the blackboard and the participants agree that the first statement would fit the empirical situation if the right-hand apple is taken away.

In this episode, the relationship between a pictorial representation and mathematics became explicitly ambiguous. The student Matthias and other students offered arguments that diverged from the teacher's expectation. The teacher's initial attempts to direct all the students' attention to counting the apples failed. A dispute began about the validity of the mathematization. Through the process of negotiation, the participants recognized that the mathematization depended on the subject's focus of attention: Was one interested in apples or in worms? Different presuppositions resulted in alternative mathematical statements.

The classroom culture, its regularities, and its evolution

The emergence of intersubjective meanings and the constitution of the classroom culture

Over time, the negotiation of meaning forms commitments between the participants and stable expectations from the individual's point of view. Meanings that were previously constituted explicitly remain tacit. Cobb (1990: 211–213) has called this the institutionalization of knowledge. Studying smooth interactions in everyday life, ethnomethodologists point out that knowledge is confirmed to be shared and to be given by descriptive "accounting practices":

The stories, that people are continually telling, are descriptive accounts. . . . To construct an account is to make an object or event (past or present) observable and understandable to oneself or to someone else. To make an object or event observable and understandable is to endow it with the status of an intersubjective object. (Leiter 1980: 161–162)

Applying this theoretical consideration to mathematics classrooms, we can state that the participants' joint indications that something was an empirical representation of a specific mathematical statement *make* it a representation of the mathematical statement.

Every true-to-life empirical situation (given as a story, picture, text, etc.) can be mathematized in various ways, depending on one's interest. Therefore, an empirical situation is not a representation of a specific mathematical relationship per se. The participants have to interpret the empirical situation so that it can be understood jointly as that representation. Through the process of negotiation, the teacher and the students endow the empirical situation with the status of intersubjectivity. In the episode just described, at first, the empirical situation did not function as the intended representation, and intersubjectivity was jeopardized. Intersubjective meanings of the empirical phenomena emerged when the teacher and the students came to an agreement.

The ambiguity of a particular object is reduced by relating its meaning to a context that is taken-to-be-shared. At the same time, the taken-to-be-shared context, the classroom culture, is confirmed by the negotiation of the meaning of the particular object. Thus, classroom culture and the particular meanings elaborate each other. Ethnomethodologists call this relationship between a particular meaning and culture "reflexivity" (Leiter 1980: 138–156). From the ethnomethodological point of view, meanings are not considered as given by the classroom culture that, like an environment, would exist independently of the negotiation of meaning; instead, classroom culture is viewed as a microculture, as a dynamic system that is continually being constituted.

For example, first graders usually realize that apples, colored blocks, chips, and so forth, are used differently in the mathematics classroom than at home. The members of the classroom ascribe mathematical meanings to the objects. At the same time, the meaning of what is called "math" becomes clearer to the first graders: One has to count chips, to assign apples to numerical symbols, to discover the teacher's interpretation of objects and events, and so forth. The microculture makes meanings in particular interactions understandable; at the same time, the microculture exists in and through these very interactions.

If the ethnographer analyzes classroom life as an alien culture, she or he can be surprised by what is taken for granted by the members of this classroom culture. However, in the treadmill of everyday life, the participants would say that they know what mathematics and classroom practice really are. In everyday classroom situations, teacher and students often constitute the context routinely, without conflict and without awareness of their ongoing accomplishments. So, in the participants' experience, the context can seem to be pregiven. In everyday classroom practice, teacher and students assume that the context is known, although, from the observer's point of view, it is taken-to-be-shared, vague, and implicit.

The pattern of direct mathematization

Because of the permanent ambiguity, teachers want assurance, relief, implicit orientation, and reliability. "We like to settle down like in a familiar nest, the nest of everyday life" (duBois-Reymond and Söll 1974: 13, translated). Considering the problem of mathematizing, the "nest of everyday life" can be described by reconstructing a specific pattern of interaction. Two cases will be presented to illustrate this pattern. The first example is a fictitious discourse, whereas the second presents a real discourse.

Earlier in this chapter, a picture was presented of a swimming pool. If the teacher takes the picture as self-explanatory and is irritated by students' divergent offers, she or he could be tempted to direct the discourse toward the intended mathematical statement, which may be $9 - 3 = 6$. The teacher would not necessarily view such reactions as narrowing and may be convinced that she or he is helping students to understand.

- How many persons are in the picture? Students' answer: 9.
- How do we have to calculate if some go away? Answer: Minus.
- How many persons left the pool? Answer: 3.
- The result: How many persons remain in the pool? Answer: 6.

Through this procedure, the picture becomes a specific arithmetical task, and a "number sentence" represents the solution. Details of the picture become clearly related to mathematical signs. The sequence of questions occurs concurrently with writing the sentence. This procedure is an example of a pattern of interaction; it will be termed the pattern of "direct mathematization." Particular empirical phenomena are related directly to mathematical signs; the sequence of questions and answers establishes the close correspondence step by step. Neither empirical nor mathematical coherence is addressed in its own right.

Producing the mathematical statements expected by the teacher, the students do not necessarily interpret the picture in terms of numerical part—whole relations. It is possible that students learn how to relate parts of stereotyped pictures to particular numerals step by step, without thinking about the relationship among numerals. This procedure can prompt them to produce even wrong mathematical statements, although every step seems to be reasonable. For example, the teacher may replace the last question by "How many persons are playing with a ball in the pool?" Answering all the teacher's questions correctly, the students may produce the mathematically incorrect statement $9 - 3 = 5$. Thus, the smooth participation in such a pattern of interaction does not necessarily contribute to mathematical thinking.

In the next episode, 8 weeks after starting school, first graders are experiencing their first story problem (Neth and Voigt: 99–102):

Teacher: When I looked out the window this morning, two black thrushes were sitting in the meadow.
Mario: What are thrushes?
Teacher: Black birds. Which number occurred there?
Student: 2.
Teacher: And, then, a little sparrow came and landed on the grass.
Student: 3.

In a general subject lesson, Mario's question could have initiated a long discussion about birds. But, during the first weeks of school, the mathematics teacher took care to ensure that the students learned what are the specifics of a mathematics lesson. For example, the numbers of animals were more important than the species.

The students' numerical replies to the next questions met the teacher's expectations. After that, the teacher introduced a cat into the story, and the students' actions diverged:

Teacher: 5 were on the grass. And then a black cat came along and 2 flew away.
Student: Because they got frightened.
Teacher: Yes. And the other 3 saw the cat as well. And they flew up into the next tree.
Student: This one? *(pointing out of the classroom window)*
Teacher: Yes. How many were still on the grass?
Student: How many flew up into the tree?
Teacher: No, no. Then you can't hear what I'm saying. You have to be very quiet. I won't say it again.
Student: 2.
Student: 1.
Teacher: Let's go back to when the cat came.

At this point, the customs of traditional mathematics classroom culture are still missing. The utterance "because they got frightened" indicated that the student thought his way into the story and identified himself with the birds. Also, the fictitious character of the story was not taken for granted. Was it the same tree as the one that they could all see outside? In classrooms, it is usual for the teacher to ask, the students to respond, and the teacher to evaluate the responses (Mehan 1979). In our case, this role structure does not seem to be stable. The teacher disciplined the students. With regard to the problem of mathematization, the important point is that a specific convention had not yet been established, that is, in a training sequence of calculations, the category in question "birds on the grass" had not changed.

After the bird story, the teacher told a story about ducks.

Teacher: Another little number story. 4 ducks are swimming on the pond, and 1 duck dives under and looks for something to eat.

Nadine: Then there are only 3.

Teacher: A child comes along and throws a stone into the water. And then 2 fly away.

Student: That leaves 2.

Teacher: 1 duck, 1 duck is still swimming on the pond.

Why did Nadine answer that there were two ducks left? Perhaps she miscalculated or miscounted. Or she mathematized the empirical situation differently. Perhaps she also included the duck that had dived under the water. Or she considered that the ducks that had flown away from the child were the remaining ones. In a classroom discourse that can be reconstructed according to the pattern of direct mathematization, it is typical that alternative mathematical interpretations do not become explicit. So, the students' thinking remains hidden, whereas the solution intended by the teacher is built up step by step.

Comparing the beginning of the classroom discourse with the later period, we can reconstruct the emergence of routines. Especially during the closing of the classroom episode, the interaction proceeds smoothly.

Teacher: 1 duck, 1 duck is still swimming on the pond. And then the other duck surfaces. Ella.

Ella: 2.

Teacher: And the two that flew away see that nothing has happened, and they come back.

Student: 4.

Teacher: Very nice.

During the course of their interactions, the participants seemed to adjust their activities to one another and to develop routines. Whereas, during the bird story, the teacher had to ask "How many?" in the duck story, the participants took the question for granted. In the duck story, the participants did not speak of fear, the teacher did not introduce a specific species. This reduced the possibility of diverging perspectives. The students' utterances became shorter, until only a number was given, and this was taken as an indication of mathematical competence. The participants seemed, on the surface, to share the mathematization of the story problem; the conflict "2 are left," was resolved in passing.

The evolution of the classroom culture

The theoretical consideration of the classroom culture as constituted by the participants' sense-making processes has the consequence that an improvement in classroom culture can be achieved by changing the participants' understandings of classroom processes. In contrast to this, theoretical considerations of the classroom culture as an environment of classroom processes suggest changing the physical conditions of classroom life, such as the tasks or the syllabus. Of course, both approaches complement each other. Because the former is rather uncommon, it is emphasized in this study. The next two cases exemplify this changing of classroom culture.

The teacher intended to discuss a geometrical problem and, at the same time, to orient the students toward saving energy (Kämmerer 1987). The teacher set a problem: How much energy is lost when the diameter of the cooking pot is half the diameter of the hotplate? The teacher wanted the students to compare the geometrical areas of the hotplate and the bottom of the pot. From this perspective, approximately 75% of the energy is wasted.

Four girls were observed during their group work. At first, the students spontaneously assumed that between 25% and 50% of the energy was lost. Then, they compared the diameters, and they arrived at the result of a 50% loss. Later, the teacher joined the group, and she intervened in the students' discussion.

Teacher: The pot does not just consist of a line.
Elke: Then we have to determine the circle, or the perimeter, or something.
Regine: Oh no!

The students express their disappointment about their failure.

Students: Goodness me, goodness me!
Michaela: Don't panic!
Elke: The area of a circle is r squared times π.
Daniela: But we should determine the perimeter.
Elke: The area, the area!
Daniela: Okay, okay!

During the subsequent group work, the student Elke asserts her social authority over the others. Finally, the students solve the problem in the way intended by the teacher.

After the lesson, the teacher looked at a videotape and reflected on the lesson. She was disappointed about the quality of argumentation among the students, and she gained a critical attitude toward their solution. Some days later, she visited the school kitchen together with several students and they conducted an experiment. They heated water under different conditions and they measured the consumption of energy. In fact, the students' first assumption provided a better fit to this real-life situation than the geometrical comparison of areas.

Through these experiences, the teacher reorganized her thinking about the relation of mathematics to empirical phenomena. She began to distinguish between inner-mathematical reasons and empirical appropriateness. Accordingly, in classroom situations, she began to focus her attention on processes of modeling, even in seemingly simple text problems.

In this classroom, the problem of mathematization became an explicit topic. The evolving culture of this mathematics classroom differed from the common one in which empirical situations are taken as pure embodiments of intended mathematical relationships. The participants' sense-making processes and their actions contributed to this evolution of the microculture. The customs of school mathematics were broken up. In the teacher's experience, the cooking pot problem became a paradigm case (Kämmerer 1987).

The evolution of this classroom culture was supported by the teacher's reflection. She participated in a teacher training course in which teachers presented videotapes of their own teaching and analyzed the tapes together with the author (Voigt 1991). The teachers were encouraged to develop their understanding of classroom processes and to change their routines in order to improve the quality of negotiation.

The next example of a change in classroom culture is taken from the classroom presented earlier in which apples and worms were counted. After the lesson, the teacher discussed the lesson with the author. The teacher was unhappy, and she suspected that the scene had gone wrong because the task had been ambiguous. The observer objected that every empirical situation could be mathematized differently and that a one-to-one relationship of mathematical signs and empirical phenomena would not be intellectually honest. Nevertheless, during the discussion between the observer and the teacher, the teacher initially did not interpret the scene as an intellectual challenge.

However, the episode had a long-term effect on the teacher. More often than before, she began to realize that many of her tasks were ambiguous and that often the students' strange solutions should not be evaluated as wrong or as diverging from the task, but only as diverging from the teacher's own limited interpretation of the task. After such classroom situations, the teacher walked to the observer at the back of the classroom and remarked that she had posed an "apple-and-worm task" once more. The teacher also began to take an interest in mathematical problems that explicitly required students to look for alternative interpretations of empirical phenomena.

It is now time to summarize the theoretical ideas in the previous sections. In the classroom situation, objects are ambiguous. Because of the differences between their perspectives, the teacher and the students must negotiate mathematical meanings of the objects. Through explicit or implicit negotiation, the particular meaning and the classroom culture inform each other. On the behavioral level, potential conflicts are minimized through thematic patterns of interaction. Through these patterns, the teacher and the students routinely arrive at mathematical meanings taken-to-be-shared without realizing all the alternative interpretations of the phenomena. In everyday classroom life, there is a risk that these processes will degenerate into proceduralized rules and rituals, so that the students will be unable to differentiate between mathematical inferences and the conventions of mathematization. However, at the microlevel, the regularities are not pregiven. Instead, negotiation can be improvised at every moment, although the stability of the regularities is supported by stereotyped textbook problems, by tests, by the persistence of tradition, and by the need for reliability.

Conclusions

Difference instead of deficit of knowledge

From the interactionist point of view, the negotiation between teacher and students is a fascinating unit of analysis because of the participants' different perspectives. The teacher represents mathematical claims as well as the tradition of mathematics education, whereas the student has a different background knowledge. The negotiation of meaning is a necessary condition for learning when the students' understandings differ from the understanding the teacher wants the students to gain. This difference that is not necessarily a deficit characterizes discourse in the classroom. This is particularly evident when students erroneously assume that they understand an ambiguous topic mathematically in line with the teacher's intentions. Describing this situation, Griffin and Mehan (1981) have used a trenchant phrase: "Thus, instead of making entries on a blank slate, teaching in school seems to be involved in erasing entries from a too full slate" (p. 212).

Nevertheless, the student is not a minor partner at all. In several of the classroom episodes presented, the students' perspectives were reasonable, although they did not meet the teacher's expectations; and, at times, they even managed finally to convince the teacher. The students' contributions to the classroom culture and to the mathematical theme have to be viewed as an essential aspect of the theory of classroom processes. The students contribute to the emerging microculture, especially to the evolving theme.

Pattern of interaction and learning

The previous critique of the pattern of direct mathematization can be contrasted with another theoretical position that views the teacher's guidance and the patterns of interaction as necessary elements of classroom culture. Forcing the students to participate in the patterns of interaction has been understood as contributing to their cognitive development. Referring to Bruner (1976), Krummheuer (1991) has analyzed mathematics classrooms by reconstructing specific patterns of interaction, "formats of argumentation," that were initiated by the teacher and that implied conventions that became obligatory for the students. Krummheuer states that the more often students participated successfully in these formats, the more often the teacher let them take on responsibility. Eventually, students gained the ability to argue independently. Krummheuer applied this theoretical idea in order to reconstruct how materials and pictures of materials become *means* for representing mathematical relationships. Students learned to use pictures in conventionalized ways in order to justify the correctness of calculations. Insofar as they did, the materials and pictures of materials were not taken as objects to be interpreted; they were not interpreted as tasks to be solved, but as means to reflect on mathematical relationships.

But it would be problematic to conclude that the teacher has to force students to participate in a pattern, such as the pattern of direct mathematization, as if the students were only able to take conscious control over the mathematization process at some future date. This idea of "scaffolding" (Bruner 1985) is problematic. The reason for this is based on a general educational claim. *Bildung* is a major claim in the German tradition of thinking about education. In 1783, Immanuel Kant criticized what the individual considers to be habitual or natural. The individual should act on rational grounds in the individual's mind without relying on the guidance of others and on conventions. In my understanding, the prominent objective of mathematics education is not that the students produce solutions to problems, but that they do this with insight and by reasonable thinking. That which, in fact, does not make any difference on the behavioral level should be an important subjective difference. Do students act as expected because they want to please the teacher, or because they draw conclusions in order to solve a problem in their experiential world?

The emphasis of interactionism and ethnomethodology on the negotiation of meaning seems to be a mediating perspective between two other positions (Voigt 1992). Roughly described, collectivism views learning as the introduction of a person into a pregiven culture. Individualism views learning as structured by the subject's attempts to resolve what the subject finds problematic in the world of her or his experience. Instead of asking how objective knowledge becomes internalized by a person or how subjective knowledge develops toward truth, interactionism and ethnomethodology study how intersubjectivity is achieved in the negotiation of meanings between persons. To that extent, interaction and learning are appreciated as intrinsically related to each other.

Patterns of interaction are not rules people have to follow; the patterns are accomplished by the participants, even unconsciously. Therefore, they may change at any moment: "So, what we have here are neither automatic rituals – repeated endlessly and mechanically, nor instantaneous creations, – emerging uniqely upon each occasion of interaction. These are negotiated conventions – spontaneous improvisations on basic patterns of interaction" (Griffin and Mehan 1981: 205). Stressing the spontaneous improvisations of the classroom microculture can be helpful, if we reconstruct and assess the common classroom interaction as being too ritualized.

By way of direct mathematization, the participants establish conventions on how to interpret empirical phenomena. On the one hand, conventions can be helpful, if students take empirical phenomena as a means of representation in order to develop their numerical competence. On the other hand, conventions that have become taken for granted can be a burden to the learner. The next classroom episode provides an example.

At the beginning of the school year, first graders had experienced that numbers always stood for the amounts of several things or persons. Some months later, the teacher presented several columns filled with different stickers, each column consisting of uniform stickers. At the top of each column, the price of one sticker in the column was entered (e.g., "4 cents"). In order to construct addition problems, the teacher required the students to purchase several stickers. But the students protested vehemently. They demanded that the numbers at the top of the columns had to be changed. They argued that every written number had to represent the quantity of stickers in the corresponding column. Although the teacher explained her intention, a student extorted a longer and more exciting negotiation of meaning.

Another well-known effect of the direct relation between numbers and empirical phenomena usually becomes problematic when negative numbers are introduced (cf. Brousseau 1983; Schubring 1988; Steinbring 1994b). The learner can experience the statement that -4 is "more" than -5 as inconsistent with the argument that 4 apples is less than 5 apples. Also, addition of negative numbers does not signify "increase," although, in the elementary school, adding is made plausible as increasing an amount.

One could assume that by participating in the pattern of direct mathematization, the so-called weaker student might gain a sense of understanding of the mathematical signs. However, if we suspect that the weaker student's problem is not the inability to reason mathematically but difficulties in interpreting empirical phenomena according to the many conventions taken for granted by the teacher, the pattern of direct mathematization does not support the student's mathematical progress. Then, the explicit negotiation of meaning is helpful when the relationship between empirical situations and mathematical statements is flexible and when the student's spontaneous mathematization is taken into account.

Acknowledgments

Several points discussed in the chapter were developed during discussions with numerous colleagues at the Institute for Didactics of Mathematics (IDM, Germany) as well as with Paul Cobb, Terry Wood, and Erna Yackel (United States).

The research reported in this chapter was partly supported by the Spencer Foundation. The opinions expressed do not necessarily reflect the views of the foundation.

References

Bauersfeld, H. (1982). Analysen zur Kommunikation im Mathematikunterricht. In H. Bauersfeld et al. (Eds.), *Analysen zum Unterrichtshandeln* (pp. 1–40). Köln: Aulis.

Bauersfeld, H. (1988). Interaction, construction, and knowledge: Alternative perspectives for mathematics education. In T. Cooney and D. Grouws (Eds.), *Effective mathematics teaching* (pp. 27–46). Reston,VA: NCTM.

Bauersfeld, H., Krummheuer, G., and Voigt, J. (1988). Interactional theory of learning and teaching mathematics and related microethnographical studies. In H.-G Steiner and A. Vermandel (Eds.), *Foundations and methodology of the discipline mathematics education* (pp. 174–188). Antwerp: University of Antwerp.

Blumer, H. (1969). *Symbolic interactionism: Perspective and method.* Englewood Cliffs, NJ: Prentice-Hall.

Brousseau, G. (1983). Les obstacles éistémologiques et les problèmes en mathématiques. *Recherches en Didactique des Mathématiques, 2,* 164–197.

Bruner, J. (1976). Early rule structure: The case of peek-a-boo. In R. Harré, (Ed.), *Life sentences.* London: Wiley.

Bruner, J. (1985). Vygotsky: A historical and conceptual perspective. In J. V. Wertsch (Ed.), *Culture, communication, and cognition: Vygotskian perspectives.* Cambridge: Cambridge University Press.

Cobb, P. (1990). Multiple perspectives. In L. P Steffe and T. Wood (Eds.), *Transforming children's mathematical education: International perspectives* (pp. 200–215). Hillsdale, NJ: Erlbaum.

Cobb, P., Wood, T., and Yackel, E. (1991). A constructivist approach to second grade mathematics. In E. von Glasersfeld (Ed.), *Constructivism in mathematics education* (pp. 157–176). Dordrecht: Kluwer.

duBois-Reymond, M., and Söll, B. (1974). *Neuköllner Schulbuch, 2. Band.* Frankfurt/M: Suhrkamp.

Garfinkel, H. (1967). *Studies in ethnomethodology.* Englewood Cliffs, NJ: Prentice-Hall.

Goffman, E. (1974). *Frame analysis: An essay on the organization of experience.* Cambridge, MA: Harvard University Press.

Griesel, H. (1975). Stand und Tendenzen der Fachdidaktik Mathematik in der Bundesrepublik Deutschland. *Zeitschrift für Pädagogik, 21,* 19–31.

Griffin, P., and Mehan, H. (1981). Sense and ritual in classroom discourse. In F. Coulmas (Ed.), *Conversational routine* (pp. 187–214). The Hague: Mouton.

Jungwirth, H. (1991). Interaction and gender – findings of a microethnographical approach to classroom discourse. *Educational Studies in Mathematics, 22,* 263–284.

Kämmerer, E. (1987). Energieverschwendung beim Kochen – Die Odyssee einer "praktischen" Aufgabe. In W. Münzinger and E. Liebau (Eds.), *Proben auf's Exempel – Praktisches Lernen in Mathematik und Naturwissenschaften* (pp. 154–169). Weinheim: Beltz.

Krummheuer, G. (1983). *Algebraische Termumformungen in der Sekundarstufe I – Abschlußbericht eines Forschungsprojektes, Materialien und Studien Bd. 31.* Bielefeld: University of Bielefeld, IDM.

Krummheuer, G. (1991). Argumentationsformate im Mathematikunterricht. In H. Maier, and J. Voigt (Eds.), *Interpretative Unterrichtsforschung* (pp. 57–78). Köln: Aulis.

Leiter, K. (1980). *A primer on ethnomethodology*. New York: Oxford University Press.

Maier, H., and Voigt, J. (1989). Die entwickelnde Lehrerfrage im Mathematikunterricht, Teil I. *Mathematica Didactica*, 12(1), 23–55.

Maier, H., and Voigt, J. (1992). Teaching styles in mathematics education. *Zentralblatt für Didaktik der Mathematik*, 7, 249–253.

McNeal, M. G. (1991). *The social context of mathematical development*. Doctoral dissertation. West Lafayette, IN: Purdue University.

Mehan, H. (1979). *Learning lessons*. Cambridge, MA: Harvard University Press.

Neth, A., and Voigt, J. (1991). Lebensweltliche Inszenierungen. Die Aushandlung schulmathematischer Bedeutungen an Sachaufgaben. In H. Maier and J. Voigt (Eds.), *Interpretative Unterrichtsforschung* (pp. 79–116). Köln: Aulis.

Oehl, W. (1962). *Der Rechenunterricht in der Grundschule*. Hannover: Schroedel.

Paulos, J. A. (1988). *Innumeracy. Mathematical illiteracy and its consequences*. New York: Hill & Wang.

Schipper, W., and Hülshoff, A. (1984). Wie anschaulich sind Veranschaulichungshilfen? Zur Addition und Subtraktion im Zahlenraum bis 10. *Grundschule*, 16(4), 54–56.

Schubring, G. (1988). Historische Begriffsentwicklung und Lernprozeß aus der Sicht neuerer mathematikdidaktischer Konzeptionen (Fehler, obstacles, Transposition). *Zentralblatt für Didaktik der Mathematik*, 20, 138–148.

Steinbring, H. (1989). Routine and meaning in the mathematics classroom. *For the Learning of Mathematics*, 9, 24–33.

Steinbring, H. (1991). Mathematics in teaching processes: The disparity between teacher and student knowledge. *Recherches en Didactique des Mathématiques*, 11(1), 65–108.

Steinbring, H. (1994a). Frosch, Känguruh und Zehnerübergang – Epistemologische Probleme beim Verstehen von Rechenstrategien im Mathematikunterricht der Grundschule. In H. Maier and J. Voigt (Eds.), *Verstehen und Verständigung. Arbeiten zur interpretativen Unterrichtsforschung* (pp. 182–217). Köln: Aulis.

Steinbring, H. (1994b). The relation between social and conceptual conventions in everyday mathematics teaching. In L. Bazzini and H.-G. Steiner (Eds.), *Proceedings of the Second Italian – German Bilateral Symposium on Didactics of Mathematics*,. Materialien und Studien, Vol. 39 (pp. 369–383). Bielefeld: IDM.

Steiner, H. G. (1969). Examples of exercises in mathematization: An extension of the theory of voting bodies. *Educational Studies in Mathematics,* 1, 247–257.

Struve, H. (1990). Analysis of didactical developments on the basis of rational reconstructions. In *Proceedings of the 2nd Bratislava International Symposium on Mathematics Education (BISME – 2).* Bratislava, 99–119.

Voigt, J. (1985). Pattern and routines in classroom interaction. *Recherches en Didactique des Mathématiques,* 6(1), 69–118.

Voigt, J. (1990). The microethnographical investigation of the interactive constitution of mathematical meaning. In *Proceedings of the 2nd Bratislava International Symposium on Mathematics Education (BISME – 2),* Bratislava, 120–143.

Voigt, J. (1991). Interaktionsanalysen in der Lehrerfortbildung. *Zentralblatt für Didaktik der Mathematik,* 5, 161–168.

Voigt, J. (1992). *Negotiation of mathematical meaning in classroom processes: Social interaction and learning mathematics.* Paper presented at ICME – 7, Québec.

Voigt, J. (1993). Unterschiedliche Deutungen bildlicher Darstellungen zwischen Lehrerin und Schülern. In J. H. Lorenz (Ed.), *Mathematik und Anschauung* (pp. 147–166). Köln: Aulis.

von Glasersfeld, E. (1987). *Wissen, Sprache und Wirklichkeit. Arbeiten zum radikalen Konstruktivismus.* Braunschweig:Vieweg.

8 The missing link: Social and cultural aspects in social constructivist theories

UTE WASCHESCIO

In the present chapter, I would like to focus on the concept of social interaction in some social constructivist theories of learning. I am referring here mainly to the approaches of Jörg Voigt and Paul Cobb. I shall discuss the concept as used in these theories, and I shall try to show that it does not provide an explanation for the occurrence of learning and development in the classroom. Neither is it able to explain why learning occurs at all, nor can it explain the contents and the direction of development. The former is due to the fact that social constructivism lacks a psychological foundation, and the latter is due to the explicit and systematic exclusion of the cultural–historical dimension.

I shall then try to develop an alternative view of the observed interactive processes between students and teachers, one that acknowledges psychological as well as cultural aspects, drawing on the developmental psychological theory of Vygotsky.

Social interaction and development

The question of the relation between social processes and cognitive development has traditionally not been a central topic in constructivist theories of learning, especially those in the tradition of von Glasersfeld. On the contrary, these approaches have emphasized the constructive activity of the individual conceptualized as being more or less independent of social and cultural influences. For some years, however, researchers in the constructivist tradition have been increasingly concerned with these issues, a trend that, among others, manifests itself in the use of a concept like "negotiation." The reason for this change of perspective can be found at least partly in the fact that there is an increasing amount of empirical data suggesting that social interaction may have a strong influence on the contents and processes of learning.

Cobb (1995b), for example, cites several investigations that "indicate that small-group interactions can give rise to learning opportunities that do not typically arise in traditional classroom interactions" (p. 26). Equally, observations on the level of classroom interaction seem to suggest that different forms of discourse produce different results in terms of how people learn and what is learned. Bauersfeld (1978), for example, describes the potential negative effects of an interaction pattern that he calls "funnel-pattern." Voigt (this volume) discusses the differential effects of the pattern of "direct mathematization" and more open forms of interaction. Cobb, following Richards (1991), distinguishes between traditional classroom interaction and "inquiry mathematics." From a different theoretical viewpoint, Brown, Collins, and Duguid (1989) analyze how deeply the common classroom discourse influences the contents of learning. Equally, Bartolini Bussi (this volume) reports how the confrontation with differing viewpoints in the course of classroom discussion promotes children's learning.

These are only a few examples of investigations and theoretical considerations that point to the important impact of social processes on learning. The results of these studies are in need of explanation and thus constitute a challenge to constructivist theory. The challenge not only consists of the task of integrating the empirical observations into the theoretical framework of constructivism, but also concerns the controversy between constructivism and sociocultural theories. Researchers in the sociocultural tradition view studies that indicate the importance of social interaction to the learning process as strong support for their theoretical perspective. Researchers in the constructivist tradition have reacted to this challenge in a series of publications discussing the differences and similarities between the two competing approaches. I am interested in whether the concept of social interaction, as used in constructivist theories, has any explanatory value for the development and learning of mathematical concepts in the classroom.

In the following, I shall concentrate primarily on the approaches proposed by Voigt (this volume) and Cobb (see Cobb 1995a, 1995b; Cobb and Yackel this volume). Voigt focuses on the concept of negotiation as a central mechanism for the development of mathematical meaning in the classroom. Cobb discusses the impact of both social processes and cultural artifacts on conceptual development. Both authors try to integrate social processes into the epistemological framework of constructivism. A discussion on whether this is possible without abandoning the epistemological basis of constructivism is interesting but would go beyond the scope of this chapter. Ernest (1994), for example, sees no problems in subsuming under the constructivist label such different theoretical approaches as information-processing theories, radical constructivism, and even neo-Vygotskian approaches. Cobb, on the other hand, clearly distinguishes between constructivist theories and sociocultural approaches. I shall begin with a discussion of Voigt' s conception of negotiation and I shall refer mainly to his contribution in this volume.

Voigt describes how the mathematical meaning of empirical phenomena is negotiated within the classroom. He draws on concepts of an interactional theory and on ethnomethodology, claiming that this perspective takes a mediating position between so-called collectivist and individualist theories of learning. Collectivism is said to view learning as "the introduction of a person into a pregiven culture" (Voigt this volume: 216), whereas individualism "views learning as structured by the subject's attempts to resolve what the subject finds problematic in the world of her or his experience" (Voigt this volume: 216). Voigt feels that, according to his approach, interaction and learning are conceptualized as intrinsically related to each other.

It is exactly this point, the relation between interaction and learning, that I find problematic in Voigt's approach. I do not see how the proposed model offers a conceptual tool that allows us to view interaction and learning as being intrinsically related. In my view, the reason for this lies in Voigt's way of distinguishing between two levels of meaning, namely, an individual and a social one, which are strictly separated from each other. In this context, Voigt speaks of "closed systems" (personal communication). To clarify this point, I shall describe, albeit in a very abbreviated form, Voigt's conception of individual and social meaning, and their interrelationship.

On the individual level, meaning is conceptualized as the result of individual sense-making processes. It is the individual's attempt to cope with the different elements of a task. When the individual student is confronted with an empirical object or event, she or he will use her or his background knowledge to create an interpretive context. In the realm of this context, the individual student makes sense of the object or event and ascribes mathematical meaning to it. This process, called mathematizing or mathematical modeling, results in individual mathematical meaning constructed by the individual student. At the same time, this process is influenced by the student's expectations regarding what is an accepted answer. If the student's way of mathematical modeling is rejected by the teacher, she or he will modify this construction in line with the assumptions she or he has about the teacher's intentions.

This process continues until the student gains the impression of sharing with the teacher a meaning of the problem at hand. This is a kind of adaptational process in which students gradually develop their mathematical constructions. Thus, the process of mathematical modelling is determined by

- individual background knowledge of the student or teacher
- mathematical structure
- constraints of the empirical facts
- assumptions about the teacher's expectations

In this way, each student constructs her or his own individual mathematical meanings.

In the course of classroom interaction, another process of meaning construction is added to this model of individual sense-making. On this interactional level, meaning is conceptualized as the result of negotiation. Since background knowledge differs between students and certainly between the teacher and students, it can be assumed that different mathematization processes occur. These probably result in different mathematical meanings. These different meanings may or may not conflict with each other. In order to prevent or to solve conflicts, a process of negotiation of meaning takes place. As a result of this process, a kind of official or collective meaning is established: a meaning taken-to-be-shared.

It is important to note that this negotiated meaning is only taken to be shared. It is not necessary that students and teacher really share knowledge. They only act as if they do so. These interactionally established meanings do not become a part of the individual student's cognitive repertoire. As Voigt says, "Meanings taken-to-be-shared do not indicate a partial match of individuals' constructions. A meaning taken-to-be-shared is not a cognitive element; it exists at the level of interaction" (Voigt this volume: 203).

In fact, these two levels of meaning, the individual and the social, are conceptualized by Voigt as closed systems. The question then is how cognitive development and learning can be intrinsically related to social interaction. What kind of relationship is possible between two closed systems, or, asked the other way round, how closed can these systems actually be? Within Voigt's model, this relationship remains rather obscure and contradictory. If we take the individual and the social as two different systems, then the process of negotiation can be seen as a kind of interface between the two. It is the place where the social and the individual meet. For the individual student, negotiation leads to the adaptational process described, a continual reconsideration of her or his constructions. This process, however, is not specified further. Especially, it is not stated how students build up their assumptions about the teacher's expectations, how these assumptions relate to students' mathematical constructions, or in which way a teacher's rejection of a construction leads to its modification.

On the social level, negotiation produces meanings taken-to-be-shared. These, however, neither achieve the status of a cognitive element nor constitute a social element in the sense of a commonly shared structure. It is social only insofar as it emerges from social interaction, but the product of this interaction remains asocial; it is just taken to be shared. The socially constructed meanings are not the concepts that are learned, since, to say it once more, they do not, according to Voigt, constitute a cognitive element and thus do not become a part of the individual student's conceptual structure. It can be concluded from this that meanings taken-to-be-shared are not a matter of learning.

The question of how social processes contribute to individual conceptual development is left unanswered. But this is an essential question, since it concerns the possibility of learning.

If we are interested in the way students build up mathematical knowledge and in the development of mathematical thinking, we need to link interactional processes back to the individual's conceptual structure. Meaning may well be the product of social interaction, but in order to achieve the status of knowledge, it has to become part of the individual. Otherwise we cannot speak of learning.

If we really were to assume that negotiated meanings do not influence or change individual meanings and thereby the conceptual structure of the individual, we would have to offer an alternative explanation of how the conceptual structure of the child develops. There must be some other mechanism at work than the process of negotiation of meaning. However, this mechanism is not specified further.

Another question we would be confronted with is, What is the sense of negotiating if its products do not contribute to learning? Its only legitimacy seems to be its capacity to prevent time-consuming conflicts in the classroom.

So where is the social factor in this theory of social interaction and learning, and how does interaction contribute to learning? Voigt (1984, 1995) speaks of norms and routines that develop during the course of classroom interaction. Similar to meanings taken-to-be-shared, these also seem to emerge exclusively from classroom interaction. However, terms like "norm" and "routine" imply a certain continuity over time. It would be senseless to speak of a norm if this norm had to be constituted anew each time students and teachers got together in the classroom. So it seems plausible to assume that norms and routines are a part of the students' cognitive repertoire. The same should be true of students' assumptions about the teacher's expectations. These, however, are largely dependent on structures that far exceed that which is negotiated in the actual classroom interaction: the mathematical sign system, concepts, rules of the mathematical discourse, artifacts like books or instructional devices, and so forth.

Voigt describes a pattern of interaction that often does not lead to mathematical thinking: the pattern of direct mathematization. In my view, this is an example of a situation in which processes on an interactional level are not, or are only insufficiently, linked to individual processes. In patterns of direct mathematization, meanings constituted on an interactional level remain external to the child. But for all instances in which mathematical thinking develops, in which learning takes place, I would claim that interactional processes have deeply affected and changed processes on the intrapsychological level. Meanings, constructed on the interactional level, have become internal. They have become a matter of learning.

So I agree with Voigt that processes of negotiation of meaning can be very valuable for the development of mathematical thinking, but only under the condition that negotiated or intersubjective meanings become internal and thereby part of the conceptual structure of the individual student.

In my view, Voigt provides a very precise description of what is generally going on in the classroom. But this description ends at a point at which a theory of learning would start. The relation between interaction and learning remains unspecified as well as the underlying concept of learning. If we see learning as the development of knowledge, as a change in the conceptual structure of the student, the question arises, How does the process of negotiation influence learning? How does the student's conceptual structure change during negotiation? What is missing is the link between social processes and internal processes, and this link has to be provided by a psychological theory of learning and development. It should be noted that the term "psychological," as it is used here, is not meant as a synonym for "individual," "cognitive," or "internal," but, on the contrary, encompasses intrapsychological as well as interpsychological processes. It should also be noted that these processes are viewed as being dialectically related in that no one can be described or explained in isolation from the other. I shall elaborate this point further later.

The initial question whether the concept of negotiation, as it is used by Voigt, can account for conceptual development has to be answered negatively. The two levels of meaning – the individual and the social – are not connected reflexively. In the attempt to construct a reflexive relation between them, two solitary systems emerged, with the social remaining external and individuals constructing meanings that remain asocial.

Paul Cobb's conception of the function of social processes for conceptual development seems to be slightly different. Whereas Voigt seems to deny that meanings taken-to-be-shared have any influence on the development of students' individual concepts, Cobb (1995b) concedes that

the students' constructions have an intrinsically social aspect in that they are both constrained by the group's taken-as-shared basis for communication and contribute to its further development. (p. 26)

He criticizes the neo-Piagetian perspective by saying that

although social interaction is considered to stimulate individual cognitive development, it is not viewed as integral to either this constructive process or to its products, increasingly sophisticated mathematical conceptions. (p. 25)

Cobb sees his approach as taking a mediating position between neo-Piagetian and Vygotskian perspectives. In neo-Piagetian approaches, he sees an overemphasis on individual autonomous constructions. Vygotskian perspectives, on the other hand, are seen as subordinating individual cognition to interpersonal or social relations. Cobb claims that his approach "acknowledges the importance of both cognitive and social processes without subordinating one to the other" (Cobb 1995b: 25).

In order to investigate the reflexivity between social and cognitive aspects of development, Cobb, in collaboration with Erna Yackel and Terry Wood, conducted four case studies. In these, four pairs of second graders cooperated to solve arithmetical problems of addition and subtraction of two-digit numbers. The children's mathematical activity was observed over a 10-week period. The focus of the analyses lay on cognitive as well as on sociological constructs. The cognitive constructs referred to children's construction of increasingly sophisticated conceptions of 10. Sociological aspects referred to the small-group interaction patterns, especially to relations of authority within a small group. The investigation aimed at further clarifying the relation between the different types of social interaction and the learning opportunities that arose from these different interaction patterns.The teacher intervened during small-group work in order to establish social norms of interaction that were considered to be important for productive relationships between the children. These norms included

explaining one's mathematical thinking to the partner; listening to, and attempting to make sense of, the partner's explanations; challenging explanations that did not seem reasonable; justifying interpretations and solutions in response to the partner's challenges; and agreeing on an answer and, ideally, a solution method. (Cobb 1995b: 104)

The analyses indicated that the teacher was quite successful in establishing these norms in the small groups as well as in the classroom.

Concerning the question on how social interactional and cognitive processes are related, the results indicated that different types of interaction varied in their productivity. Cobb distinguishes between univocal and multivocal interactions and between direct and indirect collaboration. The term "univocal" refers to situations in which the perspective of one child dominates, and this "child explains his or her solution and the partner attempts to make sense of the explanations" (Cobb 1995b: 42). Univocal interactions were frequently found to be unproductive and were seen as not giving rise to learning opportunities. The term "multivocal" refers to situations in which "both children attempt to advance their perspectives by explicating their own thinking and challenging that of the partner" (p. 42). Multivocal interactions were usually found to be productive and therefore were judged as giving rise to learning opportunities.

Concerning the distinction between direct and indirect collaboration, the results indicated that the latter was frequently productive, whereas the former usually was not. The term "direct collaboration" was used when the children "explicitly coordinated their attempts to solve a task" (p. 42). Indirect collaboration refers to situations in which "one or both children think aloud while apparently solving the task independently" (p. 42).

The productivity of an interaction pattern was judged in terms of the progress the children made in constructing increasingly sophisticated place-value conceptions. Progress was judged according to three aspects:

1. The quality of the units of 10 and 1 the children created, especially the transition from the construction of numerical composite units of 10 to the construction of abstract composite units of 10.
2. The process of creating these units. Here, the progress was seen in the transition from image-supported to image-independent solutions.
3. The possible developmental history of and metaphors implicit in particular solutions. This third aspect concerns the distinction between counting-based solutions and collection-based solutions.

From a close analysis of both cognitive and sociological constructs and the interplay between these two, Cobb (1995b) obtains the following results: The observed interaction patterns seem to be constrained by the situated cognitive capabilities that the children bring into the small-group relationship. Thus, a small-group interaction in which one child is the mathematical authority might be characterized by univocal explanations of this child to her or his partner. The particular kind of interaction pattern sets the constraints for the occurrence of learning opportunities. In the above-mentioned example, the particular pattern of univocal explanations, which itself is the result of how the children's cognitive capabilities are distributed in the group, would typically not give rise to learning opportunities.

Cobb claims that his social constructivist approach is characterized by a reflexive relation between individual and social processes. This means social processes should reflect individual constructions and these, in turn, should reflect social interactional processes. The one direction, namely, the reflection of individual constructions by social processes, is quite obvious. The situated cognitive capabilities of the children and thereby their individual constructions enter into the small-group interaction and set the constraints of this interaction.

But what about the other direction? How do children's individual constructions reflect social interaction processes? According to this model, the different kinds of relationships among the observed children set the constraints for the occurrence of learning opportunities. In other words, some interaction patterns can give rise to conceptual change.

Concerning the reflexivity of social and cognitive aspects, Cobb seems to be very cautious in drawing conclusions from the results of the case studies. For him

the central notion that characterizes the relationship between the cognitive and social aspects of small-group activity is that of constraints. (Cobb 1995b: 103)

This is in accordance with the statement cited earlier that

students' constructions have an intrinsically social aspect in that they are both constrained by the group's taken-as-shared basis for communication and contribute to its further development. (Cobb 1995b: 26)

Concerning the question in which way individual constructions reflect social relations, it becomes obvious that, for Cobb, it is basically the constraints of an interaction that are reflected by children's constructions. The most he concedes is that the particular type of interaction may have a facilitating effect on constructing mathematical concepts. It is important to note that, for Cobb, the small-group interaction is not the origin of learning; it just may or may not provide learning opportunities. He explicitly states that, in contrast to neo-Vygotskian approaches that locate the origin of conceptual change in social interaction, social constructivist theories locate the origin within the individual.

Such an interpretation raises essentially the same questions as the model proposed by Voigt. If the origin of conceptual change is located primarily, or exclusively, within the individual, what then is the function of social interaction? Why is it that particular kinds of interaction, for example, those that social constructivists subsume under the label "inquiry mathematics," are more productive than traditional forms of mathematical activity? Stating a constraining or facilitating function which fosters the child's constructional activity does not answer the question as long as the mechanisms by which social processes perform these functions are not specified further. Looking at the interaction patterns that were judged as giving rise to learning opportunities, these are multivocal interactions and situations of indirect collaboration. In both types of productive interaction, multiple perspectives are involved; that is, the children are confronted with a view that differs from their own. This observation is in line with Bartolini Bussi's discussion of Bakhtin's notion of multiple voices. According to this approach, the confrontation with different perspectives, or voices, is an important source of conceptual development. Inherent in this view, however, is the acceptance of internalization processes as a decisive mechanism of change. Rejecting the notion of internalization makes it difficult to explain the relation between particular interaction patterns and learning.

Cobb's interpretation of his results leaves several blind spots that cannot be filled by social constructivist theory. Why do some interaction patterns lead to learning opportunities, and how can these learning opportunities be characterized? Following a Vygotskian approach, one would argue that the particular interaction pattern *is* the learning opportunity, and the form and content of this interaction constitute the form and content of conceptual development. Of course, such an interpretation is also based on the assumption that the internalization of social processes is the origin and the basic mechanism of development. Since this assumption is rejected by constructivists, other explanations of the origin and mechanisms of conceptual change would have to be offered. I shall discuss this point in more detail.

Although the quality of children's constructions clearly reflects the quality of the interaction, Cobb nonetheless locates the origin of conceptual development within the individual. In fact, for Cobb, the location of the source of change is the decisive difference between his approach and neo-Vygotskian perspectives. So, the main function of social interaction in social constructivist theories is that of providing opportunities for the child to develop further her or his autonomous individual constructions. Concerning the question how individual constructions reflect social processes, the most that can be said is that they reflect the constraints set by the interaction.

Cultural aspects of development

In their discussion of the reflexive relation between individual and social processes, Cobb and Voigt restrict their view to microsocial processes, be they small-group interactions or classroom processes. Only on this level are social processes seen as contributing to the development of children's mathematical concepts, and even there – as shown previously – only in a narrowly constrained sense. Cultural aspects are rarely discussed in social constructivist theories and are usually not seen as contributing to conceptual development or as being an integral part of the construction process or the constructed concepts. This neglect of the cultural dimension is based on central theoretical assumptions of constructivism whereby individual autonomous construction is the origin of conceptual development and precedes symbolization (Cobb 1995a: 383). This stands in sharp contrast to sociocultural theories that locate the origin of conceptual development in social processes and view cultural tools and sign systems as constitutive of development. According to Vygotsky, the conceptual structure that a child builds up is dependent on the tools of thinking available to this child (Vygotsky 1978).

Cobb discusses these differences between the two theoretical approaches, especially the question where the origin of cognitive development has to be located. In the realm of his case studies, he investigates the effects of a cultural tool on conceptual change.

During problem solving, the pairs of children had the opportunity to work with a hundreds board. The hundreds board is used as an instructional device to help children in constructing place-value numeration concepts. The investigation was based on the following assumptions: According to sociocultural approaches, cultural tools are conceptualized as carriers of meaning. The hundreds board, as an example of a cultural tool, carries, according to these theories, within it the meaning of the place-value concept. When children are confronted with this tool and are given the opportunity to work with it, they should, if these theories are valid, appropriate this meaning. This should result in the acquisition of an advanced conception of place value. Furthermore, the appropriated concept should be the same for all children who worked with the hundreds board.

Results showed that some children used the hundreds board and others did not, and the way in which it was used varied among the children. It was concluded that the hundreds board had different meanings for different children, and that these meanings depended on the level of sophistication of the underlying concepts the children had developed until then. It was further observed that only those children who had relatively advanced mathematical concepts were able to use and profited from the hundreds board. Children with less advanced concepts either could not use the hundreds board or used it in a rather restricted way. And, most important, they did not seem to profit from the use of the hundreds board.

From this observation, Cobb develops his main argument against the Vygotskian view that cognitive development originates in the appropriation of cultural tools and against the view that cultural tools are carriers of meaning. He states that the "children's use of the hundreds board did not support their construction of increasingly sophisticated conceptions of ten" (Cobb 1995a: 375f.), and further, "It instead appears that the children's efficient use of the hundreds board was made possible by their construction of increasingly sophisticated place-value conceptions" (p. 376). In other words, it is not cultural tools that allow and support cognitive development; on the contrary, it is cognitive development that enables children to use cultural tools. Cobb further concludes, first, that cultural tools are not carriers of meaning, and, second, that the origin of concept development does not lie within cultural tools and sign systems but within the individual child.

In my view, the conclusions Cobb draws from his observations are not justified, because they are based on invalid assumptions about the nature and function of cultural tools in Vygotskian approaches.

In delineating the differences between constructivist and sociocultural approaches, Cobb creates a dichotomy that I find to be – at least partly – artificial. Cobb claims that, in contrast to sociocultural theories, "constructivists characterize mathematical learning as a process of conceptual reorganization" (1995a: 364). The alternative to conceptual reorganization offered by Vygotskian approaches is said to be characterized by such processes as appropriation and enculturation. Meaning is seen as originating either in individual construction or in culturally developed tools and sign systems. By means of internalization, individuals appropriate the meaning that these tools and sign systems carry.

Such a description, however, denies the complexity of the appropriation process as described by Vygotsky (1987). In chapter 6, Development of scientific concepts, of his *Thinking and speech,* Vygotsky argues vehemently against the view that "scientific concepts do not have their own internal history, that they do not undergo a process of development in the true sense of the word"; that they are "simply learned or received in completed form through the processes of understanding, learning, and comprehension"; that they are "adopted by the child in completed form from the domain of adult thinking" (p. 169). Instead, Vygotsky views concept development as a "complex and true act of thinking" (p. 175). Further, he rejects the view that that which is "new to development arises from without" and that "the child's characteristics have no constructive, positive, progressive, or formative role in the history of his mental development" (p. 175). Again, he emphasizes that

scientific concepts are not simply acquired or memorized by the child and assimilated by his memory but arise and are formed through an extraordinary effort of his own thought. (p. 176)

Another important point in Vygotsky's theory of concept development is the aspect of restructuring:

Consciousness develops as a whole. With each new stage in its development, its internal structure – the system of connections among its parts – changes. (p. 187)

Thus, according to Vygotsky, the appropriation of scientific concepts is in no way a simple matter of internalizing formerly external systems of meaning. On the contrary, it is a developmental and constructional process. Scientific concepts undergo a development within the child, and this development is characterized by a highly complex interaction between already existing concepts and new concepts, on the one hand, and by the individual's active constructional activity on the other. Moreover, the appropriation of scientific concepts is accompanied by a reorganization of the existing conceptual structure of the child. This conceptualization of the appropriation process integrates the seemingly contradictory concepts of internalization and construction by postulating a reflexive, dialectical relation between them.

Following Vygotsky, cultural tools and sign systems are not carriers of meaning in the sense Cobb uses this term; that is, they do not carry their meaning to children. Instead, for any individual, cultural tools derive their meaning from his or her constructional activity. Since this constructional activity is characterized by an interaction with already existing concepts, construction processes vary among individuals. The fact that the children in the case studies used the hundreds board in different ways does not contradict a Vygotskian approach.

The observation that the children gave different meanings to the hundreds board and that these meanings were dependent on the conceptions of place value the children had developed until then is in accordance with Vygotsky's notion of the zone of proximal development. Children are only able to appropriate those concepts that are not too far beyond their existing knowledge. And they tend to perceive concepts according to their level of conceptual development. Cobb notes the similarity between his observations and Vygotsky's approach (he refers to Vygotsky's example of how people with different levels of expertise perceive a chess board) and acknowledges that Vygotsky was interested in qualitative developmental change. Cobb sees the point that separates the sociocultural from the constructivist perspective as the way in which these changes are accounted for. In contrast to Vygotskian approaches, which locate the origin of change in external sign systems, constructivist approaches focus on individual children's cognitive constructions. Here again, a dichotomy is created that is partly artificial. As argued, the fact that the origin of higher psychological processes is seen in "extracerebral" sign systems does not exclude the possibility of construction.

Finally, it can be questioned whether the hundreds board can legitimately be called a cultural tool. Cobb himself points to the fact that it may be seen as an instructional device that may or may not facilitate the appropriation of a culturally specific tool, namely, our numeration system (Cobb 1995a: 379–380). The view of our numeration system as a conceptual tool is much more in line with Vygotskian conceptions of cultural tools. From this perspective, the results of the case studies do not contradict a sociocultural approach. On the contrary, they are difficult to explain within a constructivist paradigm.

There are three aspects which, in my view, are especially in need of explanation: first, the norms of interaction; second, authority and interaction; and, third, the direction of development. Cobb reports that the teacher who took part in the investigation made a great effort to introduce certain norms into the classroom as well as into the small-group interaction. As described previously, these included listening to others; trying to make sense of and eventually challenging their explanations; explaining one's own thoughts; justifying them in response to challenges; and finally agreeing on an answer and, ideally, a solution method.

The process of establishing these norms lasted several months, and the teacher frequently intervened in the small-group interaction to make sure that the children followed the rules. The fact that it takes effort and time to establish certain norms of interaction makes it clear that these are neither constructed individually nor naturally occurring. Instead, it seems plausible that they constitute a powerful tool that has developed historically. The observation that interaction patterns characterized by these norms are more productive than others in helping children build up advanced conceptual structures suggests that these formerly external tools of communicative exchange have become internal tools of thinking.

The second point concerns the question in which way the situated cognitive capabilities of the children lead to a certain interaction pattern. Cobb suggests that it is mainly the distribution of authority within a group that sets the constraints of the interaction. Authority, however, is not a cognitive construct but a social–psychological one. If one child in the group is the mathematical authority and the interaction is characterized by univocal explanations, then this is not due to the advanced conceptual structure of this child but to the meaning the children give to this distribution of knowledge. And again, this meaning is probably not constructed individually but the result of the children's experiences with authority and knowledge.

The third point concerns the direction of development. Cobb observes – and expects – a certain development of place-value numeration concepts. He speaks, for example, of more or less sophisticated conceptions of 10. What does sophistication mean here? According to Cobb, the direction of development leads, among other routes, from concrete to abstract composite units of 10 and from image-supported to image-independent arithmetical problem solving. It leads from counting by 1s to counting by 10s. It is well known that children in different cultures develop different numeration concepts (see, e.g., Saxe 1991). How can this be explained in constructivist terms? Concerning this question, Cobb points to the different research interests of constructivists on the one side and sociocultural theorists on the other. Vygotskian approaches focus on differences between cultures, and, in order to explain these differences, refer to cultural tools and sign systems that vary across cultures and thus produce varying systems of meaning. Constructivists, on the other hand, focus on individual differences within a certain culture that cannot be explained by referring to cultural tools since, following this argument, the same cultural tool should not produce different concepts or meanings. As discussed, this view of cultural tools has to be rejected. Individual differences do not contradict Vygotskian theory. Besides, these differences are located within a certain realm that is set by the cultural tools and sign systems of a specific culture. The variances Cobb observes in his case studies may be great, but they are all differences in constructing a culture-specific place-value numeration system.

In my view, these epistemological differences center around the concept that, for Cobb, marks the limitation of Vygotskian theory:

The limitation of adopting solely a sociocultural perspective becomes apparent when we consider one of the central theoretical constructs that Vygotsky used to account for development, that of internalization. (Cobb 1995a: 380)

He sees the specific contribution of the constructivist perspective as providing an account of the appropriation process "without recourse to the notion of internalization" (p. 382).

The rejection of the notion of internalization leaves very few possibilities to account for conceptual development. Constructivists often refer to the work of Piaget as a psychological theoretical foundation (see, e.g., von Glasersfeld 1982; Confrey 1994). Piaget's theory, however, is dialectical in nature. The individual's internal structures develop as the result of an interaction between the cognizing subject and the external objective world. The formal aspects of this interaction are abstracted and interiorized (Piaget 1948, 1974; Furth 1969). According to Piaget, development is characterized by two complementary processes: assimilation and accommodation. Assimilation concerns the tendency of the organism to incorporate the external world according to the existing internal structures. In other words, the cognizing subject tends to interpret external objects and events in terms of already established meanings.

This tendency is emphasized by the constructivist perspective. Cobb, for example, cites Blumer's (1969) argument that people respond to signs and symbols in terms of the meanings that they have for them. Cobb's argument concerning cultural tools points in the same direction.

The hundreds board is structured by the children in accordance with their already existing conceptual structures. Only children with relatively sophisticated place-value conceptions, that is, those children who could create composite units of 10 in an image-independent manner, were able to use the hundreds board in such a way that they counted by 10s and 1s. Thus, children tend to transfer their internal conceptual structure to external objects or events. In Piaget's terms, they assimilate the external world to their internal operational structures.

Constructivists lay great emphasis on the assimilative aspects of development. For Piaget, however, assimilation is only half the story. The other half is characterized by the process of accommodation. Accommodation is the "answer" of internal structures to disturbances in the assimilative process. It is the tendency of the organism to change its structures according to the experienced structure of external objects or events. Thus, there is a constant interplay between these two tendencies aiming toward a state of equilibrium between internal and external structures. This interplay, in Piaget's terms the process of equilibration, makes development possible. If subjects always interpreted things (signs, objects, events, concepts) in terms of the meanings they have for them, as Cobb's quotation of Blumer suggests, development would stand still. At the same time, it gives development its direction in such a way that it is determined by a dialectical relation between internal and external structures. This part of Piaget's theory is not just neglected by constructivists; it is rejected explicitly.

Central to Piaget's theory is the notion of interiorization. As the child acts on her or his outer environment, the equilibration process described results in an interiorization of certain, namely, the formal, aspects of the outcome of these actions. Following Piaget, the origin of conceptual development does not lie within the individual but in the dialectical relationship between internal and external aspects. If the central limitation of Vygotskian theory is the notion of internalization, then this should also be the central limitation of Piagetian theory.

Arguments against the Vygotskian notion of internalization often seem to be based on the tacit assumption that the underlying epistemological basis is a representational theory of development whereby the external world is somehow mirrored by the internal structures (see, e.g., Cobb and Yackel this volume; Cobb, Yackel, and Wood 1992). External objects or concepts are passively incorporated by the individual without transformation. The individual makes no active contribution to building up her or his conceptual structures. These theories leave no room for the notion of active individual construction. Such a view of internalization is reflected in Cobb's argument against the notion of cultural tools as carriers of meaning. As discussed, a representational view of development does not apply to Vygotsky. Instead, the internalization process as described by Vygotsky involves an interaction between the subject's existing concepts and new concepts. This interaction involves the active construction of new concepts by the child, and the developmental process is characterized by conceptual reorganization (for a detailed analysis of the relation between representation and construction, see Seeger this volume).

In both theories, Piagetian and Vygotskian, interaction plays a central role, and in both theories, certain aspects of this interaction are internalized. For Piaget, it is the interaction of the cognizing subject with her or his objective reality. The internalization of the formal aspects of this interaction is the condition for the internal conceptual structures to develop. Social interaction, however, is secondary, since, according to Piaget, development is a unidirectional process and the resulting cognitive structures are universal. They are not subject to any form of negotiation. The structures of knowledge are not taken-as-shared; they are shared. Although there are tendencies to interpret Piaget's theory in such a way that social interaction is seen as an integral part of conceptual development, these approaches do not touch the basic epistemological assumptions of the theory, namely, the dialectical relation of internal and external aspects and the notion of internalization (see, e.g., Youniss 1994; Chapman 1988).

According to Vygotsky, it is social interaction and the interaction between the subject and cultural tools and sign systems that are internalized – by way of active construction – and that form the basis of conceptual development.

Thus, in both theories, the notion of interaction is linked closely to the notion of internalization, and, in both cases, internalization is the *conditio sine qua non* of conceptual reorganization and development.

Social constructivist theories, on the other side, seem to lay great stress on social interaction, but, at the same time, reject the notion of internalization. So the question remains how constructivists conceptualize the relation between social interaction and conceptual development. It is hard to understand how social interaction should become an integral part of the constructed concepts, as Cobb claims, without conceding that parts of this interaction are internalized. Likewise, it is hard to see how conceptual development, which implies a change in meaning, should come about when the subject interprets objects, signs, and concepts solely in terms of already existing systems of meaning. Finally, by locating the origin of change exclusively within the individual, it is hard to explain why development takes a certain direction.

Constructivists try to resolve this dilemma with recourse to the concept of negotiation. In the process of negotiation, a socially accepted meaning is established that is taken as shared by the members of a certain group, be it a pair of children working together, a classroom, a scientific community, or a society as a whole. Thus, negotiation may explain why communication and cooperation between different individuals is possible. However, as I have tried to show, without accepting some kind of internalization process, it cannot explain individual conceptual development. Taking a close look at individual constructions, as constructivists do, provides a valuable description of the development of these constructions, but it says little or nothing about the mechanisms of change.

Sometimes it is suggested that the two competing theories, constructivism on the one side and Vygotskian approaches on the other, may complement each other (Cobb 1995a; Confrey 1995; Ernest 1994). Confrey, for example, states that sociocultural approaches are looking at wider society while constructivism is concerned more with the individual. Cobb states that both theories address different issues. Together they might provide a more complete picture of development.

Such statements are based on the assumption that constructivist approaches take cognitive psychological as well as sociological aspects into account, whereas Vygotskian approaches focus solely on the latter. Moreover, it is said that Vygotsky subordinates psychological aspects to sociological ones. Besides the fact that the social constructivist conception of sociological aspects is limited to face-to-face interactions in the classroom, which does not seem to be an adequate model for human communities (cf. Latour 1996), this distinction also seems to be artificial. It suggests a separation between cognitive, psychological, and individual processes on the one side and sociological, cultural, and interactional processes on the other. It fosters the view that cognitive psychological processes are internal whereas external processes are sociological; that is, they are noncognitive, nonpsychological. By citing Vygotsky's famous genetic law of cultural development whereby internal higher mental functions formerly were external, it seems only plausible to conclude that Vygotsky subordinates psychological to sociological aspects. However, the distinction Vygotsky makes between external and internal processes is that between inner-psychological and interpsychological ones. Thus, the main distinction between constructivist and Vygotskian approaches is that Vygotsky locates psychological processes not just within individuals. Psychological processes also take place between individuals. This is crucial, since it shows that the difference between the two conflicting approaches is not just a matter of focus. It rather concerns the origin of conceptual change and thereby the possibilities for a better understanding of teaching–learning processes in the classroom. It is therefore not justified to create a contrast in which individual and cognitive aspects of development are labeled psychological whereas interindividual and cultural aspects are labeled sociological, and then ascribe the first label to constructivism and the second to Vygotskian approaches. Indeed, the central claim of Vygotskian theory is that these levels – cultural, interpsychological, and inner-psychological – cannot be separated. It is this claim that provides the theory's revolutionary potential. To state that sociocultural approaches are valuable if they confine themselves to the description and explanation of cultural differences is nothing other than to reject Vygotskian theory altogether.

The more serious argument against the view of two complementary theories, however, lies in the weaknesses of constructivism itself. The claim to cover the inner-psychological aspects of conceptual development cannot be sustained. Constructivists reject the view of internalization as a central mechanism of development but fail to offer an alternative explanation. The introduction of the concept of negotiation does not solve this dilemma. Instead, it raises the question as to what is the sense of looking at social interaction when nothing of it is internalized anyway.

References

Bauersfeld, H. (1978). Kommunikationsmuster im Mathematikunterricht. Eine Analyse am Beispiel der Handlungsverengung durch Antworterwartung. In H. Bauersfeld (Ed.), *Fallstudien und Analysen zum Mathematikunterricht*. Hannover: Schroedel.

Blumer, H. (1969). *Symbolic interactionism: Perspectives and method*. Englewood Cliffs, NJ: Prentice-Hall.

Brown, J.S., Collins, A., and Duguid, P. (1989). Situated cognition and the culture of learning. *Educational Researcher,* 18 (1), 32–42.

Chapman, M. (1988). Contextuality and directionality of cognitive development. *Human Development,* 31, 92–106.

Cobb, P. (1995a). Cultural tools and mathematical learning: A case study. *Journal for Research in Mathematics Education,* 26 (4), 362–385.

Cobb, P. (1995b). Mathematical learning and small group interaction: Four case studies. In P. Cobb and H. Bauersfeld (Eds.), *Emergence of mathe-*

matical meaning: Interaction in classroom cultures. Hillsdale, NJ: Erlbaum.

Cobb, P, Yackel, E., and Wood, T. (1992). A constructivist alternative to the representational view of the mind in mathematics education. *Journal for Research in Mathematics Education,* 23 (1), 2–33.

Confrey, J. (1994). A theory of intellectual development. *For the Learning of Mathematics,* 14 (3), 2–8.

Confrey, J. (1995). A theory of intellectual development. *For the Learning of Mathematics,* 15 (1), 38–48.

Ernest, P. (1994). Constructivism: Which form provides the most adequate theory of mathematics learning? *Journal für Mathematik-Didaktik,* 15 (3/4), 327–342.

Furth, H. (1969). *Piaget and knowledge: Theoretical foundations.* Englewood Cliffs, NJ: Prentice-Hall.

Latour, B. (1996). On interobjectivity. *Mind, Culture, and Activity,* 3(4), 228–245.

Piaget, J. (1948). *Die Psychologie der Intelligenz.* Zürich: Rasche.

Piaget, J. (1974). *Biologie und Erkenntnis.* Frankfurt am Main: Fischer.

Richards, J. (1991). Mathematical discussions. In E. von Glasersfeld (Ed.), *Radical constructivism in mathematics education.* Dordrecht: Kluwer.

Saxe, G. B. (1991). *Culture and cognitive development: Studies in mathematical understanding.* Hillsdale, NJ: Lawrence Erlbaum.

Voigt J. (1984). *Interaktionsmuster und Routinen im Mathematikunterricht.* Weinheim: Beltz.

Voigt, J. (1995). Thematic pattern of interaction and sociomathematical norms. In P. Cobb and H. Bauersfeld (Eds.), *The emergence of mathematical meaning: Interaction in classroom cultures.* Hillsdale, NJ: Lawrence Erlbaum.

von Glasersfeld, E. (1982). An interpretation of Piaget's constructivism. *Revue Internationale de Philosophie,* 142 (3), 612–635.

Vygotsky, L.S. (1978). *Mind in society: The development of higher psychological processes.* Cambridge, MA: Harvard University Press.

Vygotsky, L.S. (1987). Thinking and speech. In R.W. Rieber and A.S. Carton (Eds.), *The collected works of L.S. Vygotsky,* (Vol. 1, Problems of general psychology, pp. 38–285). New York: Plenum Press.

Youniss, J. (1994). *Soziale Konstruktion und psychische Entwicklung.* Frankfurt am Main: Suhrkamp.

Part III

Epistemology and classroom culture

9 The culture of the mathematics classroom and the relations between personal and public knowledge: An epistemological perspective

PAUL ERNEST

This chapter is a contribution to an exploration of the culture of the mathematics classroom. It adopts an epistemological perspective, which immediately raises a number of questions.[1]

- What can be learned from an epistemological perspective on the culture of the mathematics classroom?
- Given that the central concern of epistemology is knowledge, what role does mathematical knowledge in its various manifestations play in the culture of the classroom?
- How can the specific character of mathematical knowledge be described?
- Does it change during the transition from the discipline into the classroom?
- What is the relationship between personal and public knowledge of mathematics?[2]
- What is the relationship between the cultures of school mathematics and research mathematics?
- How does mathematical knowledge give rise to epistemological obstacles?
- Which specific epistemological structures of, and processes related to, school-mathematical knowledge are generated by the everyday culture of the mathematics classroom?
- What is the relationship between "structural necessities" and "social conventions" of mathematical knowledge in the educational interaction?

These are some of the deep questions raised by an epistemological perspective on the culture of the mathematics classroom. Clearly in this chapter it is not possible to do more than clarify some of the questions, raise some of the issues, and sketch some of the directions from which answers might come. However I do offer the sketch of a theoretical model of mathematical knowledge genesis and warranting that provides a unified position from which such questions can be addressed. The types of issues that I discuss thus include the nature and constitution of knowledge, its justification and assessment, developmental perspectives of knowledge, the social basis of knowledge-acceptance, and the conversational interactions that shape both personal and public knowledge of mathematics.

In order to begin my account right at the focus point of this collection, namely, the culture of the mathematics classroom, consider the following cameos, which provide a concrete starting point for the discussion that follows.

Cameo 1

A teenage girl sits at a desk at the front of a present day mathematics classroom. In front of her are an open school mathematics text, an exercise book, an electronic calculator, and a set of mathematics instruments and pens. She is working in the exercise book at a routine trigonometry task. She finishes it and calls the teacher over. The teacher leans over her desk and reads through the work. She modifies some of the work with a red pen, says "That's the way to do it," and writes "Good work" at the bottom of the page. The student begins to work at the next trigonometry exercise in the text.[3]

Cameo 2

A mathematician is sitting at her desk with an open pad of paper in front of her. She is writing mathematical symbols with a thick black pen on the page in front of her. Around her are strewn pages of notes, diagrams, and reference books, and a computer sits humming on the corner of her desk. She finishes her writing, turns back to the first page of her pad, and starts to type the text and symbols into the computer. A few days later, she sends four copies of her completed typescript to the editor of a mathematics journal. The editor sends a copy to each of two mathematicians who are specialists in the same field. The first referee carefully reads the typescript, annotating it. Then she writes a report to the editor indicating a weak point in the proof and one or two typing errors. On the basis of the referees reports, the editor writes to the mathematician that she is happy to accept the paper subject to certain specific improvements being made.

These cameos epitomize some of the similarities and differences between school mathematics and research mathematics, that is, between the development of personal knowledge of mathematics and public

Table 9.1. *Differences between school and research mathematics*

School mathematics	Research mathematics
Studied by all children	Done by a tiny group of adults
Learning of existing knowledge	Creation of new knowledge
A preparatory learning activity, the acquisition of skills and knowledge	Based on the preparatory activity: deploying its skills and knowledge
Tasks imposed by teacher/text	Problems self-posed or selected
Comprises short-term tasks	Consists of long-term problems
Results in short written "answer"	Results in longer written outcome
Criteria for acceptance imposed	Criteria for acceptance shared
Interactions face-to-face	Interactions at a distance
Matters only to the learner	Becomes part of public knowledge

mathematical knowledge. Some of the differences and similarities are shown in Tables 9.1. and 9.2.

School and research mathematics constitute different discursive practices and different sites or contexts for the production (or reproduction) of mathematical knowledge. It is often recognized that there are significant differences between them. These are most evident in the genesis of mathematical knowledge, i.e., during its production or reproduction. But there are also significant similarities between them, as Table 9.2 shows. These are most striking during the warranting of mathematical knowledge. To explore these areas of similarity and difference, and in particular, the parallels between the warranting of school and research mathematics, it is necessary to look a little more deeply at the nature of mathematical knowledge.

Table 9.2. *Similarities between school and research mathematics*

School mathematics	Research mathematics
Based on school texts	Based on mathematics texts
Answers to school problems	Answers to disciplinary problems
Submitted to teacher	Submitted to editor
Produced and validated by dialectical pupil--teacher interaction	Produced and validated by dialectical mathematician--editor interaction
Acceptance depends on presenting text in conventional form	Acceptance depends on presenting text in conventional form
Acceptance depends on judgment of teacher	Acceptance depends on judgment of editor/referees
Judgment based on shared criteria and values of school mathematics culture	Judgment based on shared criteria and values of research mathematics culture

Reconceptualizing mathematical knowledge as situated

Traditionally, knowledge has been understood as a collection of validated propositions, which in the case of mathematics means mainly a set of theorems with proofs. However, philosophers such as Wittgenstein, Ryle, Polanyi, and Kuhn argue that not all knowledge can be made explicit.

Ryle (1949) distinguishes between an individual's "knowing that" concerning explicit knowledge of propositions and "knowing how" referring to practical knowledge, skills, or dispositions. Polanyi (1959) terms the latter "tacit" or personal knowledge, and Kuhn (1970) explicitly includes a tacit component in his account of scientific knowledge (for simplicity I shall identify "tacit knowledge" and "know-how" as the implicit part of "personal knowledge").

The motivation for including tacit "know-how" as well as propositional knowledge as part of mathematical knowledge is that it takes human understanding, activity, and experience to make or justify mathematics; in short, mathematical know-how is needed. Most personal knowledge in mathematics is not of the explicit propositional sort, but consists of tacit knowledge of methods, approaches, and procedures, which can be applied in new situations and to problems. Much that is accepted as mathematical knowledge consists of being able to carry out certain symbolic or conceptual procedures or operations. To know the addition algorithm, proof by induction, or definite integrals is to be able to carry out the operations involved, not merely to be able to state certain propositions. Thus what an individual knows in mathematics, in addition to publicly statable propositional knowledge, includes her mathematical "know-how."

In mathematics education, this has been recognized for some time. Skemp (1976) and Mellin-Olsen distinguish instrumental and relational understanding, and the former category can be understood as concerning tacit knowledge of mathematical algorithms and procedures. (There is also an implicit evaluation attached, to the effect that the knowledge is not richly connected with other knowledge.) More recently Hiebert (1986) and others have distinguished conceptual and procedural knowledge of mathematics, offering a further parallel in the domain of psychology of mathematics education.

Kuhn (1970) argues that scientists share knowledge in many forms, including statements, beliefs, values, techniques, and concrete problem solutions or exemplars. Thus it is

concrete puzzle solutions which, employed as models or examples, can replace explicit rules as a basis for the solution of the remaining puzzles of normal science. (Kuhn 1970: 175)

By it [exemplars] I mean, initially the concrete problem-solutions that students encounter from the start of their scientific education, whether in laboratories, on examinations, or at the ends of chapters in science texts. To these shared examples should, however, be added at least some of the technical problem-solutions found in the periodical literature. (Kuhn 1970: 187)

Philip Kitcher extends Kuhn's model of scientific knowledge to mathematics in his account of a "mathematical practice." This serves as a rich model of mathematical knowledge which includes both explicit propositional and tacit knowledge components. He includes

five components: a language, a set of accepted statements, a set of accepted reasonings, a set of questions selected as important, and a set of meta-mathematical views (including standards for proof and definition and claims about the scope and structure of mathematics). (Kitcher 1984: 163)

To extend and elaborate Kitcher's model, the first of these five components is the underlying language of mathematics: a mathematical sublanguage of English, German, or another language, supplemented with specialized mathematical symbolism and meanings. This comes equipped with an extensive range of mathematical objects, including mathematical symbols, notations, diagrams, terms, concepts, definitions, axioms, statements, analogies, problems, explanations, methods, proofs, theories, and texts.

The third component is the set of accepted reasonings, including the justificatory warrants or proofs accepted by the mathematical community as correct. It includes informal and traditional proofs, which although not accepted as adequate proofs are regarded as having some valuable content. It also includes accepted problem solutions, comprising analyses and computations.

Last, there is the set of metamathematical views, which includes

at least the following issues: (i) standards of proof [and definition]; (ii) the scope of mathematics; (iii) the order of mathematical disciplines; (iv) the relative value of particular types of inquiry. (Kitcher 1984: 189)

The first part is a set of standards, the norms and criteria that the mathematical community expect proofs and definitions to satisfy if they are to be acceptable to that community. Kitcher claims that it is not possible for the standards for proof and definition in mathematics to be made fully explicit. Exemplary problems, solutions, definitions, and proofs serve as a central part of the accepted norms and criteria that definitions, proofs, and mathematical knowledge are expected to satisfy. Like Kuhn, he argues that proof standards may be exemplified in texts taken as a paradigm for proof (e.g., Euclid's Elements), rather than in explicit statements. In this, he is in agreement with fallibilists who deny that mathematical proof standards are absolute (Davis and Hersh 1980; Lakatos 1976; Tymoczko 1986; Ernest 1991).

From the point of view of the acquisition of personal knowledge of mathematics, that is, from an educational perspective, this broadening of the concept of knowledge is very significant. Not only does it mean that mathematical knowledge extends beyond the explicit to include a tacit dimension, but that beyond the abstract and general knowledge of the results, methods, and language of mathematics there is an important concrete knowledge of mathematics. This includes knowledge of instances and exemplars of problems, situations, calculations, arguments and proofs, applications, and so on. Knowledge of particulars as well as of generalities thus plays an important part in mathematics. This has been recognized in some mathematics education research. For example, Schoenfeld (1985; 1992) in his research on mathematical problem solving, argues that experience of past problems leads to an expanding knowledgebase which underpins successful problem solving. Too often we ignore the concrete in mathematics education and view the general outcomes of learning as the only ones that matter.

One of the outcomes of Kuhn's, Kitcher's and others' views of knowledge as shared among members of a culture or community is that it can no longer be regarded as exclusively sited in an individual's mind. This is the dual of another misconception, that public knowledge is external to humanity.[4] Not surprisingly, these views arise from a dualistic view of the mind and body. The view that personal knowledge is contained in the mind is an assumption that underpins many learning theories of today, both cognitivist and constructivist, for many such theories take knowledge to be organized as mental structures, schemata, etc.

However, there is a modern tradition that rejects this view. Philosophically, the alternative view can be found in Wittgenstein (1953), who argues that knowledge and meaning reside in various "language games" embedded in organized social "forms of life." This view suggests that mathematical knowledge items differ in significance and meaning according to the context. From the perspective of the mathematics classroom, it can be said that there are already extensive developments in psychology of mathematics education which recognize this, including various forms of "situated cognition" of Lave and Wenger (1991), Vygotsky (1978), Saxe (1991), Walkerdine (1988), and others. Typical of these views is the notion that some parts of personal mathematical knowledge relate to shared social activities, namely, the contexts of acquisition and use. In particular, tacit aspects of mathematical knowledge are often elicited in their contexts of origination as part of an automatic component of recognition and engagement with the situations. Research on the transfer of learning suggests that particular knowledge and tacit knowledge do not transfer well from the context of acquisition, whereas general and explicit knowledge is more susceptible to transfer (Kilpatrick 1992).

Justification and acceptance of mathematical knowledge

An important issue is that knowledge *qua* knowledge must have a warrant or form of justification. Tacit knowledge can only be legitimately termed knowledge if there are grounds for asserting it or if it is justified. Of course, since the knowledge is tacit, then so too its justification must, at least in part, be tacit, on pain of contradiction. Thus the validity of some tacit knowledge will be demonstrated implicitly by the individual's successful participation in some social activity or form of life. Not in all cases, however, need the justification be tacit. For example, an individual's tacit knowledge of the English or German language is likely to be justified and validated by conformity to the public norms of correct grammar, meaning, and language use (as related to the context of use). Thus a speaker's production of a sufficiently broad range of utterances appropriately in context can serve as a warrant for that speaker's knowledge of English. This position fits with the view of knowledge in Wittgenstein's (1953) later work, in which to know the meaning of a word or text is to be able to engage in the appropriate language games and forms of life (i.e., to "use" it acceptably). Practical "know-how" is also likely to be validated by public performance and demonstration. To know language is to be able to use it to communicate (Hamlyn 1978). As Ayer says "To have knowledge is to have the power to give a successful performance" (Ayer 1956: 9). Such a validation is to all intents and purposes equivalent to the testing of scientific theories in terms of their predictions. It is an empirical, predictive warrant. It is a weaker warrant than a mathematical proof, for no finite number of performances can exhaust all possible outcomes of tacit knowledge as a disposition (Ryle 1949), just as no finite number of observations can ever exhaust the observational content of a scientific law or theory. Thus tacit knowledge of mathematics can be defended as warranted knowledge provided that it is supported by some form of justification or validation (possibly tacit) of which the knower or another judge of competence is, or can become, aware. What an individual knows in mathematics, in addition to publicly statable propositional knowledge, includes her mathematical "know-how."

This view fits well with the practices of assessment of knowledge and understanding in mathematics education, for requiring full explicit statements of propositional knowledge as evidence of personal knowledge is generally regarded as a very low level of assessment. The recall of such explicit verbal statements is typically placed at the lowest level of educational taxonomies of knowledge such as that of Bloom et al. (1956). Of course to demonstrate that verbal statements are a part of personal knowledge, as opposed to personal belief, necessitates the knower's demonstrating her possession of a warrant for that knowledge, typically a proof in mathematics. To be able to produce a warrant for an item of knowledge, and to explain why it is a satisfactory warrant, is a higher-level skill (perhaps at the level of "evaluation" in Bloom's taxonomy).

Although Bloom's taxonomy cannot by any means be taken as ultimate, as the profound criticisms by Freudenthal (1978) and others show, it does serve to show something that is near universally accepted in Western education, namely, that the reproductive recall of verbal knowledge is rated lower than the application of personal knowledge.

At the highest level of Bloom's taxonomy is evaluation: the ability to critically evaluate the knowledge productions of self and others. This is clearly a necessary skill for the teacher of mathematics. It is also a necessary skill for the research mathematician, not only as a means to judge the mathematical knowledge productions of others, as in Cameo 2, but as a skill that the mathematician internalizes and applies to her own work.

This section illustrates an unremarked parallel between the justification of mathematical knowledge and the assessment of mathematical learning. In the case of tacit mathematical knowledge these are identical, namely, the production of behaviors providing evidence for the possession of tacit personal knowledge. Of course there are differences of emphasis: epistemology focuses on the logic and principles, whereas educational assessment is concerned with techniques and specifics.

Although traditionally it has been thought that the acceptance of mathematical knowledge depends on logically correct proof, there is growing recognition that proofs do not follow the explicit rules of mathematical logic, and that acceptance is instead a fundamentally social act (Lakatos 1976; Kitcher 1984; Tymoczko 1986). A social view of proof is also shared by a growing number of mathematicians and philosophers, such as Manin, Beth and Piaget, Wilder, and Rorty. From this perspective, the structure of a mathematical proof is a means to its epistemological end of providing a persuasive justification, a warrant for a mathematical proposition. To fulfill this function, a mathematical proof must satisfy the appropriate community, that of mathematicians, that it follows the currently accepted if largely tacit criteria for a mathematical proof.

A proof becomes a proof after the social act of "accepting it as a proof." This is as true of mathematics as it is of physics, linguistics and biology. (Manin 1977: 48)

If, however, we think of "rational certainty" as a matter of victory in argument rather than of relation to an object known, we shall look toward our interlocutors rather than to our faculties for the explanation of the phenomenon. If we think of our certainty about the Pythagorean Theorem as our confidence, based on experience with arguments on such matters, that nobody will find an objection to the premises from which we infer it, then we shall not seek to explain it by the relation of reason to triangularity. Our certainty will be a matter of conversation between persons, rather than an interaction with nonhuman reality. (Rorty 1979: 156–157)

Thus, the deployment of informed professional judgment is what underpins the acceptance of public mathematical knowledge, not the satisfaction of inflexible logical rules.

Likewise in mathematics education, the teacher's decision to accept mathematical answers in a student's work depends on the teacher's professional judgment, not on some inflexible rules of what is correct and incorrect. As in the case of public mathematical knowledge, such judgments relate to the shared criteria, practices, and culture of the mathematics education and mathematics communities. This insight raises the issue of what criteria are involved in professional judgments in such communities. My claim is that there are a variety of rhetorical styles that learners are expected to master during their years of schooling and that judgments as to acceptability have as much to do with rhetorical style as with content.

The rhetoric of mathematics and the classroom

Recently, there has been a growth of a rhetoric of science movement, concerned to describe the stylistic forms used by scientists to persuade others of the validity of their knowledge claims (Latour, Knorr-Cetina, Woolgar). Instead of being used pejoratively, the word "rhetoric" is used to indicate that style is inseparable from content in scientific texts.

Scholarship uses argument, and argument uses rhetoric. The "rhetoric" is not mere ornament or manipulation or trickery. It is rhetoric in the ancient sense of persuasive discourse. In matters from mathematical proof to literary criticism, scholars write rhetorically. (Nelson, Megill, and McCloskey 1987: 3–4)

At base, "rhetoric is about persuasion" (Simons 1989: 2), and logic and proof provide the strongest rational means of persuasion available to humankind. As in the other sciences, the rhetoric of mathematics plays an essential role in maintaining its epistemological claims (Rotman 1988; 1993). Rhetorical form plays a basic part in the expression and acceptance of all mathematical knowledge (Ernest 1997). However, to persuade mathematical critics is not to fool them into accepting unworthy mathematical knowledge; it is to convince them that the actual proofs tendered in mathematical practice are worthy. Both the content and style of texts play a key role in the warranting of mathematical knowledge, and both are judged with reference to experience of a mathematical tradition, rather than to any specific explicit criteria.

In fact, there are varying accepted rhetorical styles for different mathematical communities and subspecialties. Knuth (1985) compared page 100 in nine mathematical texts from different subspecialties and found very significant differences in style and content. There is no uniform style for research mathematics (nor indeed for school mathematics), just as there are no wholly uniform criteria for the acceptance of mathematical knowledge.

Elsewhere I have specified some of the general stylistic criteria that mathematical knowledge representations or texts are required to satisfy (Ernest 1997). Some of the criteria of rhetorical style which apply to school mathematics are as follows. In order to be acceptable, a mathematical text should do the following:

- Use a restricted technical language and standard notation
- Use spare, minimal overall forms of expression
- Use certain forms of spatial organisation of symbols, figures, and text on the page
- Avoid deixis (pronouns or spatiotemporal locators)
- Employ standard methods of computation, transformation, or proof

These criteria are of course far from arbitrary. They depersonalize, objectify, and standardize the discourse and focus on the abstract and linguistic objects of mathematics alone. They serve an important epistemological function, both in delimiting the subject matter and (simultaneously) in persuading the reader that what is said is appropriately standard and objective.

During most of their mathematics learning career from 5 to 16 years and beyond, students work on textual or symbolically presented tasks. They carry these out, in the main, by writing a sequence of texts (including figures, literal and symbolic inscriptions, etc.), ultimately arriving, if successful, at a terminal text, "the answer." Over the course of their compulsory education, students carry out many such tasks. A typical British schoolchild will attempt thousands of mathematical activities in his or her statutory school career. In performing these mathematical activities, a student is learning

1. the content of school mathematics: the symbols, concepts, conven-

tions, definitions, symbolic procedures, and linguistic presentations of mathematical knowledge
2. modes of communication: written, iconic, and oral modes; modes of representation; and rhetorical forms, including rhetorical styles for both written and spoken mathematics

Thus, for example, in written teacher–pupil "dialogue" pupils submit texts (written work on set tasks) to the teacher, who responds in a stylized way to its content and form (ticks and crosses, marks awarded represented as fractions, crossings out, brief written comments, etc.) as exemplified in Cameo 1.

In the main, a school mathematics task is a text presented by someone in authority (the teacher), specifying a starting point and indicating a goal state: where the transformation of signs is meant to lead. A completed mathematical task is a sequential transformation of, say, n signs inscribed by the learner, implicitly derived by n-1 transformations. Sometimes this sequence consists of the elaboration of a single piece of text (e.g., the performing of three-digit column addition). Sometimes it involves a sequence of distinct inscriptions (e.g., addition of two fraction names with distinct denominators, such as $1/3 + 2/7 = [1 \times 7]/[3 \times 7] + [2 \times 3]/[7 \times 3] = 7/21 + 6/21 = 13/21$). In each case, carrying out the task usually involves a sequence of transformations of text. In addition to the "answer," the rhetorical mode of representation of these transformations is the major focus for negotiation between learner and teacher. Thus in the case of three-digit column addition, the learner will often be expected to write the "sum" on three lines, and to indicate any units regrouped as 10s. In more extended, sequentially represented tasks, the learner will often be expected to label the final answer as such (Ernest 1993a). Another feature noted by Ernest (1993a) is the disparity between the learner's carrying out of a mathematical task and representing it as a text. One of the lessons of the rhetoric of mathematics is that the text produced as answer to a mathematics task does not exactly match the learner's process of deriving the answer. The written text is in fact a "rational reconstruction" (Lakatos 1978) of the derivation of the answer, and the disparity is usually determined by the rhetorical demands of the context. (This is better illustrated in science education, where there is a general awareness that student records of experiments are not personal accounts, but conform to an objectivized rhetoric which requires headings such as "apparatus," "method," "results," etc., in a strictly regulated account.)

The rhetorical style demands are not arbitrary; they represent an elementary justification of the answers derived in tasks. They provide evidence for the teacher that the intended processes and concepts are being applied. However, there are significant variations in the rhetorical demands of teachers in different contexts, confirming that they are to a large extent conventional.

The culture of the "progressive" mathematics classroom

It is worth remarking that the introduction of project or investigational work in school mathematics (i.e. "progressive" mathematics teaching) involves a major shift in rhetorical style. For instead of representing only formal mathematical algorithms and procedures, with no trace of the authorial subject, the text produced by the student may also describe the judgments and thought processes of a mathematical subject. This represents a major shift and there are often difficulties associated with it. In particular, these arise from the traditional culture and ideology of school mathematics shared by different groups, including pupils, teachers, parents, administrators, and politicians (Ernest 1991).

In view of the weight that the mathematics education community attaches to "progressive" mathematics (Cockcroft 1982; NCTM 1989), I offer an alternative scenario to illustrate the type of classroom indicated in the previous paragraph. This is particularly relevant in this chapter, because in the rhetoric of the reform movement it is often suggested that the activity of learning mathematics in the classroom should be made more like that of the research mathematician. Table 9.1 of differences suggests some of the key aspects of the traditional classroom mathematics culture that need to change. Most centrally, they concern the issue of the relation between power and knowledge, and also the individual–social distinction. Changing the culture of the mathematics classroom to overcome these differences might result in the following:

Cameo 3

A group of teenage children sit around a table at the side of a present day mathematics classroom. In front of them is a large sheet of paper on which they have drawn diagrams and symbols, which they are discussing. There are also some open mathematics books, exercise books, electronic calculators and mathematical apparatus, instruments, pencils, and colored pens. They are discussing how best to display the results of their investigations of "Happy Numbers," which they jointly chose from a selection of open-ended tasks. They finish their display and call the teacher over. She asks whether they are all satisfied with the work, and then indicates a part of the wall where it can be displayed. At the end of the lesson, the rest of the class listen, while the group take turns to explain the problem, their strategy, their results, the display, and further open questions that they generated. The class ask questions and discuss the work with the authors. Later the teacher makes a written assessment of the project and presentation, following a set of criteria for projects. She passes on her report to the group, who read and discuss her judgment and write their comments on the assessment form. Finally, the teacher discusses her judgment and the learners' responses and writes a final assessment comment and grade, based on the discussion.

A classroom like that described in Cameo 3 embodies a culture that is closer to that of the reform movement in mathematics education. The processes of problem posing and solving; of representing, conjecturing, testing, and of mathematizing in general, resemble the processes of knowledge generation applied in the context of research mathematics. Many mathematics educators argue that this is the culture that it would be most productive to emulate in school mathematics.

If this is the type of classroom that mathematics teachers and teacher educators would like to see today, one key question concerns the obstacles to change. These include the conceptions of mathematics and of the teaching and learning of mathematics of the pupils, the teachers, the administrators, and the politicians. In particular, a conversational view of the nature of activity in research mathematics is needed, as well as a more conversational style of interaction in the mathematics classroom. However, epistemologically empowering learners to have a say over the content of their learning (the task), their means of working (investigative and cooperative), and the assessment of their outcome is not just a matter of changing the culture of the mathematics classroom toward conversation. Ultimately, how we regulate and empower learners in mathematics is a political issue.

As indicated, one of the important changes in such a shift would be that of mathematical rhetoric: the criteria of acceptance for a student-produced mathematical text. Of course the irony is that by bringing the contexts of knowledge genesis in school and research mathematics closer together, the contexts of knowledge justification move further apart. For the rhetorical shift away from an impersonal, standard code to a more personal account of mathematical investigation is a shift away from the rhetoric of professional mathematics. It is widely recognized that the contexts of genesis and justification in research mathematics differ widely. Hersh (1988) has likened the difference to that between the front and back of a restaurant. Rotman (1988) uses the simile of the stage and backstage in the theater. It may be that for educational reasons we prefer to let the ethos of the mathematics classroom match that of the context of the genesis of mathematical knowledge rather than that of justification, in the culture of research mathematics.

Obstacles and revolutions in the development of mathematical knowledge

A major impediment to change in school mathematics, I have claimed, are the existing expectations, ideologies, and epistemologies of learners, teachers, and the public. These deeply embedded preconceptions about knowledge and learning in their obstructive role might be termed an epistemological obstacle. This notion is significant across the domains of both personal and public mathematical knowledge, and in the philosophy of mathematics.

One of the major shifts in the philosophy of mathematics is the inclusion of the phenomena from the history of mathematics (Kitcher 1984; Lakatos 1976). The significance of this shift is that it shows the development and change of mathematical knowledge, including the concepts, methods, results, and metamathematical beliefs and values. Some authors argue that some of these changes are dramatic enough to be termed revolutions in mathematics, and that they come about when there are major shifts in the metaphysics, methodology, and underlying epistemological paradigm of mathematics (Dunmore 1992; Wilder 1981); thus revolutions occur when unspoken conventions and assumptions of mathematicians are overturned. Often, only by making some of these unspoken assumptions explicit can hidden conceptual obstacles be overcome. Bachelard (1951) invented the concept of an "epistemological obstacle," which prevents progress and development in the history of ideas.[5]

Bachelard regards the common-sense mind's reliance on images as a breeding ground for epistemological obstacles. However, epistemological obstacles may also arise from successful scientific work that has outlived its value. . . . The views and attitudes that constitute epistemological obstacles are often not explicitly formulated by those they constrain but rather operate at the level of implicit assumptions or cognitive or perceptual habits. (Gutting 1990: 135)

Bachelard writes of revolutions as "ruptures" in thought when an old style of thinking which is acting as an epistemological obstacle is swept away.

In the domain of personal knowledge of mathematics it is also important to adopt a developmental perspective. It is necessary to see that the school-based learning of mathematics is an essential forerunner of any mathematician's thought. Of course for most students, mathematical study does not lead to this end. But major changes in mathematical thinking and knowledge occur in all students between the age of 5 and 16 or 18. Some of these changes reflect continuities and the progressive development of mathematical knowledge. Some reflect discontinuities which needed to overcome epistemological obstacles in personal knowledge of mathematics.

Among the features of mathematical knowledge shared by the development of both school mathematics and mathematical knowledge in history are the change in and increasing complexity of mathematical concepts. For example, the concept of number proceeds from counting, and whole numbers, which become objects that can be operated on, and then to fractional numbers, integers, algebraic numbers, real numbers, complex numbers, and so on. In each transition, many features are retained, and some are relinquished. Thus the fact for natural numbers $(n > 1)$ that multiplication is a procedure that produces larger numbers ceases to be true for rational numbers or integers. Because of such conceptual changes in mathematics, as well as the learning of new knowledge, unlearning to overcome epistemological obstacles is also needed. Thus in the restructuring of knowledge components obstacles arise particularly when tacit components are restructured, at the level of both personal and public knowledge of mathematics. What is not often remarked, certainly in the classroom, is that some of the developments described here involve major shifts in the meanings of given terms and concepts. The same + (addition) sign is used to mean many different things, and such ambiguities are usually not recognized explicitly. Perhaps we should tell our students when we are "extending" (i.e., changing) meanings.

Finally, to reiterate, epistemological obstacles also exist in the domain of institutional ideology and change, especially in the domain of mathematics education. Although my goal in this chapter is descriptive, rather than prescriptive, at the back of mathematics education concerns always lies a desire to improve practice. The epistemological obstacles on the route to such progress (fully acknowledging that any claims of progress or good practice are not only value laden but fraught with risk) must never be underestimated.

Parallels and interactions between personal and public knowledge: A social constructivist account

One of the central parallels I wish to draw between personal and public knowledge of mathematics concerns the type of interaction that leads to the development and warranting of each, namely, human conversation. An increasing number of researchers in philosophy and the social sciences have adopted conversation as epistemologically basic.[6]

The original form of conversation can also be seen in derivative forms in which the correspondence between question and answer is obscured. Letters, for example, are an interesting transitional phenomenon: a kind of written conversation, that, as it were, stretches out the movement of talking at cross purposes before seeing each other's point. (Gadamer 1965: 332)

Mathematics can be regarded as conversational because of the dialectical mechanism of both personal and public knowledge generation and warranting in mathematics, with its ebb and flow, its assertion and counterassertion (Lakatos 1976; Ernest 1991, 1994). Elsewhere I have given an account of the interrelationship and mutual development of personal and public knowledge of mathematics, according to social constructivism (Ernest 1991). In brief, some salient features of this interpretation are as follows.

Becoming a mathematician, including achieving and completing an apprenticeship stage, involves an extended period of training to acquire the necessary knowledge and skills. This is followed during apprenticeship by participation in the institutions of mathematics, to acquire the knowledge, culture, and values of the mathematics subcommunity (Davis and Hersh 1980; Tymoczko 1986). The training requires interaction with other mathematicians, and with information technology artifacts (books, papers, software, etc.). Over an extended period this results in personal knowledge of mathematics and the norms of mathematical content and rhetoric, as well as of the social institution of mathematics.

Thus over a period of time, as a consequence of participation in social practices, individuals develop personal knowledge of mathematics, in both propositional and tacit form. This knowledge is manifested in linguistic behaviors and symbolic productions in "conversations" situated in a variety of different social contexts. The full range of contexts includes those of (a) professional academic or research mathematics; (b) mathematics education, both within formal institutional settings and outside them; and (c) a whole variety of other contexts, such as those of everyday, recreational, technical, or academic applications and uses of mathematics. (These will not be discussed further here, but are significant for the relation between mathematics and other forms of culture; see Dowling 1988.)

Within the context of professional research mathematics, individuals use their personal knowledge (a) to construct mathematical knowledge claims (possibly jointly with others) and (b) to participate in the dialectical process of criticism and warranting of others' mathematical knowledge claims. In each case, the individual mathematician's symbolic productions are (or are part of) one of the voices in the warranting conversation.

Within the contexts of mathematics education individuals use their personal knowledge of mathematics and mathematics education to direct and control mathematics learning conversations (a) to present mathematical knowledge to learners directly or indirectly (i.e., teaching) and (b) to participate in the dialectical process of criticism and warranting of others' mathematical knowledge claims (i.e., assessment).

Formal dialectical conversational exchanges (based on Lakatos's "Logic of Mathematical Discovery") located in the social institutional centers of mathematics are the basis of the process whereby published knowledge claims become warranted knowledge of mathematics. These depend on the criticism of mathematical claims and proofs, based on criteria which are shared within a significant section of an academic mathematical community. The successful application of the Logic is sufficient for acceptance as public mathematical knowledge, although it always remains open to challenge (and knowledge claims can of course be rejected). Beyond acceptance, the extent to which mathematical knowledge is used in the varying contexts described, especially those of mathematics, itself is a vitally important factor (Ernest 1997).

Individuals (learners) develop personal knowledge of language, mathematics, and logic through prolonged participation in socially situated conversations of varying types. In the context of mathematics education (in or out of formal institutional settings) other individuals (teachers) structure mathematical conversations on the basis of their own knowledge and texts in order to communicate mathematical knowledge to learners.

However the public representation of socially accepted mathematical knowledge within a teaching–learning conversation (including its textual variants) is necessary but not sufficient for such knowledge to become the personal mathematical knowledge of an individual learner. Sustained two-way participation in such conversations is also necessary to generate, test, correct, and validate personal mathematical knowledge. Such knowledge has idiosyncratic features, but its "mesh" with that of others is brought about by personal interaction and shared participation in forms of life.

The acquisition and use of subjective knowledge of mathematics by individuals are irrevocably interwoven. For only through utterance and performance are the individual construals made public and confronted with alternatives, extensions, corrections, or corroboration. Continual participation in dialogue (even in various stylized and institutionalized forms in education, for example) is essential for the construction of subjective mathematical knowledge, if it is to "mesh" ("fit") with the subjective knowledge of others (or rather with its manifestations, their utterances). Ultimately, such interactions are what allows an individual's personal knowledge of mathematics to be regarded as an interiorization of objective knowledge. Sustained two-way participation in such conversations is also necessary to generate, test, correct, and validate subjective representations of mathematical knowledge.

This account can be represented as a cyclic process that alternates between academic and school contexts, a process cycle in which public and personal mathematical knowledge recreate each other (Figure 9.1.). In this cycle the mathematical productions of individuals are made public. If they are potential contributions to accepted knowledge, they are subjected to a generalised logic of mathematical discovery (Lakatos 1976; Ernest 1991, 1997), and, having survived mathematical criticism and reformulation, are accepted and become a part of public mathematical knowledge. The cycle continues with this mathematical knowledge being reconstructed within individuals as personal mathematical knowledge, as a result of access to public representations of mathematical knowledge and interaction and dialogue with other individuals. On the basis of this personal mathematical knowledge, individuals are able to engage in further mathematical productions, which may be novel creations, or everyday applications, completing one full revolution of the cycle.

In this cycle what travels is embodied, either as a text or as a person. Symbolic representations of would-be mathematical knowledge travel in the academic domain, with accepted versions joining the stock of

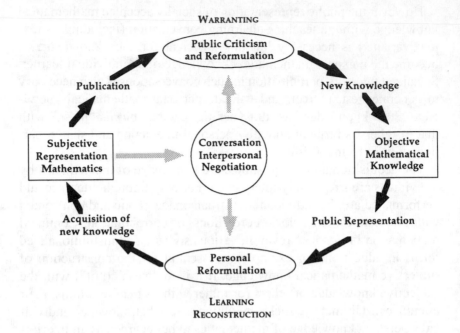

WARRANTING

Public Criticism
and Reformulation

Publication

New Knowledge

Subjective
Representation
Mathematics

Conversation
Interpersonal
Negotiation

Objective
Mathematical
Knowledge

Acquisition of
new knowledge

Public Representation

Personal
Reformulation

LEARNING
RECONSTRUCTION

THE SOCIAL CONSTRUCTION OF MATHEMATICAL KNOWLEDGE
The Role of Conversation / Interpersonal Negotiation

Figure 9.1. The creative/reproductive cycle of mathematics.

public mathematical knowledge. Some of these are recontextualized in the school context as symbolic representations of school mathematical knowledge and are presented to learners in teaching/learning conversations. Interactions in this context give rise to knowledgeable persons (accompanied by textual certifications of their personal mathematical knowledge). These persons can themselves travel and enter the school or academic context as either teacher or mathematician and participate in the respective warranting conversations.

Conclusion

In this chapter I have adopted an epistemological perspective on the culture of the mathematics classroom. I have explored the relationship that exists between personal and public knowledge of mathematics and between the cultures of school and research mathematics. I appreciate that the social constructivist account I have given is a radical departure from traditional conceptions of epistemology. For instead of focusing on logic, it takes as basic the social processes whereby mathematical knowledge is created, warranted and learned. Taking conversation as epistemologically basic in this way regrounds both public and personal mathematical knowledge in socially situated acts of human knowing and communication. It rejects the Cartesian dualism of mind versus body, and knowledge versus the world. Mind and knowledge are viewed as physically embodied, and a part of the same world in which the learning and teaching of mathematics take place.

In the chapter I have sought to bring out both parallels and differences between the cultures of school and research mathematics.[7] Perhaps my strongest claims are twofold: first, that these two domains are irrevocably wound together in the cycle I describe, through which they sustain and create/recreate each other; second, that from an epistemological perspective the warranting of knowledge is of prime importance, and that both types of knowledge receive their warrants in a parallel way through social mechanisms (and their underlying institutions). In my view, in both domains of knowledge there is an area of research needing extensive further theoretical development and empirical study: the textual, rhetorical nature of mathematical knowledge (which, it must not be forgotten, is inseparable from the social basis of knowledge acceptance). Such concerns illustrate how epistemology, psychology, sociology, history, semiotics, and virtually all other such perspectives, once regarded as distinct domains, are necessarily interpenetrating overlapping aspects of the culture of the mathematics classroom. The other chapters in this collection serve to reinforce this point and illustrate it from different directions.

Notes

1. "Epistemology" in this chapter refers to the theory of knowledge, with concerns including the genesis, structure, and warranting of knowledge. Too often discussions of "epistemology" in education neglect knowledge justification, which in philosophy is understood to be the central feature of epistemology.
2. Earlier steps toward such an inquiry can be found in Ernest (1991, 1992).
3. This example is intended to be representative of the most widespread traditional practices in mathematics classrooms in anglophone countries. Those who consider it inauspicious in the (claimed) current climate of reform may be reassured by the progressive cameo 3.
4. Accepting that there is tacit, embodied knowledge leads to a rejection of this view.

5. Ironically, the cultural dominance of the Euclidean paradigm of mathematics as absolute and objective truth is itself an epistemological obstacle obstructing the acceptance of a historicocultural view of mathematics.

6. This includes Vygotsky, Bakhtin, Mead, Habermas, Pask, Gergen, Shotter, Harré, Rorty, Gadamer, and Wittgenstein and is central to Ernest (1997).

7. Of course there are many significant issues that I have not had the space to raise. For example, if the cultures of school and research mathematics are characterized as distinct discursive practices, then the problem of transfer between them becomes extremely problematic. The issues involved include the recontextualization of text-based knowledge and the repositioning of the persons involved in the different practices and problems of the different subjectivities thus elicited or generated and their relationships.

References

Ayer, A. J. (1956). *The problem of knowledge.* London: Penguin Books.

Bachelard, G. (1951). *L'activité rationaliste de la physique contemporaine.* Paris.

Bloom, B. S. et al. (1956). *Taxonomy of educational objectives 1, Cognitive domain.* New York: David McKay.

Cockroft, W.H. (Chair) (1981). *Mathematics counts.* London: Her Majesty's Stationery Office.

Davis, P. J. & Hersh, R. (1980). *The mathematical experience.* Boston: Birkhäuser.

Dowling, P. (1988). The contextualising of mathematics: Towards a theoretical map. In M. Harris (Ed.) (1991), *Schools, mathematics and work* (pp. 93–120). London: Falmer.

Dunmore, C. (1992). Meta-level revolutions in mathematics. In D.A. Gillies (Ed.), *Revolutions in mathematics* (pp. 209–225). Oxford: Clarendon Press.

Ernest, P. (1991). *The philosophy of mathematics education.* London: Falmer.

Ernest, P. (1992). The relationship between objective and subjective knowledge of mathematics. In F. Seeger and H. Steinbring (Eds.), *The dialogue between theory and practice in mathematics education: Overcoming the broadcast metaphor* (pp. 33–48). Materialien und Studien, Band 38. Bielefeld (Germany): University of Bielefeld: IDM.

Ernest, P. (1993a). Mathematical activity and rhetoric: Towards a social constructivist account. In N. Nohda (Ed.), *Proceedings of PME-17.* Tsukuba (Japan): University of Tsukuba.

Ernest, P. (1994). The dialectical nature of mathematics. In P. Ernest (Ed.), *Mathematics, education and philosophy: An international perspective* (pp. 33–48). London: Falmer.

Ernest, P. (1997). *Social constructivism as a philosophy of mathematics.* Albany, NY: SUNY Press.

Freudenthal, H. (1978). *Weeding and sowing: Prelude to a science of mathematical education.* Dordrecht: Reidel.

Gadamer, H. G. (1965). *Truth and method.* (Trans. by W. Glen-Doepler). London: Sheed and Ward, 1979.

Gutting, G. (1990). Continental philosophy of and the history of Science. In R.C. Olby, G.N. Cantor, J.R.R. Christie & M.J.S. Hodge (Eds.), *Companion to the history of modern science* (pp. 127–147). London: Routledge.

Hamlyn, D. W. (1978). *Experience and the growth of understanding.* London: Routledge & Kegan Paul.

Hersh, R. (1988). Mathematics has a front and a back. Paper presented at the *Sixth International Congress of Mathematics Education.* Budapest, July 27–August 4.

Hiebert, J. (Ed.) (1986). *Conceptual and procedural knowledge: the case of mathematics.* Hillsdale, NJ: Erlbaum.

Kilpatrick, J. (1992). A history of research in mathematics education. In D.A. Grouws (Ed.), *Handbook of research on mathematics teaching and learning* (pp. 3–38). New York: Macmillan.

Kitcher, P. (1984). *The nature of mathematical knowledge.* New York: Oxford University Press.

Knuth, D. E. (1985). Algorithmic thinking and mathematical thinking. *American Mathematical Monthly, 92,* 170–181.

Kuhn, T. S. (1970). *The structure of scientific revolutions.* Chicago: Chicago University Press (2nd ed.).

Lakatos, I. (1976). *Proofs and refutations.* Cambridge: Cambridge University Press.

Lakatos, I. (1978). *The methodology of scientific research programmes (Philosophical Papers Volume 1).* Cambridge: Cambridge University Press.

Lave, J. & Wenger, E. (1991). *Situated learning: Legitimate peripheral participation.* Cambridge: Cambridge University Press.

Manin, Y. I. (1977). *A course in mathematical logic.* New York: Springer.

NCTM (1989). *Curriculum and evaluation standards for school mathematics.* Reston, VA: National Council of Teachers of Mathematics.

Nelson, J., Megill, A. & McCloskey, D. (Eds.) (1987). *The rhetoric of the human sciences.* Madison, WI.: University of Wisconsin Press.

Polanyi, M. (1958). *Personal knowledge.* London: Routledge & Kegan Paul.

Rorty, R. (1979). *Philosophy and the mirror of nature.* Princeton, NJ: Princeton University Press.

Rotman, B. (1988). Towards a semiotics of mathematics. *Semiotica, 72* (1/2), 1–35.

Rotman, B. (1993). *Ad infinitum the ghost in Turing's machine: Taking god out of mathematics and putting the body back in.* Stanford, CA: Stanford University Press.

Ryle, G. (1949). *The concept of mind.* London: Hutchinson.

Saxe, G. B. (1991). *Culture and cognitive development: Studies in mathematical understanding.* Hillsdale, NJ: Erlbaum.

Schoenfeld, A. (1985). *Mathematical problem solving.* New York: Academic Press.

Schoenfeld, A. (1992) Learning to think mathematically. In D.A. Grouws (Ed.), *Handbook of research on mathematics teaching and learning* (pp. 334–370). New York: Macmillan.

Simons, H. (Ed.) (1989). *Rhetoric in the human sciences.* London: Sage.

Skemp, R. R. (1976). Relational understanding and instrumental understanding. *Mathematics Teaching,* No. 77, 20–26.

Tymoczko, T. (Ed.) (1986). *New directions in the philosophy of mathematics.* Boston: Birkhäuser.

Vygotsky, L. (1978). *Mind in society.* Cambridge, MA: Harvard University Press.

Walkerdine, V. (1988). *The mastery of reason.* London: Routledge.

Wilder, R. L. (1981). *Mathematics as a cultural system.* Oxford: Pergamon Press.

Wittgenstein, L. (1953) *Philosophical investigations* (trans. G. E. M. Anscombe). Oxford: Basil Blackwell.

10 Problems of transfer of classroom mathematical knowledge to practical situations

JEFF EVANS

In this chapter, I want to consider the problematical relationship between the mathematics learned in the mathematics classroom and the "math" which is used in practical situations outside the school.[1] I shall examine traditional and "alternative" views of the relationship between the two kinds of mathematics – according to their conceptions of the context of mathematical thinking, and of the relationship between cognition and context. I set out the features that need to be considered in a more satisfactory account of thinking-in-context. I then expand on this notion of contexted cognition and illustrate its use in research on numerate thinking and affect among adult students, using in-depth consideration of a case study. Finally, I put forward some conclusions about the ways that the problem of transfer should be considered by mathematics educators.

Two views of mathematics and transfer

The "transfer" of learning can be considered to refer in general to the use of ideas and knowledge from one context in another. It often is used particularly for the "application" of knowledge from academic contexts to work or everyday activities. This is clearly a central problem of training, and also for most conceptions of education. It is especially important for mathematics. Of all the subjects taught and learned at school, it is one of those claimed to have the widest applicability outside school. However, at the same time, it is perhaps the subject in which the classroom at school (as opposed to outside activities) provides the context for the greatest part of children's learning and problem solving – though not for all of it.[2]

The views on numerate thinking and on the possibilities of transfer from school to outside contexts discussed in the mathematics education community can be classified into two broad approaches for the discussion here. The first approach, which includes traditional views favoring the use of general learning objectives with little or no indication of the context of performance, most ethnomathematics research[3] and "revised traditional" views such as those of the Cockcroft Committee Report (1982) or Sewell (1981), shares several important ideas. The problem or "task" and the mathematical thinking involved in performing it are seen as able to be separated or abstracted from the context. Many tasks can then be seen as *essentially* mathematical, and hence it is possible to talk about "the same mathematical task" across several different contexts. Therefore it is possible to expect that the "transfer of learning," e.g., from school to everyday situations, should be relatively unproblematical – at least, in principle.

However, recently, there have been a number of concerns expressed about these views. To begin with, in practice, subjects often seem to fail to accomplish transfer. Some choose to blame the students, or their teachers, but there are more fundamental problems. In these views, the context of a problem can be seen to be underemphasized, in comparison with its mathematical "essence." It is also underspecified: Indeed, in much of the mathematics education literature up to the Cockcroft Committee Report (1982) in the United Kingdom and beyond, the context of a problem was generally considered to be given by its wording. This approach allowed one to see mathematics as "practical" in the way that Cockcroft did, or to stress the importance of "functional numeracy," without being very specific about the context of this mathematics, or that numeracy.

The set of approaches I am calling "alternative" includes the work of those U.S. researchers who have acknowledged the activity theorists (Cole, Scribner, Lave, Saxe), and that of poststructuralists such as Walkerdine. It diverges in several fundamental ways. The whole argument is that the individual (or his/her thinking) cannot be neatly separated from the context, nor can the task/activity be neatly separated from the context (with the latter as "background"). Thus this approach argues that cognition and performance are context-specific, in a fundamental sense.

Therefore, rather than arguing that calculating "1000 − 70" in school is "the same task mathematically" as, say, making change for 70 *lire* from a 1000 *lire* note in a street market, these alternative researchers would claim there is a *discontinuity* between school math problems and what I call numerate problems in everyday life. This means that traditional hopes of finding general solutions to the problem of the transfer of learning between contexts are undermined.

Theoretical discussions of transfer

Jean Lave makes a number of criticisms of the concept of transfer, and of transfer experiments. The "problems" presented in the latter are considered as objective and factual, because they are constructed by experimenters, rather than by subjects, and the experimenters preformulate the correct or appropriate solutions. This has two consequences: A subject who does not take on the problems or who does not produce the appropriate solution is deemed to have "failed," and the researcher is unable to study the methods – potentially fruitful and interesting – which the subject actually uses to deal with certain problems. Therefore, such research fails to describe much observable problem-solving activity.

On the issue of transfer, Geoffrey Saxe (1991a, 1991b) is more positive, and less polemical, than Lave (Pea 1990). He is interested in "the interplay of form and function across cultural practices." For example, he describes how a seller (aged 13, fifth grade completed) carries out the pricing function in selling candy – determining the wholesale unit price by trying out several possible values, each time using a standard school multiplication algorithm. He then calculates his profit per candy bar, and hence per box of 50 – using school algorithms for subtraction and multiplication (1991a: 61; though the research of Nunes, Schliemann, and Carraher (1993) has taught us to look for decomposition and repeated addition as possible alternative methods). Thus he "transfers" school algorithms in order to address problems from within the selling practice. Saxe conceives of transfer as "an extended process of repeated constructions . . . of appropriation and specialisation" – rather than as an "immediate generalisation or alignment of prior knowledge to a new functional context" (1991b: 235). Like Lave, he emphasizes the importance of repeated attempts for transfer to take place, the importance of social (or pedagogic) supports, and some depth of knowing the specialized knowledge forms involved in solving practice-linked problems, in order for them to be appropriated and transformed.

Valerie Walkerdine gives a poststructuralist approach to language, emphasizing fluid relations of signification in order to analyze the ways that meaning is constituted and interpreted (see also Henriques et al. 1984; Taylor 1990). For her, the context of any process of social action is constituted, or "highlighted," by the practices which *position* the subjects acting, by the practices in play in the setting. These practices and the related discourses are the basis for the subjects' making sense of what is happening, of formulation and thinking about problems, of expectations, e.g., as to what they ought to do. Social differences such as gender and social class are related to the positioning of a subject within a particular practice.

We can think of discourses as sets of ideas, goals, values, and techniques, "competing ways of giving meaning to the world and of organising social institutions and processes" (Weedon 1987: 35). I argue that the bases for the culture of the mathematics classroom are the different discourses – or we might say, the various discursive practices (in order to emphasize their basis in language and practice) – at play in the classroom. These discursive practices constitute (provide the basis for) the crucial aspects (goals, social relations, material resources) of the classroom culture. These practices are historically located (Foucault 1977; Henriques et al. 1984) and socially specific (Walkerdine 1990) in that they position subjects differently. These points will be illustrated.

An activity cannot be prejudged as *essentially* "mathematics in practice," since it can be described from multiple points of view, including that of practitioners, as well as that of mathematics education researchers. What is needed is to seek to describe the numerate aspects of a practice – through attention to particular signifiers in chains of meaning.

A relevant example of the issues related to transfer is provided by Walkerdine, who describes a "shopping game," played in a primary school (1988: Ch. 7). There a boy made "errors" in his sums because he did not realize that, in the game, one was allowed – indeed, one was *required* by the rules, made to ensure the game's pedagogic effectiveness – to start afresh with a new 10 pence after each purchase. Though the child *called up* – that is, identified the task as – practical shopping, through which he "made sense" of the apparent demands of the task, he nonetheless made errors because he was positioned in, and *regulated by,* the pedagogic shopping game.

This example shows why the transfer of learning can be such a difficult problem.[4] While some aspects of everyday shopping practice were also useful in the game – say, remembering the familiar result that "when you have 10 pence and buy something worth 9 pence, you will have 1 pence left," other aspects of shopping – for example, the knowledge of the requirement of giving up money to obtain a good – were not "included" in the discourse of the school shopping game. Also, importantly, the goals and purposes were quite different in the two practices. Thus, Walkerdine argues that activity within one discourse (say, playing the card game whist) will help with school math in those, and only those, aspects of the game which both are contained in school math and enter into similar relations of signification (Walkerdine 1988: 115 ff.).

The fact that the particular discourse called up provides the basis for the subject's examining a problem and thinking about it means that cognition will be specific to the discourse called up. The same is true for affect. Walkerdine has been virtually alone among the researchers discussed so far in emphasizing the importance of the relations between cognition and affect: "Meanings are not just intellectual" (Walkerdine and "Girls and Mathematics Unit" 1989: 52).

In the approach used here, affect is seen as a charge attached to particular signifiers that make up *chains of signification*. This charge can flow from one signifier to another, by *displacement*. In general, insights from psychoanalysis can allow us a fuller consideration of the affective (Walkerdine 1988; Evans and Tsatsaroni 1996); see the case study described later.

The examples given show that the specific meanings of a word, a gesture – or any other signifier – depend on the specific discourse through which the signifier is read. The discourse(s) or language(s) in use are systems of meaning which can be analyzed by considering relations of signification, and devices such as metaphor and metonymy. This gives a systematic quality to Walkerdine's discussion of language as discourse.

Toward a more developed conception of the context

The conception of the context of mathematical thinking and problem solving used here draws on a range of alternative theorists (sociocultural and poststructuralist).

To begin with, Lave (1988) has emphasized the importance of (1) the setting, in the form of the physical layout of supermarkets (as arenas), or the physical characteristics of packaged foods, in explaining how subjects make decisions in shopping or dieting practices. Others point to the importance of (2) material and institutional resources. Thus Scribner (1984) and Maier (1980) emphasize the role of available technology of computation in molding numerate thinking. And Saxe (1991a) points to the institutional basis of the local candy-selling practice, including the artifacts involved in the packing of candy bars, and the prevailing patterns of inflation. But there is more to the context of thinking. Although many writers (including some traditional ones) correctly emphasize the *wording* of numerate problems, they tend to try to describe the context simply by naming it, on the basis of a simple reading of such wordings. Against this, those working on an alternative conception using poststructuralist ideas emphasize the importance of using the tools of analysis available in linguistics, in order to read (3) the discourse or discourses in which the subject is positioned (discussed earlier). We need to acknowledge the importance of (4) social relations. Saxe (1991a, 1991b) emphasizes social interaction as a resource in learning, but this does not sufficiently emphasize the ways that subjects are regulated in particular practices such as school math, which relate to power. Walkerdine (1988, 1990; Walkerdine et al. 1989) emphasizes the importance of (5) social differences, related to gender and social class (see also Evans 1995).

In my project, I wanted to avoid the tendencies toward excessive determinism that I found in the idea of subjects' simply being positioned by practices. Thus I argue that in a given setting subjects in general *are positioned by* the practices which are at play in the setting, *and* that a particular subject will *call up* a specific practice (or mix of practices) which may differ from those called up by other subjects, and which will provide the context for that subject's thinking and affect in that setting (Evans 1993).

Thus, in my approach, the context is understood as a result of the processes of being positioned, and of calling up – that is as *positioning in practice(s)*. Each practice produces positions which subjects take up; in some contexts/practices, the availability of a particular position may depend on the subject's gender, age, etc. (cf. Hollway 1989). For example, in the activity of feeding children, the child is positioned as demanding ("more"); the parent is positioned as having to regulate the child's consumption; the form of this regulation is likely to depend on the parent's income and social class. For a discussion of how positioning in such practices might provide the context for children's thinking in one topic of primary school mathematics, see Walkerdine and "Girls and Mathematics Unit" (1989: 52–53).

In general, the subject's positioning will depend on the interplay of a number of factors, including the following,

- language and the discursive features of the situation
- social differences in, say, class or gender terms
- the subject's "investments of desire" (Hollway 1989)

In this particular project, the positioning of the students in the interview depended also on the setting of the research, and on the social relations of the researcher and the subjects within it. Indeed, a subject will generally be multiply positioned, in some mixture of discourses (Henriques et al. 1984: 117).

In studying numerate thinking during problem solving, then, a number of issues arise:

- What are the specific discourses "available" in a particular situation?
- What determines the subject's positioning at any point?
- What indicators are there for the subject's positioning?

These issues will be discussed in the next section.

Use of these ideas in research

The setting of the research was a U.K. polytechnic with a relatively high proportion of "mature" students (over 21 years of age, returning to study after some years of work or childcare). Over 3 years, new entrants to two degree courses were asked to complete a questionnaire that included items about previous numerate experiences, and a "performance" scale, followed immediately by a version of the Mathematics Anxiety Rating Scale (Richardson and Suinn 1972). At the end of their first year, a subsample were given an interview drawing on the ethnographic tradition, (e.g., Hammersley and Atkinson 1983) and on the psychoanalytic critique of traditional field work (Hunt 1989). The aim was to study the student's thinking and affect in the interview, as well as in earlier experiences with math and numbers, in relation to her/his positionings in specific practices.

Semistructured interviews were conducted with a randomly designed sample ($n = 25$) of students at the end of their first year. The interviews had "life history" and "problem-solving" phases; in the latter, subjects were presented with a number of "practical" problems – e.g., reading graphs, deciding how much (if at all) they would tip after a restaurant meal (discussed later), deciding which bottle of tomato sauce they would buy, etc. (cf. Sewell 1981). But this interview differed in its use of *contexting questions:* When the student was first shown the "props" for the problem – e.g., a facsimile of a graph from a newspaper in Question 3 (see Figure 10.1) – before being asked anything "mathematical" – s/he was asked, "Does this remind you of anything you currently do?" And after discussing the question "Does this remind you of any earlier experiences?" subjects' responses to these questions were to help me judge the context of their thinking about the problem, in the setting of the interview.

In this situation, each subject was positioned as a student on the B.A. Social Science degree course at the polytechnic. At the same time, a student's basis for attending the interview itself was that s/he had been

This graph shows the price of gold varied in one day's trading in London. Which part of the graph shows where the price was rising fastest? What was the lowest price that day?

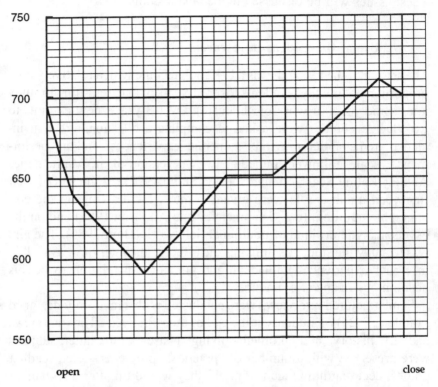

open   close

Figure 10.1. Question 3 in the interview: The London gold price – January 23, 1980 (in dollars per fine ounce) (from Evans, 1993, based on Sewell, 1981).

"chosen" (in most cases) in a random sampling exercise. Thus the two discourses which were available to provide the overt possibilities for the positioning of the subject in the interview setting were academic math (AM), with positions teacher/student; and research interviewing (RI), with positions researcher/respondent.

To the extent that RI, rather than AM, was called up by the subject, s/he would be able to call up ways of thinking, emotions, etc., from further nonacademic practices, with numerate aspects I called "practical math." For Question 3, I expected these to be practices of "business math," with several configurations of related positions possible: Manager/employee; submanager/higher manager; journalist/reader of the "business pages" of the newspaper. Each position in practice will support different ways of thinking and feeling, including different kinds of numerate or mathematical thinking.

In order to judge which practice the subject called up, I drew on various indicators (Walkerdine 1988: Ch. 3; Evans 1993):

1. the "script": e.g., how the interview, problems, etc., were introduced – as "research," "views," "numbers," rather than "test," "math," etc.[5]
2. unscripted aspects of the researcher's performance
3. the subject's talk, especially the responses to the contexting questions
4. reflexive accounts: e.g., whether I had been in the position of math teacher to each student, before the interview; and
5. the setting of the interview: where it was located (in my office at college) and the meaning of this to participants, and other features (the arrangement of furniture, the use of a tape recorder and where it was placed, etc.)

Sometimes it was relatively straightforward to make a judgment as to the subject's predominant positioning at a particular point in the interview. Sometimes it was not at all easy to decide. In addition, the subject will generally have a multiple – or interdiscursive – positioning in some "mix" of discourses – in this case, of the academic mathematics and the research interviewing discourses. And the resultant positioning may be fluid and changing. Hence the task is to describe this positioning during crucial episodes of the interview.

Two "qualitative" approaches for analysing the interviews were used. First, a cross-subject approach, based on that of Miles and Huberman (1984), aimed to consider, in a comparative way, the results from all of the interviews; for examples of this type of approach, see Evans (1994, 1995) and Evans and Tsatsaroni (1994, 1996). Second, each interview was considered as a case study of a particular subject's thinking and affect.

A case study

Donald (a pseudonym) was male, 47 years old at entry, with an O-level (age 16 qualification) equivalent in math. The background of his parents was working class; his own was middle class, as he had worked in the money markets in London. He was now on the Town Planning Track.

In response to Question 3 (a graph showing how the price of gold varied over 1 day's trading in London – see Figure 10.1), he begins by calling up business discourses, or what might be called money-market math:

JE: Does that remind you of anything that you do these days, or you've done recently?

S: Er, some of the work we done in Phase One [the first two terms of the College course], but if you ask me straight out of my head, what it reminds me of – I worked once with a credit company and we had charts on the wall, trying to galvanise each of us to do better than the other (JE: uh huh), and these soddin' things were always there and we seemed to be slaves to the charts. . . . [2 lines] . . . I found it impossible to ignore them, even though you know that they're just getting you at it. . . . [2 lines] . . . That's what that reminds me of – a bad feeling in a way – I felt that a human being was being judged by that bit of paper. (Transcript: 8–9; time in brackets indicate a pause, numbers of lines in brackets indicates omission from the transcript)

Here we notice that Donald is reminded both of his "college math" and of his earlier practice of managing a sales team – but it is the latter business practices to which he actually seems to give priority in calling up.

Next in the same episode, he mentions college math, then seems to link "work math" with it.

JE: . . . Does it remind you of Phase One [the first semester of his course]?

S: Yeah, well, we done some of the questions like this, and er, the run over the rise and that kind of thing. . . . [5 sec.] . . . trends, I suppose if you were judging a trend . . . [2 lines] . . . I find good, I like the fact I can do a chart now (JE : uh huh), but . . . I couldn't sit down and do it straight away . . . [2 lines] . . . With math I have to go back to the basic things all the time. (Transcript: 10–11)

Here he uses the language of college math, describing the gradient as "run" over "rise." He then shifts into work discourses, as evidenced by his use of the terms "trend" (rather than "gradient") and "chart" (rather than "graph"), which were not used in the college teaching. We now need to consider whether what appears to be an ability to translate elements between discourses will help in performance. In this second quotation, he has described the gradient as "run" ($X2 - X1$) over "rise" ($Y2 - Y1$) – whereas it is the inverse! At this stage it is difficult to know whether this is due to a slip or to a more basic misconception.

Next I ask specific questions about the graph.

JE: Right, okay, may I ask you which part of the graph shows where the price
     was rising fastest?

S: If I was to make an instant decision, I'd say that one, but obviously want to
     make it on a count of the line, wouldn't I? (JE: You'd? . . .) I'd count
     a line (JE: Uh huh) as it goes up . . . [25 sec.] . . . eleven over six and
     ten over six, so that one's right – in the first one [i.e., before lunch].

JE: . . . [2 lines] And um, what was the lowest price that day?

S: This one here – five hundred and eighty . . . went higher at the close, for
     some reason. (Transcript: 12)

First and second readings of the episode

Here we note that, when asked to compare the gradient of two lines, he
makes a perfectly accurate "instant decision," possibly drawing on his
work experience. However, he also feels impelled to "count a line,"
which I take to mean: Calculate the gradient by counting squares on the
graph, as in college math. There he gets the correct answer, confirming
his earlier decision based on work practices – and confirming that he
can use the formula for gradient correctly – though his calculations are
approximate. So too is his reading of the lowest price – which should be
$590, not $580. At the end of the episode, he is back in the money
market practice, as shown by his speculating about why the graph "went
higher at the close." Here he has displayed what may be a basis for
translation between the discourses of school math and what might be
called money market math.

Thus, my first reading of his misreading of the lowest point on the
graph is that it was just a "slip." It must be admitted that the photocopy
of the graph was not perfect, and also that 9 of 23 subjects slipped up
here. However, there remains something slightly surprising about his
slip on this question: We might have expected some transfer of learning
from his work practices – familiarity with "charts" – to college math –
familiarity with graphs.

Here we need to consider the context, the positioning within which he
addressed Question 3. For this problem, Donald was considered to call
up predominantly money-market math (i.e., business discourses) –
along with three other students (an ex-stockbroker, an ex-manager, and
a stockbroker's daughter). And all four made an error in reading the
lowest point on the graph. I suggest that, in money and business practices, the readings of graphs are regulated differently, since they are
made for different purposes – e.g., for rough comparisons, rather than
for precise individual readings. This view would provide the basis for a

second reading of his slip in Question 3: Addressing the problem within a "practical" discourse, rather than within academic mathematics, reduces the need for precision.

Further readings: Bringing in affect

Might Donald's slips have something to do with the range of feelings he has expressed in these two episodes?

If we look in this episode for indicators of affect, we find that Donald ranges between "good" and "bad" feelings. He again expresses mixed feelings about work later:

S: Once you're in there you do perform – you wouldn't do a bad deal in a million years, 'cause it's yourself's on the line . . . [5 lines]
JE: . . . that sounds like pressure, doesn't it . . .
S: Oh, dreadful!
JE: . . . did you feel the pressure or the anxiety?
S: Oh, very much so, yeah. . . . Sometimes I got a pain in your chest [2 lines] – you had, the form gets stuck to your hand, the tension, the sweat. . . . But once you do a good deal – somehow, it could kill you somehow, you just feel good or something, as if it's your own money. (Transcript: 14)

But he expresses confidence about his numeracy at work: "I'd no confidence with figures when I started [work] . . . sheer use made me good at them ." But now "I can read figures . . . I just had a gift for that." (Transcript: 6)

He also expresses a range of feelings about academic math. He tells of his new-found positive feelings for mathematics:

S: I found connections of something there to go from one thing to another, and I found it [math, during the second term] exciting, you know, I couldn't get bored with it at all . . . [2 lines] . . . I liked it. (Transcript: 15)

But these new feelings are still tentative. He is "not [. . .] afraid of figures, but the formulas and things still frighten me really (Transcript: 15, my emphasis). He also describes an experience of feeling a "block," when he first attempts the current math worksheet – followed by "panic:"

JE: Panic, uh huh. So when you look at a question, what happens? . . . [1 line]
S: Some kind of inferiority inside of me says I can't do it . . . My brains tell me I can do it, but something says I can't (Transcript: 17)

Thus a third reading of his slips on Question 3 is that they may relate to a chronic lack of confidence about graphs and algebra.

Throughout the interview, Donald makes a distinction between different "aspects" or "types" of mathematics which is similar to those made by a number of other subjects (see Evans 1993, 1997). He appears to distinguish between what might seem to be two different "types" of mathematics, school math and a particular kind of practical math, but these occur in different contexts and are marked by a range of different feelings. For example, "banking and figures were a job, or something, but math were there to trip me up or something," and "I feel – not afraid of figures – but the formulas and things still frighten me really" (Transcript: 15, my emphasis).

This is related to the fact that at school "these things were splashed up on the board, and nobody said what the hell the reason." For Donald, "till I get it into my head what it [i.e., a formula] means, I'm not happy" (Transcript: 22, my emphasis). He also finds the formula frightening, because "it's divorced from reality in my mind." Also, "I'd get an answer to pass an exam, but I'd no idea what it was all about. I [. . .] couldn't see the point at all – to my *real* life, you know" (Transcript: 5, his emphasis).

Thus we can see that Donald uses a deeply felt distinction – between "figures," on one hand, and "formulas and rules," on the other. This distinction relates in turn to others as follows:

- Work (math) vs. school/college math;
- Figures being "a job" vs. math being "there to trip me up";
- Being meaningful/"having a point" vs. being "divorced from reality";
- Having confidence vs. being "frightened";
- "Feel(ing) good (about numbers)" vs. "a bit of panic" about "formulas and rules."

Returning to his first slip on Question 3, which involved inverting the formula for the gradient as "run over rise," we have now seen that he is "frightened" and in "panic" over formulae in general, so it is no wonder that he suffered a "memory slip" on this one. This relates closely to the third reading, which explained the slips by his chronic lack of confidence about algebra and graphs.

Let us reconsider several episodes in the interview, using insights from psychoanalysis. Hunt (1989) suggests some guiding principles for using such insights in analyzing the interviews, here done within an ethnographic perspective:

1. Much thought and activity take place outside conscious awareness; thus, everyday life is mediated by unconscious images, fantasies, and thoughts – which sometimes appear as jokes, slips, or dreams, or are subtly disguised as rational instrumental action
2. The unconscious meanings which mediate everyday life are linked to complex webs of signification which are ultimately traceable to childhood experiences
3. Intrapsychic conflict (among desires, reason, ideals, norms) is routinely mobilized vis-à-vis external events, especially if they arouse anxiety or link to unresolved issues from childhood. (1989: 25)

A very general implication of (1) and (3) is that any product of mental activity – including interview talk – may, on deeper investigation, reveal hidden aggression; I would add suppressed anxiety, forbidden desire, and defenses against these wishes (p. 25). An important implication of (2) is that *transference,* the imposition of "archaic" (i.e., childhood) images onto everyday objects (people and situations), is a routine feature of most relationships, including fieldwork relationships.

Fantasies are a major focus of psychoanalysis. They provide a site where the subject can be in control, and Donald produces several. For example, at the end of the interview, when we return momentarily to discuss question 2 [10% of 6.65], he recalls, in a clear reference to his younger days,

S: . . . in the shop . . . my mind just would make prices up . . . [2 lines] . . . If say { inaudible } somewhere, say, reduced by 15%, I could do it in my head without thinking, almost. (Transcript: 20–21)

Thus he clearly had insistent fantasies involving making up prices and calculating discounts in shops. Could these have been on goods which the young Donald desired, but his family couldn't afford?

When he goes to work, he gets into selling money, where it sometimes feels "as if it's your own money" (in the context of feeling good after doing a good deal). This practice, and the figures, etc., may relate to deep fantasies; e.g., it is exciting to play with money. They may also be a defense against a deep anxiety about not being in control, about not having something. So far in his life, school math has failed to relieve this set of anxieties for him – in contrast to those boys attracted to the "mastery of reason" (Walkerdine 1988) – though it is interesting to note that college math has given him a taste, and he now feels that it would be "exciting" to do a math degree at the Open University (Transcript: 11).

Thus, he has different affective "investments" in school math and "formulas," from those in his money-market practices and "figures."

Returning to his second slip on question 3, it involved his reading the lowest point on the graph as $580, less than the correct value of $590. We can recall that he says that the graph reminds him of "a bad feeling," and of how he "found it impossible to ignore" the graphs showing his performance on the wall (Transcript: 7). Is it possible then that his misreading of the graph might be motivated by his desire to ignore such "charts"? Though admittedly only *very* suggestive, this fourth reading may provide the affective/psychic basis for the chronic lack of confidence about graphs, the basis of the third reading.

Conclusion

Some researchers discussed here would argue that part of the problem stems from the specific quality of both "outside" discourses and school math – as against the view that school math is more general, more straightforwardly powerful, and hence in some sense privileged. The privileging of the abstract, of school math, leads to the traditional position's being normative (Lave 1988): "Can be applied" easily becomes "should." Several of the other interview subjects illustrate how this can hinder understanding of cognition; for example, for a "best buy" question later in the interview, a middle-class young man, here called "Alan," calls up shopping practices – in which for him great value is not placed on saving a few pence, or indeed on saving money at all – and school math is not called up. In such cases, a researcher may conclude that a "mathematical" signifier is not recognized as such, whereas it may be recognized, but its mathematical meaning be undermined by competing values related to other discourses (see also Dowling 1991).

There have been responses to claims of privileged generality for school discourses like mathematics from researchers such as Scribner and Saxe. Saxe, for example, recommends that transfer be conceived as "an extended process of repeated appropriation and specialisation" – rather than as an "immediate generalisation" of prior knowledge to a new context (1991b: 235). While Saxe's work represents a weakening of cognitivist claims (toward emphasizing more specific processes), still situations are perceived in terms of their cognitive elements: Specialized knowledge forms, motivations of individuals addressing problems, and appropriation (through cognitive schemas) of the knowledge involved in the activity.

There are several differences between Saxe (and other "activity researchers") and the framework used here. First, the arguments here show that transfer would involve not only ideas, strategies, etc., but also values and feelings, carried by chains of signification. For example, it was argued that "Fiona's" (the stockbroker's daughter's) errors on question 3 may have related to the anxiety and anger associated with her positioning in family discourses as "not able to understand" her father's work in stockbroking (Evans 1993: Ch. 11). Further, the "contexting questions" used in the interviews (discussed earlier) have revealed a wealth of associations between the sorts of "mathematical" problems presented and the subject's memories and accounts of experiences which provide a context for speaking of the meanings that elements of these problems have for him or for her.

Second, the way the cognitive and the affective are linked through an emphasis on language and meaning also makes transfer depend on relations of signification in the two practices. Thus, for anything like transfer to occur, a process of "translation" across discourses would have to be accomplished through careful attention to the relating of signifiers and signifieds in particular chains of meaning. But this translation is not straightforward: It generally involves "transformation." Walkerdine also points to the possibility that differing forms of regulation, related to different positionings, in the two practices (e.g., a "game" vs. "school") may further limit possibilities of transfer (Walkerdine 1988: 114ff.). The ability of a signifier to form different signs, to take different meanings, within different practices, at once constitutes a severe limitation on the possibilities of transfer, yet also provides the basis for any such possibilities.

Thus we can see illustrations of the differences in goals and values, social relations and regulation – and especially language/signification and emotional associations – between different discursive practices which make transfer, in the sense of the application of concepts or "skills" from school math to everyday practices, highly problematical (for further examples, see Walkerdine 1988). This means that transfer, because of both the vagaries of signification and emotional charges, will be difficult to predict or control. And it may not even be "positive," even with what seems the "right" pedagogic or social support.

Walkerdine (1988: Ch. 6) does give an example of what might be seen as successful transfer, though it will be noticed that it is accomplished by moving from out-of-school to school contexts – not from school to outside, as is the aim of most transfer theories. The example involves what appears to be the harnessing of children's prior knowledge about counting objects, etc., to lead to learning about addition in school mathematics. However, she shows how the process of "translation"/"transformation" of discourses must be accomplished through careful attention to the relating of signifiers and signifieds in particular chains of meaning. Thus, "teachers manage in very subtle ways to move the children . . . by a process in which the metonymic form of the statement remains the same while the relations on the metaphoric axis are successfully transformed, until the children are left with a written metonymic statement, in which the same metaphors exist only by implication" (Walkerdine 1982: 153–154).

This discussion of transfer in the sense of (attempted) application of school math to nonschool practices is meant to clarify the problems with the traditional view, and to recommend a certain amount of scepticism in conceiving the problem.[6] However, I do not wish to be seen to be contributing counsels of despair. People clearly do transfer ideas, feelings, etc., from one context to another under all kinds of conditions – but what they transfer is not always what we as educators would like them to transfer. This chapter has aimed to help explain why.

Notes

1. This paper is one of a series from ongoing work on issues around transfer. Developments of the position outlined here have been explored in Evans (1996).
2. For a discussion of out-of-school learning in mathematics or numeracy, see Lave and Wenger (1991), Nunes, Schliemann, and Carraher (1993) and Walkerdine (1988).
3. Because the distinction between traditional and alternative views is admittedly somewhat crude, it is not helpful to try to classify all work in this area. For example, the Brazilian School, led by Nunes, Schliemann, and Carraher, are clear about challenging the Piagetian view of stages of development, which are basically abstracted from particular problem situations, with a complementary analysis of situational models (Nunes et al. 1993: Ch. 5) that has made an important contribution to the sorts of "alternative views" discussed here.
4. It is not suggested that the teacher's purpose in playing the shopping game would be to produce transferable skills (from school to shopping), nor to "harness" children's experience with shopping for pedagogic purposes – for these young children "do not really go shopping" yet (Walkerdine 1982: 150). It was to give the children an experience of action on money, or tokens, which could later be "disembedded" in the process of producing abstract mathematical knowledge.

5. The letter of invitation and the script used by me as interviewer were careful to talk about "research" and "interview" and "numbers," rather than "mathematics" or "test." This was an attempt to shift the discourse and the positioning from one having to do with academic math (AM), to that of a research interview. What I called the research interview discourse can be considered to offer a space for other discourses to be called up, in a way that the academic mathematics discourses might not. To the extent that a practice is called up other than AM, the subject will have access to ideas, methods of reasoning, "skills," and emotions from that practice. One aim of the interview, then, was, as much as possible, to create a situation with space for the subject to call up one or more discourses – just those that would be called up if the subject were to be confronted with the problem in the course of his/her everyday, out-of-college life.

6. I must briefly mention transfer in the "opposite direction" – namely, harnessing ideas or skills from nonschool practices to use in school contexts. Such harnessing is important in education at all levels, and examples exist of contexts set up in schools in an attempt to harness knowledge from pupils' nonschool activities. Harnessing is especially important in societies where recurrent education is common, or in institutions where "mature students" return to study after periods of work or childcare. However, the limits to the possibilities of transfer discussed apply in both directions.

Two examples of my attempts to use harnessing in my own teaching in ways that are based on ideas of context developed here can be mentioned. First, Evans (1989) suggests the creation of a new context for teaching social research methods and statistics, called *community research*. This context would be positioned between, and aim to draw on, both the course members' daily activities, as students and as members of various communities, and a course in social science research methods and statistics. The aim is to encourage students to bring problems from their daily activities to be addressed in the course. The meaningfulness of this created context, community research, might be expected to be *shared* to a reasonable extent, for certain groups of students – and their relationship to the academic discourses of social research methods and statistics might be seen as one of "barefoot statisticians" (Evans 1989).

The second proposal, for younger and/or more differentiated groups of students, is to seek to build up a relatively generally shared discourse around activities, in which the learners as "citizens," present and future, are highly likely to participate:

- purchase and/or growing and consumption of food and other necessities
- involvement with, and raising of, children
- paying for (perhaps building) and maintaining a dwelling and surroundings
- engagement with discussions and debates about personal, family, and

public well-being and about describing, evaluating, deciding on future
directions

A thoughtful engagement with these activities might be called *critical
citizenship* (Evans and Thorstad 1995). The "skills" necessary for its practice might provide the basis for a course offered as, say, "statistics" or
"mathematics across the curriculum" – or as "civics" or "responsible
citizenship." More work on this is needed.

References

Adda, J. (1986). Fight against academic failure in mathematics. In P. Damerow
et al. (Eds.), *Mathematics for all* (pp. 58–61). Paris: UNESCO.

Bernstein, B. (1971). On the classification and framing of educational knowledge. In M. Young (Ed.), *Knowledge and control.* London: Collier-
Macmillan.

Carraher, D. (1991). Mathematics in and out of school: A selective review of
studies from Brazil. In M. Harris (Ed.), *Schools, mathematics and work.*
Brighton: Falmer Press.

Cockcroft Committee (1982). *Mathematics counts.* London: HMSO.

Dowling, P. (1991). The contextualizing of mathematics: Towards a theoretical
map. In M. Harris (Ed.), *Schools, mathematics and work.* Brighton:
Falmer Press.

Evans, J. (1989). Mathematics for adults: Community research and the barefoot statistician. In C. Keitel et al. (Eds.), *Mathematics, education and
society,* Science and Technology Education Document Series No. 35 (pp.
65–67). Paris: UNESCO.

Evans, J. (1991). Cognition, affect, context in numerical activity among adults.
In F. Furinghetti (Ed.), *Proceedings of the Fifteenth International Conference, Psychology of Mathematics Education* (vol. II, pp. 33–39). Assisi, Italy.

Evans, J. (1993). *Adults and numeracy.* Ph.D. Thesis, University of London.

Evans, J. (1994). Quantitative and qualitative research methodologies: Rivalry
or Cooperation? In J.P. da Ponte and J.F. Matos (Eds.), *Proceedings of the
Eighteenth International Conference, Psychology of Mathematics Education (PME – 18)* (vol. II, pp. 320–27). Lisbon, Portugal, 29 July–3 August
1994.

Evans, J. (1995). Gender and mathematical thinking: Myths, discourse and
context as positioning in practices. In G. Hanna and B. Grevholm (Eds.),
*Gender and mathematics education: An ICMI study in Stiftsgarden
Akersberg, Hoor, Sweden 1993.* Lund: Lund University Press.

Evans, J. (1996). *Boundary-crossing: Another look at the possibilities for
transfer of learning in mathematics.* Paper for the Standing Conference on
Research on Social Perspectives on Mathematics Education. School of
Education, University of North London, London, June 13.

Evans, J. (in preparation). *Mathematical thinking and emotions in context: A study of adult learners and their numerate practices.* London: Falmer Press.

Evans, J., and Thorstad, I. (1995). Mathematics and numeracy in the practice of critical citizenship. In D. Coben (Ed.), *ALM – 1: Proceedings of the Inaugural Conference of Adults Learning Math – a Research Forum* (pp. 64–70). Fircroft College, Birmingham, UK, 22–24 July 1994. London: Goldsmiths College.

Evans, J., and Tsatsaroni, A. (1994). Language and "subjectivity" in the mathematics classroom. In S. Lerman (Ed.), *Cultural perspectives on the mathematics classroom* (pp. 169–190). Dordrecht, NL: Kluwer.

Evans, J., and Tsatsaroni, A. (1996). Linking the cognitive and the affective in educational research: Cognitivist, psychoanalytic and poststructuralist models. *British Educational Research Journal: Special Issue on Poststructuralism and Postmodernism,* 21(3).

Foucault, M. (1977). *Discipline and punish* (A. Sheridan, Trans.) London: Penguin.

Hammersley, M., and Atkinson, P. (1983). *Ethnography: Principles and procedures.* London: Tavistock.

Harris, M. (Ed.) (1991). *Schools, mathematics and work.* Brighton: Falmer Press.

Henriques, J., Hollway, W., Urwin, C., Venn, C., and Walkerdine, V. (1984). *Changing the subject: Psychology, social regulation and subjectivity.* London: Methuen.

Hollway, W. (1989). *Subjectivity and method in psychology: Gender, meaning and science.* London: Sage.

Hunt, J. (1989). *Psychoanalytic aspects of fieldwork.* London: Sage.

Keitel, C., Damerow, P., Bishop, A., and Gerdes, P. (Eds.) (1989). *Mathematics, education and society,* Science and Technology Education Document Series No. 35. Paris: UNESCO.

Lave, J. (1988). *Cognition in practice: Mind, mathematics and culture in everyday life.* Cambridge: Cambridge University Press.

Lave, J., and Wenger, E. (1991). *Situated learning: Legitimate Peripheral Participation.* Cambridge: Cambridge University Press.

Maier, E. (1980). Folk mathematics. *Mathematics Teaching,*93.

Miles, M., and Huberman, M. (1984). *Qualitative data analysis.* London: Sage.

Newman, D., Griffin, P., and Cole, M. (1989). *The construction zone: Working for cognitive change in school.* Cambridge: Cambridge University Press.

Noss, R., Brown, A., Dowling, P., Drake, P., Harris, M., Hoyles, C., and Mellin-Olsen, S. (Eds.). (1990). *Political dimensions of mathematics education: Action and critique; Proceedings of the First International Conference,* 1–4 April. Department of Mathematics, Statistics and Computing, Institute of Education, University of London.

Nunes, T., Schliemann, A.D., and Carraher, D.W. (1993). *Street mathematics and school mathematics.* Cambridge: Cambridge University Press.

Pea, R. (1990). Inspecting everyday mathematics: Reexamining culture–cognition relations. *Educational Researcher,* May, 28–31.

Richardson, F., and Suinn, R. (1972). The Mathematics Anxiety Rating Scale: Psychometric data. *Journal of Counselling Psychology, 19,* 551–554.

Rogoff, B., and Lave, J. (Eds.). (1984). *Everyday cognition: Its development in social context.* Cambridge, MA: Harvard University Press.

Saxe, G. (1991a). *Culture and cognitive development: Studies in mathematical understanding.* Hillsdale, NJ: Lawrence Erlbaum.

Saxe, G. (1991b). Emergent goals in everyday practices: studies in children's mathematics. In F. Furinghetti (Ed.), *Proceedings of the Fifteenth International Conference, Psychology of Mathematics Education (PME – 15)* (vol. II, pp. 320–27). Assisi, Italy.

Scribner, S. (1984). Studying working intelligence. In B. Rogoff and J. Lave (Eds.), *Everyday cognition: Its development in social context.* Cambridge, MA: Harvard University Press.

Sewell, B. (1981). *Use of mathematics by adults in everyday life.* Leicester: Advisory Council for Adult and Continuing Education.

Taylor, N. (1990). Picking up the pieces: Mathematics education in a fragmenting world. In R. Noss, A. Brown, P. Dowling, P. Drake, M. Harris, C. Hoyles, and S. Mellin-Olsen (Eds.), *Political dimensions of mathematics education: Action and critique; Proceedings of the First International Conference,* 1–4 April (pp. 235–242). Department of Mathematics, Statistics and Computing, Institute of Education, University of London.

Walkerdine, V. (1982). From context to text: A psychosemiotic approach to abstract thought. In M. Beveridge (Ed.), *Children thinking through language* (pp. 129–155). London: Edward Arnold.

Walkerdine, V. (1988). *The mastery of reason: Cognitive development and the production of rationality.* London: Routledge & Kegan Paul.

Walkerdine, V. (1990). Difference, cognition and mathematics education. *For the Learning of Mathematics, 10,* 3, 51–55.

Walkerdine, V. and "Girls and Mathematics Unit" (1989). *Counting girls out.* London: Virago.

Weedon, C. (1987). *Feminist practice and poststructuralist theory.* Oxford: Basil Blackwell.

11 Cultural perspectives on mathematics and mathematics teaching and learning

STEPHEN LERMAN

In this chapter, I will discuss the implications for mathematics teaching and learning of views of mathematics as a social construction or as the timeless description of reality. I will suggest that most mathematicians appear to adopt the latter view, but that in rhetoric at least, mathematics educators and teachers seem to argue the former. I will address the important question of how interests are served by preferred perspectives. I will then suggest that mathematics educators have not yet taken on board the potential of a view of mathematics as a social construction largely because, in the dominant neo-Piagetian tradition, a mechanism for the connection between history-and-culture and learning cannot be articulated. I will argue that discursive, or cultural, psychology offers a language for such connections and will sketch some of the shifts in perspective that are offered for the mathematics classroom and for research in mathematics education. A recurring theme will be the resources that theoretical discourses offer for analyses of knowledge as power.

Mathematics as a social construction

Western rationality is predicated upon the assumption that the nature of the universe and all its elements is determinable by scientific study and predictable using mathematical laws. Perhaps by reason of Western hegemony that image is predominant around the world, although certainly not universal. Against that background, weak sociologies of mathematics are just about acceptable and strong ones appear as radical alternatives (Bloor 1976). The former category would include studies of mathematicians and of incorrect paths in the development of mathematical ideas. The latter position would see mathematics and science merely as particular cases of the general phenomenon of the human drive to develop theories, to make generalizations from phenomena, and to search for explanations. Sociological explanations apply symmetrically to correct or incorrect mathematics, which should better be described as rejected or generally accepted mathematics. For example, Restivo (1992) argues that the development of mathematics in different times and in different places can be seen as resulting from many factors, a major one being the nature of the mathematical community.

The nature and rates of innovation in mathematics can be expected to vary in relationship to the social structure of the mathematical community, and especially in the organization of competition. (1992: 62)

I will return to Restivo's arguments in relation to mathematics as a social construction later. My main point here is that the privileging of one set of explanations over another is a historical and cultural phenomenon captured in the notion of knowledge as power, and this applies equally to views of mathematics as absolute or as fallible knowledge.

Certainty and fallibilism – the mathematician

It is now quite common in the mathematics education community to come across the claim that mathematics is a social construction (e.g., Ernest 1991). I am sure that it is still the case, however, as suggested by Hersh some years ago (1979), that most mathematicians prefer to believe that they are working with real objects whose properties they are trying to discover and establish, although on weekends, when pushed to present a justified position, they tend to fall back onto a version of formalism. In that paper and in the later book (Davis and Hersh 1981) Hersh implied a psychological motive for the platonist position of the working mathematician, in the sense of that position's being more fruitful than a fallibilist one in stimulating and sustaining creative work in mathematics. We might also offer an analysis of a different kind. The notion of mathematics as certain, as engaging with the paradigms of the signifiers "truth" and "proof," supports a privileged position for mathematics and its academics. In the common perception, and even among mathematicians themselves, mathematics has seemed to remain largely impervious to the devastating attacks both from within – the failures of the foundational studies earlier this century, Wittgenstein's sociological critique, and the growing centrality of the computer in generating unsurveyable results and proofs – and from without, in the form of poststructuralist critiques. Among academic groups, to be seen to be scientific is to gain a measure of respectability, and among scientific and aspiring-scientific groups to be seen to establish principles mathematically is to gain the greatest respectability. It is clearly not in the interests of mathematicians, or scientists, to undermine position and status by challenging certainty, proof, or truth in mathematics. Such studies and critiques are for those weekends, or they are categorized as external sociologies. "True" mathematics needs no sociology. Often it is deemed to be the work of scientists and mathematicians who have finished their productive life *within* science or mathematics and now write *about* science or mathematics.

In our own subsection of the body of working mathematicians one can see the privileging manifest. It is very comfortable to have our subject seen as vital for all children, as the gateway through which one must pass for success in adult life, as the best measure of the general intellectual ability of an individual, and as a body of essential skills for life. We may have our own debates in the mathematics education community about how little mathematical knowledge is actually required for normal successful functioning in the world (some put it as the mathematics learned in schools by the age of 11), and about our doubts of the generalizability of the intellectual skills required to be successful in school mathematics, and of course about that "ability" itself. Those discussions remain, however, within the community. We retain employment, structures, and status through the privileged position given to mathematics by society, in particular in relation to the school curriculum. This is not to suggest that we choose that privileging: simply that society in general, and governments, are not concerned with debates internal to the mathematics education community. The status of mathematics is captured in language and in the history of western culture; logic, rationality, deduction, proof and truth, and the platonic association of the mathematical forms with justice and beauty will not be changed by arguments about whether school mathematics should occupy a significant percentage of all children's studies to age 16 or 18.

Certainty and fallibilism – the mathematics educator

That mathematics may be better seen as fallibilistic is a view that has been around the mathematics education community for some years. Rather than engage with the debate about the nature of mathematics I want to examine here how that perspective supports values, beliefs, structures, and groups in mathematics education regardless of the debate. I offer four such examples:

1. Within academic (and other) communities, shifts in paradigms (or, less grandly, new trends in ideas) offer ways for new generations of researchers and writers to gain a voice. I am not suggesting that people make overt strategic moves into different domains in order to gain that voice, but that these patterns are part of the constitution of relations of knowledge/power within social groups. Identities and allegiances are formed through subgroups collaborating on furthering new approaches, and Restivo's (1992) analysis of the sociology of mathematics can of course be mirrored by a sociology of mathematics education. For example: one can argue that the development (or rediscovery – see Howson 1982) of investigational and problem-solving work in mathematics classrooms in the United Kingdom, beginning in the 1960s, was an integral part of the growth of the Association of Teachers of Mathematics in its break from the more establishment Mathematical Association (Cooper 1985); the struggle that the radical constructivists faced in the mid-1980s to establish their perspective and gain acceptability for their research contributed to the growth and the energy of those ideas; the same could be said for the increasing interest in the work of Vygotsky and other Russian psychologists and of sociocultural perspectives.

2. Early research in mathematics education drew upon traditional psychological paradigms and for the most part studied student errors and misunderstandings, or strategies for efficient learning. As more mathematically educated teachers became interested in research, mathematical perspectives, including developments in the philosophy of mathematics, began to become of interest more generally in mathematics education. In particular, the influence of Popperian and Lakatosian ideas gradually appeared in the literature (Dawson 1969; Rogers 1978; Confrey 1981; Nickson 1981; Lerman 1983). The language of uncertainty, fallibilism, certainty, and absolutism became available to writers in mathematics education. That theoretical framework offered a discourse, a discursive space, through which different interest groups could articulate their ideas. Polya's problem-solving work, picked up and developed by many (e.g., NCTM 1980; Mason, Burton, and Stacey 1982), and the literature of problemposing (e.g., Brown 1984) gained from a space in which mathematical knowledge was characterized by ways of thinking and ways of acting within a community.

3. In the late 1960s and early 1970s in the United Kingdom, secondary schools became "comprehensive," that is to say the selection process whereby the top 20% of children, determined by tests at age 11, went to so-called grammar schools, ended, and the resourcing of the education of that section of the population with the majority of funding

was shifted to all pupils. The test was abolished and almost all second-ary schools took in children by locality rather than by test results. Of course this is only part of the story; class, manifested particularly in the wealth of the locality, the reputation of the former select schools, and the competition for places in those schools, made, and continues to make, the entry much less "comprehensive." However, many schools, particularly in inner London, followed the ideal of greater social justice through to making classes of mixed ability rather than setting, and mathematics teachers were active in this (SMILE and other individu-alized learning schemes were developed as a response to mixed ability group teaching of mathematics). In keeping with other social move-ments of the times, there was a strong desire for greater democracy in the classroom, with teachers and students being seen as equal partners in investigating mathematical ideas and in learning how to think mathe-matically. The notion that mathematics was identified by the ways that mathematics was done and by specifically mathematical ways of think-ing, rather than a particular absolutist body of knowledge, reinforced that egalitarian position.

4. Finally, the language of radical constructivism, which entered the United Kingdom late in the 1980s also served the egalitarian ideas by offering a psychological rationale for the right of all learners to be seen to be creating their own mathematics for themselves, and countering a view of the teacher as authority in relation to knowledge and of teaching as the transmission of a fixed body of knowledge. Some writers (Con-frey 1985; Lerman 1989) overtly called on fallibilism and relativism to support constructivist learning theories with a philosophy of mathemat-ics which appeared compatible.

Thus, within the mathematics education community, mathematics can be claimed to be a social construction and each person is seen to (re-)create that mathematics for her/himself, with the constraint/pertur-bation of social interaction as the medium for the development and limitation of those personal creations. The most elaborate argument of the interplay between the individual construction and the public accu-mulation and sharing of mathematics, drawing on constructivism at its heart, is offered by Ernest (1991, this volume).

However, an examination of school mathematics practices and of school textbooks, at least in Britain, reveals that little has actually changed in the vast majority of classrooms. Even the investigations are aimed, for the most part, at achieving the learning of parts of the mathe-matics curriculum, in a way that does not differ from that in classrooms of 30 years ago. There may be talk of mathematics as a social construction, of notions of mathematical truth as deconstructed, and of mathematical knowledge as fallibilistic, but that revolution is not represented in significant ways in mathematics classrooms. Democratic and egalitarian tendencies and Lakatosian and constructivist frameworks are not the full story.

Juxtaposed against this account is that mentioned earlier, the power and status of an absolutist view of mathematics. The importance ascribed to mathematics in society, translated into its place in the curriculum, together with the authority which mathematical "truth," its "right-or-wrong"-ness, endows on the teacher, are powerful resources in the classroom, in the school, and in the society at large. They are not to be easily surrendered. At the micro-level, the classroom teacher is the only one who can identify mathematical understanding in an individual child. Walkerdine (1991) points to how the privileging of the status of the teacher in her/his ownership of "true understanding" manifests itself, not least of all in perpetuating gender differences. Statements such as "She may do well in tests but she doesn't really understand the mathematics; she achieves good results by hard work not mathematical ability" and "He may not get good results, but he has natural ability in mathematics and when he settles down he will do well" are familiar. What is more, the teacher is the source of confirmation of correct answers. As students learn that mathematical authority resides in the teacher, the teacher hopes that they learn that social authority does too. Of course constructivism, and radical constructivism in particular, has been and is a strong argument against this position. From this perspective the teacher can never know what is in the child's head, and knowledge can only be identified as the individual's mental constructions. However, in the growing destabilization of traditional social relationships pertaining to authority in the postmodern world, in particular in the classroom, teachers may find the adoption of radical versions of constructivism threatening.

There are also contradictions that some see in the radical constructivist position on mathematics learning (see Lerman 1996). If the mathematical knowledge that we have is a social construction, but individuals construct their own knowledge, how does that intersubjective social knowledge become intrasubjective? According to radical constructivism, social and other interactions set up perturbations, but there is no necessary or strong consequence of these interactions in terms of the individual's learning, or internalization. In this view there is no mechanism for the intersubjective to become the intrasubjective. Every internal world is potentially different from every other.

Strong sociologies of mathematics

One problem has been a lack of adequate elaborations of the social construction view of mathematics. The "unreasonable effectiveness" of mathematics, its ubiquity and power in our lives, tends to reinforce the absolutist story, in spite of our conviction otherwise, as does the sense that mathematical concepts have gradually become complete through discovery. There is a feeling that they were always complete, but it took millennia for the full story to unfold. How can one account for generality and abstraction? What drives the pure mathematician? Why, for example, did Euclid compile the Elements, which was a useless task in the sense of the technological functioning of society? For the platonist, these questions are easily answered – true mathematics needs no justification or sociology, its history is a telling of the unfolding of mathematical knowledge as it is. For the post-Weberian and post-Durkheimian sociologist the task is considerable.

Bloor (1976) and in particular Restivo (1992) are providing us with alternative stories. Restivo looks at the kinds of mathematical activity that emanated from different societies in different times and demonstrates the strong connection between the status, size, continuity, relation to religion, and patronage of the mathematically active groups and that production. Thus the mathematics of China is described as a mathematics of survival, of India as episodic, of Japan as about commercial revolution. Restivo is aware of the limitations of available translations and extant documents, but his aim is to offer possible sociological analyses of different kinds of mathematical developments, as against an explanation through a metaphysical internal teleology of mathematics itself. He points out that other descriptions of those historical periods are equally valid. He describes this as the weak case for the sociology of mathematics. The strong case requires an engagement with the production itself. Restivo argues that there is no mysticism in those mathematical objects:

For the objects with which mathematicians deal are activities of mathematicians. In building upon the operations already in existence, and making them symbolic entities upon which further operations can be performed, mathematicians are self-consciously building upon previous activities in their intellectual community. (1992: 84)

In order to illustrate this argument, consider as an example the physical action of turning a tile. That action can then become an object, so that the outcomes of a series of turns can be investigated. Groups of series of turns, cyclic or Kleinian, subsequently become objects too, and operations can be applied to the new level of abstraction.

There is no necessity to the body of mathematical knowledge which has become, through Western hegemony, the dominant approach, as Restivo's strong case demonstrates. For instance, his discussion of mathematics in Renaissance times is a story of competition between mathematicians, of rivalry, trickery, disputes and challenges. If one person had found a solution to a specific type of third degree polynomial, another would try to cap it by producing a more general solution. If one less well-known person had developed a specific algebraic method, a more well-known person would have no compunction in attempting to steal that method. Levels of abstraction and generality grow not through the interminable uncovering of mathematical truth but through the social activities of people, and they gain stability as they are subsequently built upon.

Restivo then goes on to begin an argument that mind is itself a social structure.

Selves, minds and ideas are not *merely* social *products;* nor are they *merely* socially *constructed;* they *are* social constructs. (1992: 132, emphasis in original)

In order to be able to counter the absolutist view of mathematics which is at the heart of Western thought about knowledge and reality, about the relationship between the human mind and existence, one needs accounts such as Restivo's, although he points out that his work is still only able to sketch out possible directions for study as the sociology of mathematics is at an early stage in its development.

In an interesting parallel, the British Society for the Advancement of Science debated the contribution of the sociology of scientific knowledge at its annual conference in September 1994. Eminent scientists argued that its contribution to the development and understanding of science is nil but that it has something to say about "the uses to which science can be put" (Times Higher Education Supplement Sept. 30 1994: 18). As one scientist put it (R. Dawkins, ibid: 17): "Show me a cultural relativist at 30,000 feet and I will show you a hypocrite."

Cultural, discursive psychology

Cultural studies, psycholinguistics, and sociolinguistics have had their influence in mathematics education research. It is not the case that individualistically centered accounts of mind have ignored such work. On the contrary, many people working in the area of ethnomathematics see constructivism as at the heart of their work, and writers such as Cobb, Wood, and Yackel (1993) and Wood (1994) also clearly argue that their accounts of learning take on broad issues of cultural and discursive influences. In what follows I will suggest that starting from the notion of mind as a social construct positions culture and discourse differently than constructivist accounts do. In particular it offers a way to conceive of strong connections between culture and discourse and learning. This is inconceivable within radical cosntructivism, as the following from Steffe demonstrates:

Vygotsky's notion of internalization is an observer's concept in that what the observer regards as external to the child eventually becomes in some way part of the child's knowledge. But Bickhard (in press) has pointed out that there is no explanatory model of the process. (1993: 30)

As I have argued elsewhere (Lerman 1994) there are no critical experiments, empirical or quasiempirical, that identify different arguments as alternative paradigms, nor that refute one and prove the other. The preceding discussion, concerning the nature of academic (and other) groups and theoretical debates, should also be borne in mind here. Such theoretical presentations both are polemical and at the same time provide resources for development of ideas; the two are not separable.

As the study of the mind, psychology appears to have a natural orientation toward the individual in its search for explanations of human actions and in its attempt to find ways of nurturing development. But studies in anthropology, linguistics, structuralism, and poststructuralism encourage a recognition that the activity of the individual must be looked at in social settings and that wider gaze has to be the precursor to the traditional psychological orientation. Language, in its widest sense as communication, is the limit of what one can express about the world, and that language is culturally specific and learned as one develops. Were an individual to be born and never meet with other humans, whatever propensities there were would not be realized and the individual would scarcely be human. Human consciousness is driven by communication.

For Piaget, actions are at the heart of human consciousness too, but those actions are precursors to communication:

The true course of failure in formal education is, therefore, essentially the fact that one begins with language (accompanied by drawings) instead of with real actions. (1973: 104)

A discursive psychology turns this order around. For an action to gain significance for the child, and thus to become something attached to goals, aims, and needs; to be associated with a purpose, is a social event.

Voluntary attention is not biological in its origin but a *social* act, and that it can be interpreted as the introduction of factors which are the product, not of the biological maturing of the organism but of forms of activity created in the child during his relations with adults, into the complex regulation of this selective mental activity. (Luria 1973: 262)

A theory of development which situates knowledge in the form of mental objects constructed by the individual relies on a notion of innate mental powers, such as those of discrimination and of generalization, of differentiation and of similarity. Vygotsky and his colleagues and followers argued that the baby learns to pay attention to particulars through interaction with adults. To see similarities and to class objects under one concept come about through instruction, and hence are essentially cultural. Similarly, any discussion of representation cannot ignore the social. To talk of an individual constructing a mental picture of an object, event, or concept is to shift the problem of meaning from the outside world to a supposed inner world of mental constructions, as Wittgenstein pointed out through his shift from the "Tractatus" to "Philosophical Investigations." Without the significance objects, events, and concepts gain from their function and role in social life, they are just pictures. Seeing a picture with eyes open or closed is the same. I can only use an object, event, or concept if its purpose and function are learned by me, at the simplest level, or conjectured through associations, at a more sophisticated level. Objects, events, or concepts gain their meaning through their use, the way they perform in social situations. This applies to the term "understanding," too, an issue to which I will return later, when discussing mathematics teaching and learning. Far from a mystical experience of the acquisition of some complete concept, whether platonic or socially constructed, understanding is never closed and is a process of learning when, and which, rules apply.

Do I understand the word "perhaps"? – And how do I judge whether I do? Well, something like this: I know how it's used, I can explain its use to somebody, say by describing it in made-up cases. I can describe the occasions of its use, its position in sentences, the intonation it has in speech. – Of course this only means that "I understand the word 'perhaps'," comes to the same as: "I know how it is used etc."; not that I try hard to call to mind its entire application in order to answer the question whether I understand the word. (Wittgenstein 1974: 64)

This is also the essence of the often quoted statement of Vygotsky that concepts occur first on the social plane and only secondarily on the individual plane, that is, from intersubjective to intrasubjective. There is no problem of mechanism for internalization, contrary to Steffe's argument, because the distinctions between subject and object, the internal and the external, do not exist as they do in the Piagetian view. If knowledge consists essentially of mental constructs, objects in the mind which individuals use to think, act, and speak, but which are only poorly represented by those communicative acts, then there is need for an explanation of how social meanings become individual ones. One must talk of "fit" not "match," of construal from events, of "taken as shared." However, if consciousness itself is constituted and regulated through relations with adults, so that the very processes of discriminating, of paying attention, are learned, not biologically developed, internalization *is* the mechanism through which we become human, through which consciousness itself evolves.

In short, it is not that one needs to take account of social interactions and language because it is through these that the individual constructs thoughts and concepts. It is in discourses, subjectivities, significations, and positionings that psychological phenomena actually exist (Evans and Tsatsaroni 1994). Acts of remembering, for instance, are not manifestations of hidden subjective psychological phenomena, they *are* the psychological phenomena (Harré and Gillett 1994). So too are emotions, attitudes, etc.

The study of the mind is a way of understanding the phenomena that arise when different sociocultural discourses are integrated within an identifiable human individual situated in relation to those discourses. (1994: 22)

Mathematics, and teaching and learning

I have argued that there is a growing body of literature that elaborates a strong sociology of mathematics and thus offers a resource, in a discursive sense, to mathematics education. I have also argued that discursive, or cultural, psychology provides the mechanism for situating the individual within culture and consequently overcoming the Cartesian subject/object, internal/external divide. With these resources, I will discuss three aspects of mathematics teaching and learning: development of mental processes such as voluntary attention; understanding; and mathematics as cultural artifact.

Mental Processes

Higher mental processes, such as memory and voluntary attention, are usually thought to arise in the normal development of a child and to take different forms in different stages of development. They can be exercised by activities in the classroom, but their development is a natural process, occurring at different speeds and to different degrees in children. One extreme end of these higher mental processes is exhibited by gifted children. If, however, higher mental processes are seen as resulting from communication, as social acts rather than mental ones, such as when the random movements of a baby's hand or random sounds are perceived by parents as intended and rapidly become so, then the processes of their development can be studied and consciously taught.

There has been a great deal of research in the past few decades in Russia, concerning teaching and the development of mental processes, only a small part of which has been translated. The approaches are novel to the mathematics education community outside Russia, and I will attempt to offer an illustration of that body of work to highlight its difference, and to indicate the potential that it may have in other national settings for research and teaching. I will refer to some work in progress, recently summarized in English (Veraksa and Diachenko 1994), concerning giftedness. Rather than the study of intellectual structures, with which they associate Binet, Wechsler, Guilford, and Torrance, or the development of abilities depending on personality development, with which they associate Maslow and Rogers, they are concerned with *abilities in the context of human activity.* Abilities are seen as "the systems of psychological tools existing in a culture and operations with them, which are used to orient a person in reality" (Veraksa and Diachenka 1994: 3). This approach unites the intellectual and personal approaches through the analysis and development of the forms of mediation which lead to the growth of purposeful and conscious behaviors in the child. Their work thus far has been with the preschool and early years of schooling and not subject-specific, and they have attempted to identify groups of abilities which typify the functioning of intellectual giftedness and a curriculum through which to develop those abilities, a curriculum that is appropriate for all children. In the next stage of their research they will focus on the development of higher mental structures in specific curriculum areas, commencing with mathematics.

For example, one class of structures in gifted thinking which they have identified they have called "normative–stabilizing abilities," and the interpretation of two-dimensional plans into the three-dimensional reality of the classroom, and vice versa, is seen as characteristic of that structure. The teacher places markers on a plan of the classroom and children have to find the places identified, where they discover hidden objects. They also take the other role in the game, hiding objects in the room and then marking the location on the plan for other children or the teacher to find. Another important class of structures is that of dialectical thinking. Children will analyze fairy stories for the common transformations that appear, such as from good fairy to bad fairy, and vice versa, or how a good king does something evil, which returns to him in a different form.

The important feature of this work, in the context of this chapter, is that these "gifted" abilities are seen to develop in social activities, through communication and, specifically, at the appropriate social stage of schooling, in teaching and learning. In their approach these abilities are available to all children. This orientation concerning intellectual functioning arises within a discourse which perceives that functioning as social acts, associated with purposeful acting in social life, operating with cultural tools.

Understanding

The concept of "fraction" is an elaborate one. It connects with percentages and decimals; it emerges from the inverse of multiplication of integers; and adding, subtracting, multiplying, and dividing them involve strange rules. It is defined as a/b for all integers except $b = 0$. It is also the equivalence relation over all ordered pairs of integers (a, b), where $(a, b) \int (c, d)$ iff $ad = bc$, and the set of fractions, the rational numbers, form a field. This is only part of "fraction" as it relates within mathematics in general, and the current elaborated concept takes in all these aspects and others. Does this mean that the child who only knows that $1/5$ means one part of five equal pieces and knows what to do with any $1/b$ in the same way does not understand fractions? According to Wittgenstein, and to paraphrase him, "I understand the word 'fraction'," comes to the same as: "I know how it is used etc."; not that I try hard to call to mind its entire application in order to answer the question whether I understand the word." Understanding is therefore *cumulative, fragmented,* and *incomplete.* It is *cumulative* in that it grows through learning experiences, and at every stage an individual can be said to understand, in that she or he can answer whatever is given within those particular learning experiences. But that does not necessarily mean it becomes connected, building gradually toward the supposed complete concept of the mathematician. For most students the identity of fractions, decimals, and percentages is far from transparent. They may be able to use each in different situations, applying appropriate algorithms in each. Learning is goal-oriented, and the dominant goal for children in schools is to get the right answers to the questions set, in the shortest possible time. Issues of surface learning versus deep learning, operational versus relational understanding, are perhaps researchers' categories. For learners, it is all strategic learning, in that it is always bound up with needs, aims, and goals. Of course we are all learners, in this sense, and as such thinking is always bound up with needs, aims, and goals. From this point of view, *fragmented* understanding is perfectly adequate. Thus changing from fractions to decimals to percentages becomes yet another domain of activity, not one that emerges on its own through the natural connections within a single "mathematical" notion. Finally, it is always *incomplete,* and new situations can always arise, for the teacher too. The mathematician requires and demands a "complete" understanding, but perhaps no one else does.

Mathematics as cultural artifact

Mathematical knowledge, in all its different forms, is an inheritance. As a body it can be observed on the shelves of a university library; it is absurd to deny its existence, as some do. All learning is a process of internalizing the accumulated knowledge of cultures, not least of all language. Whether and how much mathematics is to be learned in school is a complex decision, influenced by many interests, including mathematicians and mathematics teachers (as discussed previously), governments with interests of social control and (perhaps!) social development, and parents with implicit values which inhere in language. Given that a substantial quantity of such knowledge, which I take to include processes of mathematical thinking, appears in school curricula throughout the world, how might we engage with the implications of a strong position on mathematics as social construction? There are two aspects of mathematics as cultural artifact that need to be addressed here:

1. From the perspective of this chapter, it makes no sense to engage only with the mathematician's perception of the outcomes of mathematical activities, outside their context. If mathematical ideas are the products of social activity, related to their time and place, then a school mathematics curriculum which ignores history presents a peculiarly truncated, rootless, meaningless image, which serves particular interests at the expense of the cultural experience of learners and teachers. I am not referring here to the cursory introduction that many British textbooks offer, such as "Napier was a Scottish mathematician." If one looks to how renaissance art or medieval literature or church architecture in Britain might be taught in schools, they would be incomprehensible seen merely as current objects. Their location in time and space is integral to their existence, although many stories can be told of each, including the functions they have enabled and fulfilled subsequently through history. Mathematical knowledge is treated quite differently, however. Algebra, for example, is an engagement with structure and generality for the teacher and is largely an accumulation of techniques and problems for the students. The mathematical thinking that we call algebraic is, within mathematics, a powerful approach to many phenomena, and some excellent materials based on tiling, on symmetries in folk art, in the form of software etc., have been developed. But this is half the story. The view of mathematics as independent of society holds sway when the mathematics curriculum treats mathematical objects, the group for example, merely as fully formed. There have been some attempts to find ways of integrating the historical development of mathematical concepts into the learning of those concepts (e.g., for teachers in Arcavi, Bruckheimer, and Ben-Zvi 1982), but I do not believe that we have much idea how this might be done in schools.

2. This is not to argue, however, that the history (histories?) of a concept in mathematics determines the order in which that concept should be learned. Cultural artifacts change language and therefore thinking. Tools become part of cultural inheritance. Consciousness shifts. The notion that ontogeny replicates phylogeny is, therefore, highly questionable. In a technological context, it would seem absurd to argue that a child will need to progress from carving in stone, to scribing on parchment, then to using chalk and slate, before using paper and pencil and finally typing on a word processor. Similarly, children see television and hear radio from their birth. Electronic calculation, whether with calculator or cash machine, has become part of lived experience. Concerning language, one would not expect a child to pass through whatever stages language went through from earliest times.

Considering teaching notions of time without involving, from the beginning, the digital form which many (privileged) children use to program the videorecorder would be absurd. Periods of time, which previously could be visually imagined as the size of an angle swept out by the hands of a clock, have to be conceived differently when the typical representation is digital.

I would argue, therefore, that epistemological obstacles, as identified from the study of the history of mathematics from extant texts, cannot inform the mathematics educator of the learning obstacles that the child will experience today. Such obstacles, in any case, arise from particular approaches to history and will differ. It is not appropriate here to go into detail concerning this issue, but a study of mathematics in an image-based culture, such as India, will be likely to reveal different apparent epistemological obstacles to those encountered in a text-based culture, and teachers may be faced with determining which would be the most suitable epistemological obstacles around which to design a mathematics curriculum for multiethnic classrooms. Single universal obstacles emerge from a historical analysis that treats mathematical objects as fully preformed, awaiting discovery.

I have argued here against an ahistorical approach to mathematics which, unwittingly, serves particular interests and supports specific power relations in the classroom and in society. I have also suggested that using historical analyses to reveal epistemological obstacles which also face students today is a hypothesis which is supported by absolutist views of mathematics and is itself a result of the notion that each individual reconstructs for herself or himself all of mathematical knowledge. Perhaps the most fruitful approach to the implications for the classroom of viewing mathematics as cultural artifact is that which informs much Italian research (e.g., Bartolini Bussi this volume and references given there), which suggests that drawing on the sociocultural milieus in which mathematical tools developed offers an epistemological complexity that other more simplistic approaches ignore and, potentially, a fully cultural epistemology.

Conclusion

In this chapter I have attempted to suggest how different discourses about the nature of mathematics support positions of status and power. I have attempted also to sketch how these discourses might appear and gain their support in the context of the mathematics education community. I have suggested that different theoretical spaces are opened up by different discourses. Given that argument I have also been concerned to suggest here that a cultural view of psychology may enable an engagement with mathematics as a cultural inheritance, in the classroom, in a way that is not offered by individualistic psychological perspectives.

References

Arcavi, A., Bruckheimer, M., and Ben-Zvi, R. (1982). Maybe a mathematics teacher can benefit from the study of the history of mathematics. *For the Learning of Mathematics,* 3(1), 30–37.

Bloor, D. (1976). *Knowledge and social imagery.* London: Routledge & Kegan Paul.

Brown, S. I. (1984). The logic of problem generation: From morality and solving to de-posing and rebellion. *For the Learning of Mathematics,* 6(1), 9–20.

Cobb, P., Wood, T., and Yackel, E. (1993). Discourse, mathematical thinking, and classroom practice. In E. A. Forman, N. Minick, and C. A. Stone (Eds.), *Contexts for learning: Sociocultural dynamics in children's development* (pp. 91–120). Oxford: Oxford University Press.

Confrey, J. (1981). *Using the clinical interview to explore students' mathematical understanding.* Institute for Research on Teaching, Science-Mathematics Teaching Centre, Michigan State University.

Confrey, J. (1985). Towards a framework for constructivist instruction. In *Proceedings of Ninth International Meeting of Group for the Psychology of Mathematics Education* (pp. 477–483). Noordwijkerhout, The Netherlands.

Cooper, B. (1985). *Renegotiating secondary school mathematics.* Lewes: Falmer.

Davis, P. J., and Hersh, R. (1981). *The mathematical experience.* Brighton: Harvester.

Dawson, A. J. (1969). *The implications of the work of Popper, Polya and Lakatos for a model of mathematics instruction.* Unpublished doctoral dissertation, University of Alberta.

Ernest, P. (1991). *The philosophy of mathematics education.* Lewes: Falmer.

Evans, J., and Tsatsaroni, A. (1994). Language and "subjectivity" in the mathematics classroom. In S. Lerman (Ed.), *Cultural perspectives on the mathematics classroom* (pp. 169–190). Dordrecht: Kluwer.

Harré, R., & Gillett, G. (1994) . *The discursive mind.* London: Sage.

Hersh, R. (1979). Some proposals for reviving the philosophy of mathematics. *Advances in Mathematics, 31*(1), 31–50.

Howson, A. G. (1982). *A history of mathematics education in England.* Cambridge: Cambridge University Press.

Lerman, S. (1983). Problem-solving or knowledge-centred: The influence of philosophy on mathematics teaching. *International Journal of Mathematical Education in Science and Technology,* 14(1), 59–66.

Lerman, S. (1989). Constructivism, mathematics and mathematics education. *Educational Studies in Mathematics,* 20, 211–223.

Lerman, S. (1994). Metaphors for mind and metaphors for teaching and learning mathematics. In *Proceedings of Eighteenth Annual Meeting of the Group for Psychology of Mathematics Education* (Vol. 3, pp 144–151). Lisbon, Portugal.

Lerman, S. (1996). Intersubjectivity in mathematics learning: A challenge to the radical constructivist paradigm? *Journal for Research in Mathematics Education,* 27(2), 133–150.

Luria, A. R. (1973). *The working brain* (B. Haigh, Trans.) Harmondsworth: Penguin.

Mason, J., Burton, L., & Stacey, K. (1982). *Thinking mathematically.* London: Addison Wesley.

NCTM (National Council of Teachers of Mathematics), (1980). *An agenda for action.* Reston, VA.

Nickson, M. (1981). *Social foundations of the mathematics curriculum.* Unpublished doctoral dissertation. University of London Institute of Education.

Piaget, J. (1973). *To understand is to invent.* New York: Crossman.

Restivo, S. (1992). *Mathematics in society and history.* Dordrecht: Kluwer.

Rogers, L. (1978). The philosophy of mathematics and the methodology of teaching mathematics. *Analysen,* 2, 63–67.

Steffe, L. (1993). *Interaction and children's mathematics.* Paper presented at American Educational Research Association, Atlanta.

Veraksa, N., and Diachenko, O. (1994). *The possibilities of the development of cognitive abilities of 3–7-year-old children.* Centre for Mathematics Education, South Bank University, London.

Walkerdine, V. (1988). *The mastery of reason.* London: Routledge.

Wittgenstein, L. (1974). *Philosophical grammar.* Oxford: Blackwell.

Wood, T. (1994). Patterns of interaction and the culture of the mathematics classroom. In S. Lerman (Ed.) *Cultural perspectives on the mathematics classroom* (pp. 149–168). Dordrecht: Kluwer.

12 Representations in the mathematics classroom: Reflections and constructions

FALK SEEGER

The present chapter uses the concept of representation in a twofold sense: As an external, primarily pictorial, graphic, diagrammatic, or notational form; and as an internal representation, as cognition. In brief, I am discussing the (external) representation of (internal) representation, and vice versa. To indicate what kind of representations I am talking about, I shall use "representations" when I am addressing the external, notational form of representation and "representation" when talking about the internal, cognitive one. Since the advent of cognitive science, the discussion on the nature of internal representation has attracted broad attention from philosophers and psychologists (see, e.g., Dennett 1978; Fodor 1975, 1981; Millikan 1993; Putnam 1988). In this chapter I shall not try to contribute to that discussion; my analysis of representations is, however, indirectly related to one of the main points of this discussion, which has been the critique of a "picture" theory of mental representation. This theory typically assumes that a representation is like what it represents. The critique of this assumption has been picked up by constructivists in the field of mathematics education and turned into a general argument against the "representational view of mind," saying that we cannot know whether there exists something to be represented, and that representation is only construction. Now, this perspective seems only to apply as long as one sees "representation" as some basically passive reflection of the outside world. In this chapter, the line of argument will be that representation is active construction and that construction without representation is empty. External and notational representations are part of this contradictory process.

Educational practice is not only characterized by its verbal patterns, its discourse, but also by its use of representations. Representations, like the part–whole diagram, the number line, the place-value system of numbers, or the hundreds board, are major resources in mathematics education and form together with textbooks an important part of the material culture of the mathematics classroom. If one tries to understand the complex situation that leads to learning in the mathematics classroom as well as to learning problems, one can hardly overlook the role played by pictorial and graphical representations, that is by pictures, schematic diagrams, and the like. An inflationary use of representations can be observed that reflects an underlying growingly methodized mathematical content (cf. Keitel, Otte, and Seeger 1980).

"Representational overkill" exerts a devastating influence, especially on those students whom it is meant to help: The "weaker" students and the students with "learning problems" (cf. Lorenz 1991; Radatz 1980, 1984). These students are supposed to benefit most from representations because it is assumed that what appears complex and complicated to them is made easier by projecting it onto the plane of the seemingly simple and obvious language of pictures, graphs, schematic diagrams, and the like.

For many students, using representations and manipulatives as a remedy to learning problems is of little help. For them, it actually creates a new problem, because these students are now facing two problems: To understand the intention of representations and to understand the world of numbers without being able to make a connection between the two. What was meant to be a remedy, thus, can create an additional problem (see, e.g., Fingerhut and Manske 1984).

Leont'ev (1978: 149) has devoted considerable analytical attention to this "lack of correspondence between the supposed and the actual content" of learning. If visual material is used, its function is to build up the internal actions of the learner: Using a display of pencils in arithmetic teaching is not meant to enrich the sensory experience of the child, but to help her or him to learn counting.

The internal actions that must be structured in the learner require an abstraction from the subject content of the picture, and the richer the content, the more difficult this is. Counting uninteresting pencils is psychologically easier for the child than counting interesting tanks. (Leont'ev 1978: 161)

In a similar vein, Dewey (1902) has criticized the "sugar coating" of scholastic content, and Davydov (1990) has criticized the "visuality principle" in primary schools according to which visuality is seen as the prime starting point for learning.

In this situation, it seems appropriate that teachers leave the general notion that one can never use too many representations and become more careful with their use, especially with pictures that are supposed to represent what is taken for granted, what is taken to be self-evident. This applies particularly to the idea that mathematical understanding could be grounded by looking at or referring to some perceptual structure with a taken-for-granted meaning.

In this chapter, the starting point is that near to nothing is self-evident with representations, diagrams, and schemas. They are tools, and, as part of our culture, one has to be introduced into their use; as they are symbolic tools, one has to learn their language. Currently, a stimulating scientific discussion on the "nature" and "culture" of representations is going on in mathematics education and developmental/educational psychology (cf., e.g., Bauersfeld 1995; Boulton-Lewis 1993; Carpenter, Moser, and Bebout 1988; Cobb 1995; Cobb, Yackel, and Wood 1992; Janvier 1987; Kaput 1987; Lorenz 1991, 1993a, 1993b; Meira 1995; Perkins and Unger 1994; Perner 1991; Voigt 1993). The present chapter is meant to contribute to that discussion by discussing the complementarity of representing and constructing that forms the basis of the cultural–historical understanding of artifact representation. Indeed, reflection on the mediated and mediational nature of thinking and learning is becoming crucial now that "new media" and new forms of representation are becoming elements of occupational and everyday life.

The chapter starts with a brief review of the history of the concept of representation. This review will arrive at the conclusion that "representation" has to be understood in terms of mediation. Understanding "representation" in terms of mediation, then, means that cultural means are inserted between representation and what is represented. The process of representing is, as it were, culturally mediated. One can say that representing and culture are mutually creating and sustaining each other. Thus, the representation of culture is also to be seen as a culture of representation.

Following that, the role of representation in the process of generalization and abstraction in learning is discussed with a presentation of the views of Dienes, Bruner, and Davydov. The following section sketches five aspects or elements of a cultural–historical approach to representation: (1) representations as artifacts; (2) the relation of perception and representation; (3) reflexive, self-referential representations; (4) representation and exploration, and (5) representation and re-mediation.

The concept of representation: A short
history of a long past

Representation is a slippery concept, and it has been for quite a long time. There are different connotations of the concept. As with many concepts that have a long past, it does not seem possible just to define what is understood as "representation" and make a fresh start. Instead, it seems reasonable to make the different meanings present, to re-present them. My account of the history of the concept largely follows Scheerer (1995) and Scholz (1995).

In ordinary language, "representation" can refer to a wide range of meanings. It is a classic of epistemology that continues to be used in phenomenology and philosophy of language in the nineteenth and twentieth centuries. In psychology and cognitive science, it has been a major concept since the "cognitive turn," and knowledge representation is a major topic in artificial intelligence and related fields. Basically, one can find the following four meanings: (1) a broad conception of representation, that is, any mental state with a specific content; (2) a narrow conception of representation as mental reproduction of a former mental state; (3) a structurally equivalent "presentation" through pictures, symbols, or signs; and (4) something "in place of" something else.

The concept of representation has been treated among others in scholastic philosophy and in the writings of Descartes. According to Cassirer (1907: 83), the turn from a "picture" theory of representation to a "symbol" theory did not occur before Leibniz articulated the insight (which also referred to a new conception of mathematics) that for a representational relation, it is in no way necessary for that which is represented to be similar to the original; a certain analogy, a "symbolic" similarity will do.

This general turn of philosophical thought was more radically continued in Peirce's semiotic approach. He defines "to represent" as

to stand for, that is, to be in such a relation to another that for certain purposes it is treated by some mind as if it were that other . . . Thus, a spokesman, deputy, attorney, agent, vicar, diagram, symptom, counter, description, concept, premise, testimony, all represent something else, in their several ways, to minds who consider them in this way. (Peirce 1931–1935: 2.273)

Peirce then notes that "mediation" is the more suitable term for representation, because "representation" would not capture what he wanted to express with the hierarchy of the different levels of semiotic functions, quality, relation, and mediation, which he also called firstness, secondness, and thirdness to prevent false associations.

With the advent of the "symbolic turn" of thinking about the relation of reality and representation mentioned by Cassirer, a "crisis of representation" began that can be seen to have become almost a hallmark of postmodern thinking. Some thinkers find it necessary to get rid of representation and the representational view of mind as soon as possible (e.g., Lakoff 1987; Lakoff and Johnson 1980; Rorty 1979). We will return to their argument later. Others, like Dennett (1978), see the necessity of a totally new conception of representation, as the following statement, which strongly echoes Peirce's fundamental insights, indicates:

What is needed is nothing less than a completely general theory of representation, with which we can explain how words, thoughts, thinkers, pictures, computers, animals, sentences, mechanisms, states, functions, nerve impulses and formal models (inter alia) can be said to represent one thing or another. It will not do to divide and conquer here – by saying that these various things do not represent in the same sense. Of course that is true, but what is important is that there is something that binds them all together, and we need a theory that can unify the variety. (Dennett 1978: 91)

It is not my task here to sketch what such a general theory could look like. I take it that the concept of mediation would play a decisive role in any such theory, as Peirce has already indicated. This mediational stance necessarily will have to infiltrate into the standard view of visual perception, to which I shall return later in this chapter.

Summing up this all too brief excursion into the history of the concept, it can be said that the direct relation between the representation and what is represented has been replaced by an indirect relation (cf. Ogden and Richards 1949) and by a mediated one (cf. Vygotsky 1978); and that at the same time the representational stance is expanded into other domains by using the idea of significative relationships and signs as mediating tools.

Representation and abstraction in the mathematics classroom

In this section, I shall present three approaches on how, in the process of abstraction, representations are supposedly meant to function in grounding learning.

Reaching generalization in the process of abstraction has long been a major focus of the mathematics classroom. One central issue has been where to start. If teaching begins with concrete materials, it seems to be a long way finally to reach abstract generalizations, especially if pictorial representations and pictures as illustrations are used. On the way, some terrible paradoxes (cf. Brousseau 1984) and double-bind situations (cf. Bateson 1972; Engeström 1987) may arise. Actually, a learner already has to have a rather complete idea about what a graph, a diagram, or a picture is supposed to illustrate, about how it is supposed to function. If a learner does not have this initial idea, representations can readily function as an additional obstacle in the learning process, additional in the sense that not understanding a given representational item adds to the difficulty of understanding the mathematical content.

Using representations in the mathematics classroom is something to be built up gradually as a cultural practice of using representations as means to meaningful mathematics (for an example, see Bartolini Bussi this volume). Part of this culture can be taught. But there is a large part that has to be nurtured. This blend of teaching and nurturing has to begin in first grade. Now, much of the current practice of using representations in the primary grades is entangled in a paradoxical, contradictory situation. This contradiction is a corollary of the main event of primary mathematics: The transformation of preschool counting and knowledge of numbers into the number concept. This transformation has to relate simultaneously to two types of entities: On the one hand, there are representations that connect the world of numbers with the world of children by using things, scenes, situations from everyday life; on the other hand, there are representations that result from an understanding of the necessity of getting into the structure of the number concept. On the one hand, we have representations of everyday artifacts; on the other hand, there are artifacts that one would not find outside a math classroom: The pictorial representations connect easily to the experiential world of the children and they are hard to relate to the number concept, whereas structural representations readily connect to the number concept but find hardly any connection to the experiential world of the children. Figure 12.1 illustrates this contradictory situation: The question marks indicate the critical connections.

In the light of the preceding paradox, I shall give a short discussion on three approaches that, more or less implicitly, try to give an answer to this issue: Dienes's approach to abstraction and generalization; Bruner's sketch of three systems of representations: The enactive, the iconic, and the symbolic; and Davydov's approach of ascending from the abstract to the concrete. In presenting and discussing these approaches, I will con-

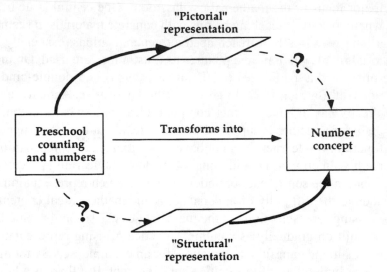

Figure 12.1. The dilemma of using representations in primary mathematics education.

centrate on aspects tangent to the topic of this chapter. My presentation, thus, is selective.

Z. P. Dienes on abstraction and generalization

Dienes has not only been one of the main protagonists of the "new math" movement in the 1960s and 1970s (see, e.g., Dienes 1960), but is also well known for his "Dienes Blocks," which are manipulatives used in the primary mathematics classroom. I decided to analyze one of Dienes's papers (Dienes 1961) because it seems interesting to learn how someone who created such a widely used means as the Dienes Blocks, giving the "hands-on approach" an important tool, was thinking about abstraction and generalization. It also attracted my interest because it would have to be a rather radical approach, given that the gap between manipulative, hands-on, and experience, on the one hand, and abstraction and generalization, on the other, could not be wider.

Dienes begins his paper by giving a definition of abstraction in the classical Aristotelian spirit:

Abstraction is defined as the process of drawing from a number of different situations something which is common to them all. Logically speaking it is an inductive process; it consists of a search for an attribute which would describe certain elements felt somehow to belong together. (Dienes 1961: 281)

The definition could also well pass as a definition of generalization, and this definition of generalization actually is used frequently and has been criticized equally frequently as insufficient and unsatisfactory (see, e.g., Davydov 1990; Vygotsky 1987; Weimer 1973; Wertheimer 1959). Dienes gives an example of this kind of generalization as gradual "impoverishment" of a concept:

For example the forming of the concept of the natural number two is an abstraction process, as it consists mainly of experiences of pairs of objects of the greatest possible diversity, all properties of such objects being ignored except that of being distinct from each other and from other objects. The essential common property of all such pairs of objects is the natural number two. For all pairs of objects encountered (elements) we form the attribute (class) of two; this is the process of abstracting the natural number two from our experience. (Dienes 1961: 282)

It is quite obvious that this definition of number as "abstracting from our experience" is more than incomplete. What would be the diversity of experience in terms of elements that would allow us to form the natural number "789" or "1.235.578"? It is quite clear that the probability that we shall have ever to count seven hundred and eighty-nine elements, let alone more, is close or equal to zero. For numbers 0FD 7, Dienes's claim is refuted by the well-known finding of "subitizing," that is, the rapid recognition of a number of objects that possibly does not involve counting (see, e.g., Kaufman et al. 1949; Mandler and Shebo 1982; von Glasersfeld 1982; and, for a discussion of recent explanatory approaches, Dehaene 1992).

Dienes's view on the advantages of his definition of abstraction[1] is captured in the following:

We might set up the hypothesis that the degree of abstraction of a concept is in direct proportion to the amount of variety of the experiences from which it has been abstracted. (Dienes 1961: 286)

According to our definition of abstraction, it is not possible to abstract from one type of experience, it is necessary to have at least two types. It is just the recognition of the same structure in two different types of experience that constitutes the process of abstraction. . . . Some people do indeed have a leaning towards abstraction. . . . Others, with a lesser leaning towards abstractions . . . will need to encounter a greater variety of situations. (Dienes 1961: 287)

These statements reveal the dangers of focusing too much on the structure without looking at the representational means and artifacts that are used in learning generalizations. The individual learners are empty-handed while being confronted with the structures to be found. They are thrown back to "experience." That being confronted with different domains may not automatically lead to generalizations but to domain-specific and situated learning seems to be a common perspective today. Dienes can interpret different forms of learning only as differences between unequally gifted individuals. Although he is the creator of an ingenious manipulative that allows the representation of number relationships in a multitude of ways, Dienes's view on generalizations can tell us nothing about their theoretical impact.

J. S. Bruner on enactive, iconic, and symbolic representation

Thirty years ago, Jerome Bruner in his "Toward a theory of instruction" (1966) made an attempt to define the central role of representation for the instructional process. At the bottom of this enterprise, we can find the idea of a mediating link between external representations, be they bodily, pictorial, or related to symbols, and the corresponding representational–cognitive outcome of learning. Bruner uses the concept of "representational system" to indicate this interest in the mediation of external and internal representation, reflecting his reading of Vygotsky (cf. Bruner 1962; for a recent critical appraisal of Bruner's view on culture and mind, see Bakhurst 1995). Bruner's analysis of three basic forms of representations in human development and learning, the enactive, the iconic, and the symbolic, has had an enormous impact on educational practice. It has been used ever since within the discourse among teachers and in teacher education as one rule that is grounding all teaching efforts: The first step in learning is to do it with real objects, manipulatives, embodiments; the second, with pictures; and the third, with symbols. This idea has been so widely appropriated that one is tempted to conclude that it must already have been an element of teachers' discourse long before Bruner came up with the names for the three systems of representation. In reviewing his attempt briefly, it will become clear that Bruner was personally not as sure about how the systems were interrelated as later users seem to have been.

Bruner used the concept of representation or system of representation in his approach to a new understanding of teaching and learning as the process to "translate experience into a model of the world" (1966: 10). Bruner identifies the three well-known ways:

The first system of translating, the enactive, is through action. The example of coaching someone how to play tennis is supposed to make clear that the coach is trying to let the trainee develop a model of playing tennis for which he or she has no imagery and no words.

The second system of representation, the iconic, depends on visual or other sensory organization – and on the use of summarizing images.

The third system, the symbolic, is a representation in words or language, with some of the typical features of symbolic systems. These features are arbitrariness and compactability: On the one hand symbols are arbitrary; they are remote from reference; on the other hand, they are so productive and generative that they are open for condensing a whole lot of information.

With regard to the relations among the three processes, Bruner is cautious to point out that this is an issue yet to be better understood:

How the transitions are effected – from enactive representation to iconic, and from both of these to symbolic – is a moot and troubled question. (1966: 14)

This caveat has obviously not attracted the same attention as the presentation of the three systems, so that Bruner's presentation was read as a three-layer hierarchy in which experience has to start necessarily with the enactive and end up with the symbolic level of representation.

Since the publication of Bruner's book, research has found repeatedly that the question of the transitions is a moot and troubled question indeed. An interesting study by Clements and Del Campo (1989) shall be presented as an example. In their study, Clements and Del Campo were giving tasks on fractions to second- to fifth-grade students. On the first trial, students had to identify certain fractions as parts of a circle or a triangle that were shaded (cf. Table 12.1). The results show that even very simple and basic representations of fractions have to be learned, and that the seemingly obvious character of these representations is not so obvious even to fourth- and fifth-grade students. On the second trial, the same students were asked to cut the same fractions out of cardboard figures or color them in with a pencil.

Bruner's theory of the succession of enactive, iconic, and symbolic representation would have predicted that these tasks would find higher percentages of correct solutions, being "enactive" and, by that attribute, "easier" than the previous tasks. In the enactive situation, however, the students had even more difficulties in producing the correct solutions (cf. Table 12.2). This does not necessarily indicate that the enactive mode is inappropriate in general. It may indicate that the test persons were not accustomed to using scissors and cardboard for fraction tasks in their usual scholastic task environment. This same explanation can, however, not be used in the case of the coloring task. This should be a

Table 12.1. *Percentages correct on four fraction recognition tasks (from a study by Clements & Del Campo, 1989)*

	Grade			
	2	**3**	**4**	**5**
⊕	38	53	67	79
(circle in thirds)	15	24	48	63
(triangle in thirds)	11	29	44	54
(triangle in quarters)	10	20	41	53

Table 12.2. *Percentages correct on four fraction construction tasks (from a study by Clements & Del Campo, 1989)*

	Grade			
	2	**3**	**4**	**5**
Use these scissors to cut exactly one-quarter of this circle	19	28	43	63
Use these scissors to cut exactly one-third of this circle	8	10	21	38
Use this pencil to color in one-third of this triangle	3	7	18	30
Use this pencil to color in one-quarter of this triangle	0	1	6	15

procedure well known to any child who has gone through the primary grades, and also well known within the environment of fraction tasks. Yet, the results are even worse with these tasks. As Clements and Del Campo do not provide any data or information on this issue, one is left to speculate on possible reasons for this finding. Their results demonstrate, however, that the "enactive" mode cannot be viewed as an automatic for grounding learning across task domains.

In a later part of the book, Bruner tries to make clear that these systems of representations are also similar insofar as they are related to three systems of skills: The enactive system of representation uses "tools for the hand," the iconic system uses "tools for the distance receptors," and the symbolic system uses "tools for the process of reflection." Again, Bruner is not very explicit about how these tools are related to representations, or how representation can be said to be tool-related. Thus, his conception of mediational systems of representation remains rather vague.

V. V. Davydov on ascending from the abstract to the concrete

In comparison to the two approaches discussed, Davydov's not only is outstanding for its philosophical and psychological grounding, but has also been recognized as a truly revolutionary and challenging one since it became known outside the Soviet Union in the early 1970s. Here was an attempt to begin elementary mathematics teaching in first and second grade with what then appeared and still today appears to be the most difficult theme: Algebra. Freudenthal's rather detailed review (1974) on Davydov's 1969 book (in English 1991) reflects the surprise and astonishment about the fact that the overall goal of the approach seemed to have been achieved.

The central feature of the approach, ascending from the abstract to the concrete as the general teaching strategy, rests on a painstaking analysis of the developmental situation of the schoolchild. School learning, according to Davydov, is not just the prolongation of preschool learning, it is a new type of activity, an activity that has theoretical concepts and theoretical knowledge as its object (see, in detail, Davydov 1988, 1990; and for an elaboration and critique, Engeström 1987, 1991). Theoretical concepts, thus, are not the end point of a long and winding road of "concrete experiences"; they are what school learning is about from the beginning. School learning, thus, is not about "induction from experience," as Dienes's approach suggested, an idea that also somehow underlies Bruner's view on systems of representation, but about deducing more particular abstractions from an initial abstraction.

In ascending from the abstract to the concrete, one of the main tasks is to find the "initial abstraction," the "germ cell":[2]

When moving toward the mastery of any academic subject, schoolchildren with the teacher's help, analyze the content of the curricular material and identify the primary general relationship in it . . . by registering in some referential form the primary general relationship that has been identified, schoolchildren thereby construct a substantive abstraction of the subject under study. . . . Then children utilize substantive abstraction and generalization consistently to deduce . . . other, more particular abstractions and to unite them in an integral (concrete) academic subject. When schoolchildren begin to make use of the primary abstraction and the primary generalization as a way of deducing and unifying other abstractions, they turn the primary mental formation into a concept that registers the "kernel" of the academic subject. This "kernel" subsequently serves the schoolchildren as a general principle whereby they can orient themselves in the entire multiplicity of factual curricular material which they are to assimilate in conceptual form via an ascent from the abstract to the concrete. (Davydov 1988: 22–23)

In finding the initial abstraction in algebra teaching in the primary grades, a schematic representation, a diagram, seems to play a major role in Davydov's approach: It is the well known part–whole diagram (Figure 12.2).

It can be conjectured from the available sources about how the teaching of algebra in the primary classroom actually proceeded that the part–whole diagram was the outstanding and prime means to cope with the difficulties in setting up algebraic equations. Figure 12.3 gives an example of how a student in one of the classes that Mikulina (1991) observed actually proceeded: The part–whole diagram and the notations used in Figure 12.3 grew out of a task to measure the amount of water from two different measuring cups, k and c, poured together in a third container, b. First, the task was done with letters; then numbers were used to replace the letters.

Though Freudenthal is very critical about the quality of the task used, about the "choice of artificial problems, contrived in order to exercise this method" (1974: 412), he simultaneously applauds the overall strategy and momentum of the approach:

What is more important is what I called . . . a sound pedagogical–psychological idea behind the experiments. I mean the idea that abstraction and generality are – in many cases – not reached by abstracting and generalizing from a large number of concrete and special cases. They are reached by one –

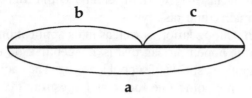

Figure 12.2. A part–whole diagram.

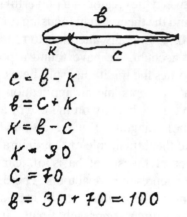

$$c = \beta - \kappa$$
$$\beta = c + \kappa$$
$$\kappa = \beta - c$$
$$\kappa = 30$$
$$c = 70$$
$$\beta = 30 + 70 = 100$$

Figure 12.3. Solving an algebraic equation with the help of a part–whole diagram (from Mikulina, 1991: 219).

paradigmatic – example, or . . . by a straightforwardly abstract and general approach. . . . The experiments convincingly show that algebra can be taught more adequately, and at an even earlier age than it is now. (Freudenthal 1974: 412)

However, as later empirical work on the part–whole diagram has shown, it cannot be treated as a general means to cope with different domains of school mathematics. For example, Wolters (1983) has shown that a rather lengthy teaching experiment involving the part–whole diagram led to positive results in some arithmetical domains, while leaving others relatively untouched: The effect on word problems, for instance, could be shown to be positive for part–whole problems, whereas it was actually negative for joining/separating and comparing problems. One could object that these results refer to arithmetic and not to algebra. This, however, is not my point here. The point is that one single representational means may work differently with different content domains, task domains, and task types.

Davydov's theory in its foundation as well as in its application places a strong emphasis on the nonempirical, that is, conceptual; on the theoretical and on the cognitive aspects of school learning. The social basis of school learning does not come into focus, as Engeström (1991) has criticized.

What kind of perspectives has the discussion in this section brought about for a deepened understanding of representations? In Dienes's account, we have found a split between the creator of a powerful manipulative and representational tool and the theoretician focusing on "structures" and "experience," while not paying any attention to representational tools at all. In Davydov's account, we have found a powerful conceptual framework that overcomes the traditional split between empirical experience and theoretical concepts and an application of that framework in algebra teaching that rests heavily on the use of a representational diagram, the part–whole diagram. But we do not find an equally penetrating analysis of the theoretical role of that diagram: The theoretical and analytical neglect of the social basis of learning is matched by a neglect of the role of representational tools. There is only the very general acknowledgment of the central role of mediation in Vygotsky's sense (Vygotsky 1978). Bruner's approach, finally, in trying to develop a theory of representational systems goes to the heart of the matter. His attempt, however, has been received primarily as confirming the traditional view of ascending from the concrete to the abstract – even though this point seems to have been the most problematic one for Bruner himself.

Elements of a cultural–historical approach to representations

This section will sketch five elements or aspects of a theory of representations that are related to the cultural–historical approach. It starts by presenting the notion of representations as artifacts. In the next section, a discussion of the constructivist critique of representation will show that an appropriate theory of perception seems to be essential for a full understanding of representations. Third, I shall discuss self-referential or reflexive representations in terms of Vygotsky's concept of "reverse action." Fourth, the notion of exploratory representations will be introduced using the metaphor of the map. Finally, the exploratory function of representations will be related to the concepts of mediation and re-mediation.

Representations as artifacts — artifacts as representations

Looking up "artifact" in the dictionary, one will find that it is a compound of the Latin word ars, in the sense of "skill," and factum, in the sense of something that is produced or done. According to Webster's Dictionary its primary meaning refers to a "characteristic product of human activity as a usually hand-made object (as a tool or ornament) representing a particular culture or stage of technological development."

The cultural–historical approach to an understanding of artifacts is not making a halt at the above definition of artifacts as material objects. It goes further by asking what exactly it is that makes an artifact "represent a particular culture or stage of technological development." How can it be that an artifactual object represents "ideal" qualities?

E. V. Ilyenkov has devoted a considerable part of his work to the "Problem of the ideal" (Bakhurst 1991). He has shown that an appropriate understanding of artifacts will lead to a solution of the riddle that the ideal can be thought of as outside human minds, as something that exists as an object. The source of this "objective" quality of ideas is not some kind of mysterious projection whereby things are endowed with "mental" attributes.[3] It is human activity. Human activity is embodied, "objectified" when an artifact is produced or manufactured:

Ideality is a characteristic of things, but not as they are defined by nature, but by labour, the transforming, form-creating activity of social beings, their aim-mediated, sensuously objective activity.

The ideal form is the form of a thing created by social human labour. Or, conversely, it is the form of labour realized . . . in the substance of nature, "embodied" in it, "alienated" in it, "realized" in it, and thereby confronting its very creator as the form of a thing or as a relation between things, which are placed in this relation (which they otherwise would not have entered) by human beings, by their labour. (Ilyenkov 1979: 157; as cited in Bakhurst 1991: 182)

Here, Ilyenkov is talking about not only a rearrangement of the physical properties of natural objects, but also the fact that by rearrangement of these physical properties, things gain significance because they represent social human activity and social human activity is represented in them. Artifacts, thus, represent human meaning and purpose. Specifically, they possess "affordance" and *Aufforderungscharakter:* There is a certain way to use an artifact, and much of child development is about being introduced to the proper use of artifacts (cf. Valsiner 1987). The significance and the meaning of the artifact are not just appropriated – even though cultural agencies like parents and schools often do act as if the meaning is something they have at their command and have only to transfer to the younger ones. The meaning of an artifact is not only appropriated – it is constructed, reconstructed, and remediated in the process of appropriation. The list of artifacts of major importance whose meaning is constantly being reconstructed and remediated would be endless. The example of language, as an artifact, may be sufficient here: Its very use is constant remediation.

A closer look at language also reveals the psychological functioning of the artifact. A classic example is given by Vygotsky:

When a human being ties a knot in her handkerchief as a reminder, she is, in essence, constructing the process of memorizing by forcing an external object to remind her of something; she transforms remembering into an external activity. This fact alone is enough to demonstrate the fundamental characteristic of the higher forms of behavior. In the elementary form something is remembered; in the higher form humans remember something. In the first case a temporary link is formed owing to the simultaneous occurrence of two stimuli that affect the organism; in the second case humans personally create a temporary link through an artificial combination of stimuli. (Vygotsky 1978: 51)

Vygotsky points out that the sign functions as an "intermediate link" (1978: 40), rendering the direct stimulus–response relation into an indirect one. With this intermediate link, humans "control their behavior from the outside" (1978: 40), because the sign "possesses the important characteristic of reverse action (that is, it operates on the individual, not on the environment)" (1978: 39). This statement also captures the essential characteristic of artifacts as representations of activity.

Artifacts are all alike insofar as they are human constructions. But, of course, they differ from each other regarding the place they occupy within an activity system. Wartofsky (1979: 202–203) gives a useful distinction among primary, secondary, and tertiary artifacts. Primary artifacts are those directly used in the production of forms of action or social praxis; secondary artifacts, roughly speaking, are representations of that same praxis used to preserve and transmit those skills that are bound to the forms of action. The preceding artifacts are "on-line," accompanying actions; whereas tertiary artifacts, the third type, "constitute a domain in which there is a free construction in the imagination of rules and operations different from those adopted for ordinary 'this-worldly' praxis" (Wartofksy 1979: 209). As one can say that tertiary artifacts are about the meaning of representation as such, about mediation as such, and that secondary artifacts are reflexive and primary artifacts are qualitative, this distinction corresponds with Peirce's distinction among firstness, secondness, and thirdness.

Engeström (1990), using Wartofsky's distinction as a starting point, has proposed to distinguish four types of artifacts, "What," "How," "Why," and "Where to" artifacts in which the first three types largely correspond to Wartofsky's classification. Engeström argues convincingly that "Where to" artifacts are needed to explain the coming into life of innovation and the restructuring of complex activity systems.

Representation and perception

This section attempts to relate the issue of representations to an ongoing discussion on different views on representation as cognition. The constructivist critique of the "representational view of mind" is relevant here for two systematic reasons: First, the notion of "construction" in itself seems to be indispensable for any attempt to conceptualize representations; and, second, the critique departs from a notion of perception. Obviously, an attempt to understand representations badly needs a theory of perception as well. My overall impression is that constructivism is heavily preoccupied with the (re)construction of the social and does not devote much attention to external things in the classroom like representations. It is all about the internal constructions' becoming a topic of negotiation in classroom discourse.

Constructivist approaches have recently embarked on a criticism of the "representational view of mind," joining some major figures in the philosophical discourse on the limits of cognitivism in a theory of mind. The arguments developed in this discussion have been used to mount an attack against psychological approaches in classroom studies as being "individualistic" and against the sociocultural approach as being "representational." The first issue will not be discussed here (cf. Waschescio this volume). I shall focus on the notion of representation that is used in constructivist arguments, and I shall try to show that their critique makes use of an understanding of perception that basically goes back to geometrical optics. In giving an account of their approach to mathematics education, Cobb and Yackel (this volume) state, "Such an approach explicitly rejects the representational view of mind and instead treats mathematics as an activity." In a paper devoted entirely to the problem of representation in cognition, Cobb, Yackel, and Wood (1992) have laid out a "constructivist alternative to the representational view of mind" that

rejects the view that mathematical meaning is inherent in external representations and instead proposes as a basic principle that the mathematical meanings given to these representations are the products of the students' interpretive activity. (Cobb, Yackel, and Wood 1992: 2)

In the following, I shall try to show that constructivism is right in shifting the emphasis from representation to activity, and that constructivism is wrong in postulating that representation can be discarded. In this attempt, a critical point seems to me that "representation" in the philosophical, psychological, and social sense has nothing to do with what Cobb and Yackel (this volume) call the "representational" approach in mathematics education.

What are the reasons that constructivism gives for discarding the concept of representation? The basis is a critique of "objectivism" that can be found in the writings of Lakoff and Johnson (1980), Lakoff (1987), and philosophers like Rorty (1979). I shall not go into the details of, for example, Lakoff and Johnson's claim that all ordinary language is metaphorical.

I shall try to discuss the general structure of the arguments raised against objectivism. Objectivism is criticized for

the assumption that meaning is primarily a characteristic of signs and concepts, that it is a kind of externally determined "cargo" carried by symbols and sentences. (Sfard 1993: 96)

Objectivism, according to its critics, holds that "meanings are disembodied, that they are received by a human mind rather than shaped by it" (Sfard 1993: 96).

A comparison between objectivist and nonobjectivist approaches reveals the necessity

to reverse the Objectivist version of the relationship between meaning and understanding: While Objectivism views understanding as somehow secondary to, and dependent on, predetermined meanings, non-Objectivism implies that it is our understanding which fills signs and notions with their particular meaning. While Objectivists regard meaning as a matter of relationship between symbols and a real world and thus as quite independent of human mind, the non-Objectivist approach suggests that there is no meaning beyond that particular sense which is conferred on the symbols through our understand-

ing. . . . *The meaning lies in the eyes of the beholder.* (Sfard 1993: 97, my
italics)

In reading this last statement, I found my hunch confirmed that constructivism in its critique of the concept of representation is still very
much entangled in what I would like to call the visual metaphor of
representation, using only the reverse of that metaphor, a mirror image
of the mirror image, so to speak. This is also substantiated by the use of
the visual metaphor that "meaning lies in the eye of the beholder."

Let me illustrate this point with some very simple diagrams. First, I
shall give an illustration of the visual metaphor of representation (Figure 12.4). You will notice that the illustration is an analogy to a common
representation of visual perception: Light waves are emitted by real
things, then refracted by the lens of the human eye, and projected upside
down onto the retina. The crossing of the two arrows indicates that
"something" is happening here, information is processed or converted,
resulting in the reversed image of the tree. Adaptation to the reversed
image is another point: Humans become accustomed to the reversed
image, it does not hamper them in seeing how trees "really" grow. This
kind of perceptual theory – even though it is quite common as an
element of folk psychology – goes back to seventeenth century perceptual theory as found in the writings of Hume, Locke, and Bishop Berke-
ley (cf. Wartofsky 1979). The constructivist version – or should I say
reversion – of this picture looks as follows (Figure 12.5): You will
notice that this picture possesses all the elements and also the structure
of the first illustration, except the direction of "projection": Meaning is
constructed individually and projected into the social realm, in which,
by some mechanism or process – namely negotiation – taken-as-shared
meaning results.

Obviously, this picture will not be accepted by constructivists as
completely representing their understanding of perception. It is, however, one way to illustrate that constructivism is not dealing systematically with the issue of the mediatedness of activity and, as a consequence, uses a rather simple, if not simplistic, theory of perception.

Marx Wartofsky's (1979) ideas about an activity-oriented approach
to the relation of representation and perception can serve as an excellent
starting point to explore this issue. In his chapter, Wartofsky states that
thinking about the nature and role of representations presupposes a clear
idea about what perception is all about. Understanding perception as a
universal faculty at the disposal of human beings is obviously not very
helpful for gaining new insights into the relationship between percep-

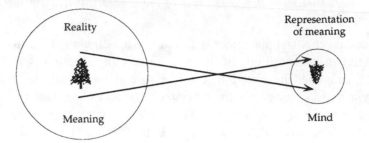

Figure 12.4. The visual metaphor of representation.

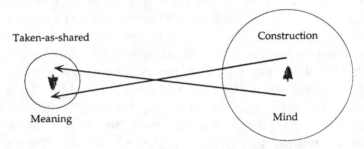

Figure 12.5. The constructivist emergence of meaning.

tion and representation. Wartofsky argues that it is not the universal, unchanging, physiological, and physical properties of perception that deserve our attention. Instead, it is the historically changing nature of perception that provides the key to an understanding of representation.

How could a possible discourse on the historically variant forms or modes of perception look, as, quite clearly, the "wetware" that allows for vision and perception has not changed significantly since the evolution of the human species? In other words, how could the historically changing forms in which the world is seen and presents itself be captured? The only way to get an idea about this change in the relation between the perceptual world and perception is to use a conceptual tool that mediates between the two, that is, to postulate that it is a specific feature of perception to be mediated by representation, and that the variation and change in modes of representation connect the historicity of perception to other historically changing forms of human practice. The constellation of the perceptual world, perception, and representation is such that in perceiving an activity system uses the tool of representation to perceive, that representation mediates perception. As it

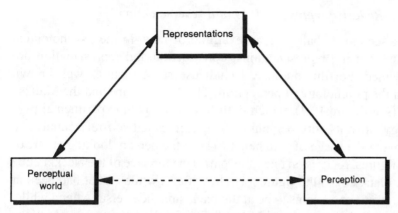

Figure 12.6. Illustrating the mediational role of representations.

were, there is no direct link between the perceptual world and perception. The above diagram illustrates the mediational role of representations. It should be kept in mind that the diagram in Figure 12.6 is stripped of all links that connect the three concepts to other concepts and other links. It is meant to represent only the essential links and relations that make up the mediational role of representations. In particular, the quality of representations as artifacts that will be discussed below is missing here. However, the diagram illustrates quite nicely that

the perceived world of the organism is in effect a map or an image of its activities: Just as, conversely, the perceptual apparatus is itself shaped to the modes of interaction by which the species survives. . . . To say that we see by way of our picturing, then, is to claim that perceptual activity is now mediated not only by the species-specific biologically evolved mechanisms of perception but by the historically changing "world" created by human practical and theoretical activity. (Wartofsky 1979: 195–196)

So far, we have seen some of the basic features of an activity-oriented approach to understanding representations: Tool aspect and the mediational aspect (for a more extended discussion on the features and principles of activity theory, cf. Asmolov 1986–1987; Cole and Engeström 1993; Davydov 1991; Engeström 1991a).

If one adopts Wartofsky's notion that "we see by way of our picturing," it follows quite clearly that representation and construction are two sides of one coin. Representation is putting more emphasis on the internalization of processes, whereas construction is underlining the externalization aspect. In learning, representation without a constructive appeal is as empty as construction that does not represent anything.

Reflexive representations and reverse action

In this section, I shall give an example of how, in the psychological process of using representations, construction and representation are intertwined. For this purpose, I shall use some pictures well known within the psychology of perception: The Necker cube and the Maltese cross (see Figure 12.7), which both have been with experimental psychology since its very beginning. They are called self-referential here (following the usage of Mitchell 1994). If the person, looking at one of those figures, focuses on one of its parts (the top face of the Necker cube or the black cross) the figure switches after a while: What had been at the forefront now seems to be at the back, and vice versa. After a while, the figure switches again with a tendency for the time during which the image remains stable to become shorter with each switch. The usual semiotic function of the sign as pointing or referring to something else that is not given in the sign or the picture is short-circuited because the image refers to itself. As a consequence, the viewer is "left to her or his own devices," so to speak, and the normally unconscious processes of perception are made conscious. The resulting impression is that of an object-image plus an object-process that flip and change according to a rate that is felt to be independent of the perceiver, but only as long as she or he deliberately continues to use a certain way of looking at the image. This mechanism is used in other cultural – especially religion- and meditation-related – artifacts like the well-known mandalas (see Figure 12.8).

The feature of self-referentiality or "reflexivity" cannot be understood without looking at how carefully these pictures have to be constructed so as to produce the intended effects. The key to an understanding of the psychological functioning of those pictures is Vygotsky's idea of "reverse action," of producing an artifact that is operating on the individual, not on the environment. As in the case of Vygotsky's (1978: 51) example of tying a knot in a handkerchief so as to remember something and thus creating a mechanism of "reverse action," these self-referential pictures use the same mechanism, that is, they are carefully constructed devices for the production of certain perceptual effects.[4]

The pictures in Figure 12.7 are "pure cases" in the sense that they demonstrate the effect of self-referentiality, and nothing else. They are in vitro artifacts. However, the mechanism of self-referentiality is not restricted to this kind of representational artifact. Self-referentiality seems rather a general feature of representations. Actually, it is possible

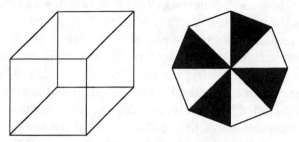

Figure 12.7. Two well-known self-referential pictures: The Necker cube (left) and the Maltese cross (right).

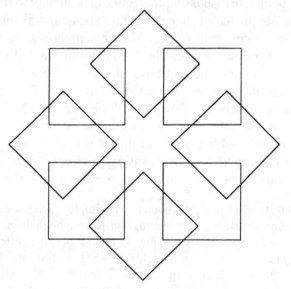

Figure 12.8. A more complicated self-referential picture: A Sufi mandala.

to think of the self-referentiality of representations as being the origin of their exploratory potential.

Representation and exploratory action: The map-territory relation

Representations reflect our activity in the world and not the world as such. One central metaphor that this claim is referring to is the map, and the map–territory relation. "The map is not the territory" is a quote from

Korzybski (1941) used by Bateson (1972) in his attempt to gain a new perspective on the mind and on representation. This metaphor refers to the fact that a message does not consist of those objects that it denotes, or, more generally, that any kind of representation is ontologically different from the objects it represents. Rather, the relation between word and object is like the relation between map and territory. One could argue that, after the "symbolic turn" in thinking about mind, thought, and representations, the "difference" and the coexistence of map and territory expressed in this metaphor capture one of the central issues of a psychological image of humans.

In an interesting chapter on psychological development, Downs and Liben (1994) discuss the role of maps in representing the environment and place in particular. In discussing this issue, they attempt to reconcile the "ontogenetic" approach of Piaget and the "sociocultural" approach of Vygotsky so as to preserve the best of both approaches.

Downs and Liben first comment on the taken-for-granted status of maps. There seems to be nothing questionable about them, because, in the common view, they represent the world "as it is." The map is a "miniaturization of the world" (Downs and Liben 1994: 167). An additional connotation that accompanies this view is that maps are transparent, some kind of looking glass on the world. They criticize this common view as wrong and advocate instead a view of maps as symbolic statements about the world that are truly works of art:

As representations, they abstract, generalize, and simplify. They are models. As products of symbol systems, they are not transparent and their meaning is not immediately available. They are opaque. As realizations, they offer one of an infinite variety of views of what the world might be like. They are opinions. As societal tools, they have been shaped by accident and by design, by ignorance and knowledge, by convention and invention. They are cultural theories of the social and spatial environment. (Downs and Liben 1994: 167)

Downs and Liben underline that maps can also be used as tools just because they offer a realization of the world. In this sense, to describe their function as "wayfinding" seems to reflect a narrow understanding. Rather, they find maps to have two complementary functions: One of substituting for direct experience and the other of amplifying direct experience with the environment. From this definition of the functional properties of maps, Downs and Liben go on to the definition of maps in ontogenesis and cultural development. Here, they see the opaqueness of maps as the decisive aspect. Maps lack transparency because, as complex symbolic inventions of society, they have "evolved into a myriad of different forms" (Downs and Liben 1994: 173). With this argument, they are criticizing Rogoff's (1994) view that tacit knowledge of maps is important for an understanding of maps. They argue that one has to learn what one can "see" with a map.

Some of the arguments in the Downs and Liben chapter, like the one that one has to "learn to see through maps," go directly to the point. Others seem to be more surface descriptions. I want to mention two points here: The substitution/amplification thesis and the claim that one of the most important aspects of maps is their lack of transparency, their opaqueness, their ambiguity.

With regard to the thesis that maps are substituting experience, on the one hand, and amplifying it, on the other, it has to be said that it is precisely their connectedness that constitutes their unique function. The map is an amplification precisely because it is substituting symbols for object, precisely because the map is not the territory. Maybe, introducing the concept of experience into the discussion on maps has led to some conceptual confusion, because it is tempting to use this word for something "genuine," "original," and "authentic" that seems to be reserved for moving through the "real" territory. If one thinks about moving around in a territory with the help of a map, however, "experience" then means the activity that is directed toward a goal, an activity that means moving around plus reading the map. As is often the case, the understanding of symbolic activity as something like a crutch, as not the real thing, on the one hand, and, on the other, the understanding of symbols as empowering, do not go together well.

The issue of the missing transparency of maps as representations is directly relevant to the discussion on representations in the classroom. It is true that maps are not transparent. But in terms of learning, the opposite of "transparent" is not "opaque" – at least not in relation to a representation used as a tool. The opposite is, rather, "ambiguous" or "exploratory." Maps, or pictorial representations in general, can fulfill their function of giving orientation just because they are different from the territory they are supposed to represent, just because they do not reflect the territory but possible trajectories of activity in that territory. It might be added that they also incorporate – as Downs and Liben (1994) suggest – theories about the environment. They might be called "cultural theories," because they sometimes also embody things one is supposed to do or not to do in a specific environment as part of the larger culture.

"Ambiguity" is another critical feature of maps and of pictorial representations in general. It is important to realize that maps and pictorial representations can only function because they are ambiguous – that is, they are expressions of the "difference" between map and territory. Ambiguity allows for multiple exploration, for exploration of alternatives – which, in a 1 to 1 model, would not be possible or would be rather pointless. The fact that there are "options" in the map that do not change the territory, but allow one to explore the range of possible action, is precisely the feature that creates the power of orientation, wayfinding, and navigation through maps.

The exploratory function of maps, as of representations in general, springs from the fact that one uses the self-referentiality of this pictorial representation in moving around within the map.

Representation and re-mediation

If one follows the line of argument developed in this chapter, representations play an important, if not a central, role in the culture of the mathematics classroom, either in sustaining a given culture or in creating a new one.

Now, in which way could representations be used to restructure not only their own function, but also the system of teaching–learning activity and the culture of the mathematics classroom? This question leads back to the self-referential and exploratory characteristic of representations discussed. Putting an emphasis on these characteristics in the classroom means to potentially break away from the tacit assumptions underlying the use of representations.

Representations are often seen as something beyond discussion, as tacitly and automatically grounding learning. This includes that representations in the classroom are often given one specific, singular meaning and one specific, singular way of use. Focusing on self-referentiality and exploration, instead, necessarily brings the question of multiple interpretations and multiple ways of use of representations to the fore: Obviously, there is no exploratory point if one does not have at least two interpretations of a given representation.

To give more room to multiple interpretations could also lead to changes in the time structure of the mathematics classroom. Here the challenge seems to be simultaneously to change to more projects that comprise larger content domains and to change the social life in the classroom. If the challenge is met, the first aspect, larger and more cohesive content, will break the dominant time structure in the classroom, a necessary condition for changing content; the second aspect will result in breaking the "persistence of the recitation pattern" (Hoetker and Ahlbrandt 1969) and lead to more discussion and argumentation in the classroom. How both challenges can be successfully met is well documented in the work of Maria Bartolini Bussi (this volume and references there).

There is another aspect that deserves to be mentioned. Self-referentiality is not only a characteristic of representations in general, it is also one basic feature of doing mathematics. Cultivating the self-referential and exploratory use of representations, thus, contributes to cultivating mathematics in the proper sense.

Self-referentiality may, however, lead to a self-contained classroom, and this in turn can easily be followed by an "encapsulation" (Engeström 1991b) of the culture of the mathematics classroom. To prevent this encapsulation, one has to look beyond the confines of the classroom, focusing not only on how the classroom is referring to itself but also how it refers to the everyday world.

A look at Figure 12.1 shows only the way from the everyday world to the school world. The loop of the number concept back to the everyday world is clearly missing. In a diagram form, the complete problem situation illustrating the perspective beyond the classroom is shown in

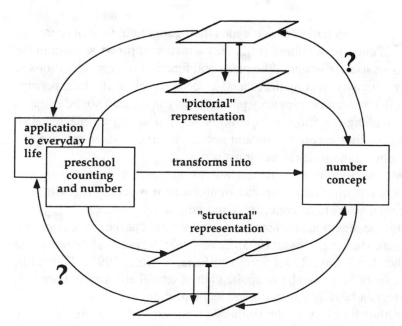

Figure 12.9. Using representations in their relationship to everyday life and the number concept.

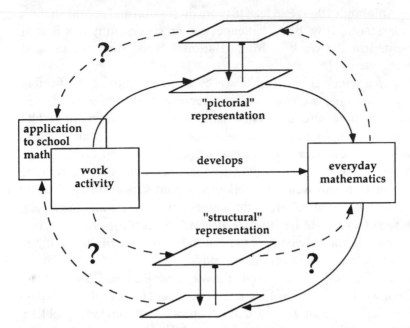

Figure 12.10. Potential relationships between everyday mathematics and school mathematics.

Figure 12.9, which illustrates the idea that changing the use of representations in classroom culture is not only a matter of processes within the classroom itself. Changing the place and function of representations in the classroom by treating the crucial aspects of their artifactual, perceptual, reflective, and exploratory potential may in the end not be enough. Without going beyond the classroom, without changing the place of schools within the community and society at large, it may turn out that all changes remain fickle and transitory.

Exploring Figure 12.9, now, leads to the reverse question of how everyday counting and everyday mathematical practice might inform the scholastic number concept. A growing body of research is documenting the ingenious arithmetic techniques used at the workplace and elsewhere (see e.g. Beach 1995; Lave 1988; Nunes, Schliemann and Carraher 1993; Saxe 1991; Scribner 1984a, b; Ueno 1995). This arithmetic cannot be said to be an application of school arithmetic. Rather, it constitutes a heavily situated knowledge of a unique kind.

If arithmetic tasks in the primary classroom refer to the everyday world, to everyday activity, this reference mostly is made up for the sake of making the tasks supposedly more interesting. Everyday activity is only taken as an exercise field where the algorithms to be learned are executed. In Figure 12.10 the potential relationships between everyday mathematics, school mathematics and representations are shown. The picture given there is largely utopian which is indicated by using mostly dashed lines. The picture shows, however, that research and development on the role of pictorial and structural representations as mediating between everyday mathematics and school mathematics seems promising (see Evans this volume). To invest work in the clarification of this issue seems all the more necessary as the notion of two worlds of mathematics and of "two cultures" of mathematics – the school mathematics and the street mathematics – may prove not to be very far reaching in the end.

Concluding remarks: Rethinking representations

In this chapter, a line of argument has been developed in favor of thinking through anew the crucial role of representations in the mathematics classroom. Representations cannot be taken for granted – and they cannot be understood as producing meaning that is taken for granted. Rather, representations can be seen as exploratory artifacts that allow the production of multiple perspectives on mathematical content. This exploratory function of representations is heavily dependent on a culture of the classroom that teaches and nurtures multiple perspectives.

Currently, representations are a core element of classroom culture. They guarantee a certain continuity and stability of that culture. As with many ingredients of a culture that give stability, they are taken as obvious and taken for granted. A perspective of change, however, has to address the question how representations can be instrumental in the process of change and which kind of representations should be chosen for that purpose. I have argued that representations that promote exploratory activity are prime candidates. To unfold the potential of this kind of artifact, two corresponding changes necessarily must cooccur; first, a change of classroom culture that supports and is supported by the exploratory use of representations; second, a change of perspective beyond the classroom walls that does not confine exploration to the classroom and the textbook.

Acknowledgments

I am grateful for encouragement, comments, and discussions from/with Ritva Engeström, Yrjö Engeström, Michael Hoffmann, Clive Kanes, Merja Kärkkäinen, Reijo Miettinen, Mircea Radu, Arne Raeithel, Kari Toikka, Hanna Toiviainen, and Liisa Torvinen on earlier versions of the chapter. I also wish to mention the great support of Ute Waschescio and Jörg Voigt as editors.

Notes

1. Dienes distinguishes abstraction from generalization: Abstraction goes from elements to classes, whereas generalization is passing from one class to another. I shall not discuss this distinction here, because I am more interested in the practical consequences of the definition of abstraction.
2. Davydov is not very explicit about whose task it is to find the initial abstraction. Is it the task of a curriculum planning unit, of educational psychology, or of the single teacher in the classroom or a team of teachers? In each case, the result would be rather different.
3. In a similar vein, Bruno Latour (1996) has pointed out in a recent paper that a "fetish" is normally held to be only a dead object animated by feelings and wishes projected onto it, only a "projection screen" for social life. He underlines that it actually expresses the quality of objectification in the complementary sense of what is produced or fabricated and what is true or valid.
4. This perspective has been demonstrated to be very well applicable to the analysis of the so-called perceptual illusions (cf. Stadler, Seeger, and Raeithel 1975; Velichkovskij 1988).

References

Asmolov, A.G. (1986–1987). Basic principles of a psychological analysis in the theory of activity. *Soviet Psychology,* 78–102.

Bakhurst, D. (1991). *Consciousness and revolution in Soviet philosophy: From the Bolsheviks to Evald Ilyenkov.* Cambridge: Cambridge University Press.

Bakhurst, D. (1995). On the social constitution of mind: Bruner, Ilyenkov, and the defense of cultural psychology. *Mind, Culture, and Activity,* 2(3), 158–171.

Bateson, G. (1972). *Steps to an ecology of mind.* New York: Ballantine Books.

Bauersfeld, H. (1995). "Language games" in the mathematics classroom: Their function and their effects. In P. Cobb and H. Bauersfeld (Eds.), *The emergence of mathematical meaning* (pp. 271–298). Hillsdale, NJ: Lawrence Erlbaum.

Beach, K. (1995). Activity as a mediator of sociocultural change and individual development: The case of school–work transition in Nepal. *Mind, Culture, and Activity,* 2(4), 285–302.

Boulton-Lewis, G.M. (1993). Young children's representations and strategies for subtraction. *British Journal of Education,* 63(3), 441–456.

Brousseau, G. (1984). The crucial role of the didactical contract in the analysis and construction of situations in teaching and learning. In H.G. Steiner (Ed.), *Theory of Mathematics Education.* IDM – Occasional Paper Nr. 54, (pp. 110–119). Bielefeld: Institut für Didaktik der Mathematik.

Bruner, J.S. (1962). Introduction. In L.S. Vygotksy, *Thought and language* (pp. v–x). Cambridge, MA: MIT Press.

Bruner, J.S. (1966) *Toward a theory of instruction.* Cambridge, MA: Belknap Press.

Carpenter, H.P., Moser, J.M., and Bebout, H.C. (1988). Representation of addition and subtraction word problems. *Journal for Research in Mathematics Education,* 19, 345–357.

Cassirer, E. (1907). *Das Erkenntnisproblem in der Philosophie und Wissenschaft der neueren Zeit* (Reprint 1971). Darmstadt: Wissenschaftliche Buchgesellschaft.

Clements, M.A., and Del Campo, G. (1989). Linking verbal knowledge, visual images, and episodes for mathematical learning. *Focus on Learning Problems in Mathematics,* 11(1), 25–33

Cobb, P. (1995). Cultural tools and mathematical learning: A case study. *Journal for Research in Mathematics Education,* 26(4), 362–285.

Cobb, P., Yackel, E., and Wood, T. (1992). A constructivist alternative to the representational view of the mind in mathematics education. *Journal for Research in Mathematics Education,* 23(1), 2–33.

Cole, M., and Engeström, Y. (1993). A cultural–historical approach to distributed cognition. In G. Salomon (Ed.), *Distributed cognitions: Psychological and educational considerations* (pp. 1–46). Cambridge: Cambridge University Press.

Cole, M., and Griffin, P. (1981). Cultural amplifiers reconsidered. In D.R. Olson (Ed.), *The social foundations of language and thought* (pp. 343–364). New York: Norton.

Davydov, V.V. (1988). Problems of developmental teaching. *Soviet Education,* 30 (Nos. 8, 9, and 10).

Davydov, V.V. (1990). *Types of generalization in instruction.* (Soviet Studies in Mathematics Education, Vol. 2). Reston, VA: National Council of Teachers of Mathematics.

Davydov, V.V. (1991). The content and unsolved problems of activity theory. *Activity Theory,* Nos. 7/8, 30–35.

Davydov, V.V. (Ed.) (1991). *Psychological abilities of primary school children in learning mathematics.* Reston, VA: National Council of Teachers of Mathematics.

Dehaene, S. (1992). Varieties of numerical abilities. *Cognition,* 44, 1–42.

Dennett, D.C. (1978). *Brainstorms. Philosophical essays on mind and psychology.* Montgomery, VT: Bradford.

Dewey, J. (1902/1959). The child and the curriculum. In M.S. Dworkin (Ed.), *Dewey on Education: Selections* (pp. 91–11). New York: Teachers College Press.

Dienes, Z.P. (1960). *Building up mathematics.* London: Hutchinson.

Dienes, Z.P. (1961). On abstraction and generalization. *Harvard Educational Review,* 31, 281–301.

Downs, R.M., and Liben, L.S. (1994). Mediating the environment: Communicating, appropriating, and developing graphic representations of place. In R. Wozniak and K.W. Fischer (Eds.), *Development in context: Acting and thinking in specific environment* (pp. 155–181). Hillsdale, NJ: Lawrence Erlbaum.

Engeström, Y. (1987). *Learning by expanding. An activity-theoretical approach to developmental research.* Helsinki: Orienta-Konsultit Oy.

Engeström, Y. (1990). *Learning, working and imagining: Twelve studies in activity theory.* Helsinki: Orienta-Konsultit Oy.

Engeström. Y. (1991a). Activity theory and individual and societal transformation. *Activity Theory,* Nos. 7/8, 6–17.

Engeström, Y. (1991b). *Non scolae sed vitae discimus:* Toward overcoming the encapsulation of school learning. *Learning and Instruction,* 1, 243–259.

Fingerhut, R., and Manske, C. (1984). *"Ich war behindert an Hand der Lehrer und Ärzte" – Protokoll einer Heilung.* Hamburg: Rowohlt.

Fodor, J.A. (1975). *The language of thought.* New York: Thomas Crowell.

Fodor, J.A. (1981). *Representations.* Cambridge, MA: MIT Press.

Freudenthal, H. (1974). Soviet research on teaching algebra at the lower grades of the elementary school. *Educational Studies in Mathematics,* 5, 391–412

Hoetker, J., and Ahlbrand, W.P.J. (1969). The persistence of the recitation. *American Educational Research Journal,* 6, 145–167.

Ilyenkov, E.V. (1977). The concept of the ideal. In *Philosophy in the USSR: Problems of dialectical materialism* (pp. 71–99). Moscow: Progress.

Janvier, C. (Ed.). (1987). *Problems of representation in the teaching and learning of mathematics.* Hillsdale, NJ: Lawrence Erlbaum.

Kaput, J.J. (1987). Toward a theory of symbol use in mathematics. In C. Janvier (Ed.), *Problems of representation in teaching and learning of mathematics* (pp. 159–196). Hillsdale, NJ: Lawrence Erlbaum.

Kaufman, E.L., Lord, M.W., Reese, T.W., and Volkmann, J. (1949). The discrimination of visual number. *American Journal of Psychology,* 62, 498–525.

Keitel, C., Otte, M., and Seeger, F. (1980). *Text–Wissen–Tätigkeit. Das Schulbuch im Mathematikunterricht.* Königstein: Scriptor.

Korzybski, A. (1941) *Science and sanity.* New York: Science Press.

Lakoff, G. (1987). *Women, fire, and dangerous things: What categories reveal about the mind.* Chicago: University of Chicago Press.

Lakoff, G., and Johnson, M. (1980). *Metaphors we live by.* Chicago: University of Chicago Press.

Latour, B. (1996). On interobjectivity. *Mind, Culture, and Activity,* 3, 228–245.

Lave, J. (1988). *Cognition in practice – Mind, mathematics and culture in everyday life.* Cambridge: Cambridge University Press.

Lave, J., and Wenger, E. (1991). *Situated learning: Legitimate peripheral participation.* Cambridge: Cambridge University Press.

Leont'ev, A.N. (1978). *Activity, consciousness, and personality.* Englewood Cliffs, NJ: Prentice-Hall.

Lorenz, J.H. (1991). *Anschauung und Veranschaulichungsmittel im Mathematikunterricht.* Götttingen: Hogrefe.

Lorenz, J.H. (1993a) (Ed.). *Mathematik und Anschauung.* Köln: Aulis.

Lorenz, J.H. (1993b). Veranschaulichungsmittel im arithmetischen Anfangsunterricht. In J.H. Lorenz (Ed.), *Mathematik und Anschauung* (pp. 122–146). Köln: Aulis.

Mac Cormac, E.R. (1985). *A cognitive theory of metaphor.* Cambridge, MA: MIT Press.

Mandler, G., and Shebo, B.J. (1982). Subitizing: An analysis of its component processes. *Journal of Experimental Psychology: General,* 111, 1–22.

Meira, L. (1995). The microevolution of mathematical representations in children's activity. *Cognition and Instruction,* 13(2), 269–313.

Mikulina, G.G. (1991). The psychological features of solving problems with letter data. In V.V. Davydov (Ed.), *Psychological abilities of primary school children in learning mathematics* (pp. 181–232). Reston, VA: National Council of Teachers of Mathematics.

Millikan, R.G. (1993). *White queen psychology and other essays for Alice.* Cambridge, MA: MIT Press.

Mitchell, W.J.T. (1994). *Picture theory: Essays on verbal and visual representation.* Chicago: University of Chicago Press.

Nunes, T., Schliemann, A.D., and Carraher, D.W. (1993). *Street mathematics and school mathematics.* Cambridge: Cambridge University Press

Ogden, C.K., and Richards, I.A. (1949). The *meaning of meaning: A study of the influence of language upon thought and the science of symbolism.* London: Routledge & Kegan Paul.

Peirce, C.S. (1931–1935). *Collected Papers of Charles Sanders Peirce* (C. Hartshorne and P. Weiß, Eds.). Cambridge, MA: Harvard University Press.

Perkins, D.N., and Unger, C. (1994). A new look in representations for mathematics and science learning. *Instructional Science,* 22, 1–37.

Perner, J. (1991). *Understanding the representational mind.* Cambridge, Mass.: MIT Press.

Putnam, H. (1988). *Representation and reality.* Cambridge, MA: MIT Press.

Radatz, H. (1980). *Fehleranalysen im Mathematikunterricht.* Braunschweig: Vieweg.

Radatz, H. (1984) Schwierigkeiten der Anwendung arithmetischen Wissens am Beispiel des Sachrechnens. In J.H. Lorenz (Ed.), *Lernschwierigkeiten – Forschung und Praxis* (pp. 17–29). Köln: Aulis.

Rogoff, B. (1990). *Apprenticeship in thinking: Cognitive development in social context.* Oxford: Oxford University Press.

Rogoff, B. (1994). Children's guided participation and participatory appropriation in sociocultural activity. In R. Wozniak and K.W. Fischer (Eds.), *Development in context: Acting and thinking in specific environment* (pp. 121–153). Hillsdale, NJ: Lawrence Erlbaum.

Rorty, R. (1979) *Philosophy and the mirror of nature.* Princeton, NJ: Princeton University Press.

Saxe, G.B. (1991). *Culture and cognitive development: Studies in mathematical understanding.* Hillsdale, NJ: Lawrence Erlbaum.

Scheerer, E. (1995). Repräsentation, IV. Psychologie und Kognitionswissenschaft. In J. Ritter (Ed.), *Historisches Wörterbuch der Philosophie* (Vol. 8, pp. 834–846). Basel: Schwabe.

Scholz, O.R. (1995). Repräsentation, 19. und 20. Jh. In J. Ritter (Ed.), *Historisches Wörterbuch der Philosophie* (Vol. 8, pp. 826–834). Basel: Schwabe.

Scribner, S. (1984a). Studying working intelligence. In B. Rogoff and J. Lave (Eds.), *Everyday cognition: Its development in social context* (pp. 9–40). Cambridge, MA: Harvard University Press.

Scribner, S. (1984b). Pricing delivery tickets: "School arithmetic" in a practical setting. *Quarterly Newsletter of the Laboratory of Comparative Human Cognition,* 6, 19–25.

Sfard, A. (1993). Reification as a birth of a metaphor: Non-objectivist's reflections on how mathematicians (and other people) understand mathematics. In P. Bero (Ed.), *3rd Bratislava International Symposium on Mathematical Education,* August 25–27, (pp. 93–118). Bratislava, Slovakia.

Stadler, M., Seeger, F., and Raeithel, A. (1975). *Psychologie der Wahrnehmung.* München: Juventa.

Ueno, N. (1995). The social construction of reality in the artifacts of numeracy for distribution and exchange in a Nepalese Bazaar. *Mind, Culture, and Activity,* 2, 240–257.

Valsiner, J. (1987). *Culture and development of children's action: A cultural–historical theory of developmental psychology.* New York: Wiley.

Velichkovskij, B. (1988). *Wissen und Handeln: Kognitive Psychologie aus tätigkeitstheoretischer Sicht.* Berlin: Deutscher Verlag der Wissenschaften.

Voigt, J. (1993). Unterschiedliche Deutungen bildlicher Darstellungen zwischen Lehrerin und Schülern. In J.H. Lorenz (Ed.), *Mathematik und Anschauung* (pp. 147–166). Köln: Aulis.

von Glasersfeld, E. (1982). Subitizing: The role of figural patterns in the development of numerical concepts. *Archives de Psychologie,* 50, 191–218.

Vygotsky, L.S. (1978). *Mind in society: The development of higher psychological processes.* Cambridge, MA: Harvard University Press.

Vygotsky, L.S. (1987). Thinking and speech. In R.W. Rieber and A.S. Carton (Eds.), *The collected works of L.S. Vygotsky,* (Vol. 1, Problems of general psychology, pp. 38–285). New York: Plenum Press.

Wartofsky, M.W. (1979). Perception, representation and the forms of action: Towards an historical epistemology. In M.W. Wartofsky, *Models: Representation and the scientific understanding* (pp. 188–210). Dordrecht: Reidel.

Weimer, W.B. (1973). Psycholinguistics and Plato's paradoxes of the Meno. *American Psychologist,* 28 (January), 15–33.

Wertheimer, M. (1959). *Productive thinking* (enlarged edition). New York: Harper.

Wolters, M. (1983). The part–whole schema and arithmetical problems. *Educational Studies in Mathematics,* 14, 127–138.

13 Mathematical understanding in classroom interaction: The interrelation of social and epistemological constraints

HEINZ STEINBRING

Understanding as the deciphering of social and epistemological signs

The notion of understanding a mathematical concept or problem plays an important role in any educational investigation of school mathematics, be it in research contexts or in the framework of curricular constructions for the teaching of mathematics. To make possible, to support, and to improve mathematical understanding seem to be central objectives of any theoretical or practical endeavor in the didactics of mathematics.

There is a lot of reflection about how to understand understanding, how to classify different types and degrees of understanding (Pirie 1988; Pirie and Kieren 1989; Schroeder 1987; Skemp 1976), and how conceptually to conceive of understanding (Maier 1988; Sierpinska 1990a, 1990b; Vollrath 1993), and many models describing ideal and everyday processes of understanding mathematics have been developed (Herscovics and Bergeron 1983; Hiebert and Carpenter 1992). Our first step in approaching the puzzling concept of mathematical understanding is to list a number of seemingly contradictory attributes that are associated with this notion.

- Understanding is conceived of as an expanding process, gradually improving the comprehension of a concept or problem step by step, and there is never an absolute understanding; on the other side, when being confronted with a new, unsolved problem that seems to be totally incomprehensible, sometimes there is sudden and complete understanding without any further doubt.
- Understanding requires one to relate the new knowledge to the knowledge already known, to integrate the unknown into the known; but on the other hand, understanding a new mathematical concept

often requires it to be comprehended in itself, without being able to derive the new concept from concepts already understood.

- Every fruitful process of understanding requires the learning subject to be active and to construct his or her own understanding; on the other hand, the understanding of fundamental mathematical ideas, concepts, and theorems cannot be reinvented completely by each student. He or she has to accept it as preconstructed.
- Understanding is an individual and personal quality of the learning subject in every mathematical learning process; on the other hand, the understanding of abstract and general mathematical ideas often needs social support and stabilization.

How can these contradictory statements about the conception of understanding be better explained and untangled? The main perspective taken in the following is on problems of understanding school mathematics *in the course of everyday teaching and learning processes.* This is an important framework, because students' learning in school is always embedded in social and content-dependent situations, subordinating the processes of understanding to specific conditions and intentions. One could argue that the school milieu creates a specific local mathematical culture that contributes to the contradictions of understanding mentioned, because, traditionally, there are the combination of individual and shared apprehension on the social side, and the construction of all new knowledge erected upon old knowledge on the content side.

From an epistemological perspective, I would like to characterize the core problem of understanding school mathematics as follows: *Students have to decipher signs and symbols!* There are two different types of signs and symbols: *mathematical signs and symbols* (ciphers, letters, variables, graphics, diagrams, visualizations, etc.) and *social signs and symbols* (in the frame of the classroom culture: hidden hints, remarks, reinforcements, confirmations, rejections, etc., made by the teacher, and comments and remarks of similar types made by other students). Being able to decipher all signs of these two types correctly would guarantee understanding. But the deciphering of social and epistemological signs and symbols in classroom culture is not an easy task. It is not a simple translation according to fixed universal rules and dictionaries. Both types of signs are intentional; they are referring to something other; and what they refer to may even change during discourse or may undergo a decisive shift of meaning.

Scientific concepts are interconnected and comprehensive systems of understanding which have been elaborated and refined in the course of social history. . . . Scientific concepts are not simply internalized, but undergo a complex transformation in their inward movement from artefacts external to the child's activity to mental processes which are interwoven with the child's intellectual functions. (Saxe 1991: 11)

For example, when introducing negative numbers through the idea of the debt model, in which an amount of debt is represented by a number of red counters (the black counters represent the positive numbers), the mathematical signs (red and black counters) are used with some kind of concrete reference; more elaborated operations in this model need the "rule of compensation" that represents the zero by a number of pairs of red and black counters. On the observable surface, the same sort of signs is used here, that is, red and black counters with concrete reference. But, in fact, simply the combination of the same number of red and black counters has drastically changed its referential function; even if traditional school mathematics pretends that these are still concrete objects and methodically they are used this way, from an epistemological perspective, they have become true mathematical symbols. A concrete zero would have been represented by "nothing," by no black or red counter; the way of representing the zero by a (unlimited) number of pairs expresses a relation representing the zero, that is, a true symbol. The concretely chosen combination of pairs *is* not the zero (as five red counters are minus 5), but is a *symbolic representation* of the zero (cf. Steinbring 1994). The mathematical sign of counters has totally changed its intentional reference.

Correspondingly the social signs introduced by the teacher into the classroom culture have no fixed and direct meaning. Instead, they have to be deciphered relative to a specific context. The introduction of the "impossible event" during a course in elementary probability, for instance, may produce referential interpretations for the students that differ from those aimed at by the teacher. Whereas the students always have in mind some kind of concrete impossibility, the teacher tries to develop the mathematical idea of the impossible event compared with the other elementary events of a sample space. And all remarks and hints (social signs) he or she is giving in the interaction have a twofold referential meaning. The teacher, for example, asks, "How would we describe this with an adjective?" provoking the student to answer, "The uncertain event." And, again, the teacher: "The uncertain one? We shall simply say: the impossible event. And now for my question: What kind of a subset is this, if I speak of an impossible event?" But for the students, this impossibility remains real and concrete; they answer directly, "That won't work at all!" All the teacher's remarks remain open for different intentional references; they cannot be made strictly unambiguous (cf. Steinbring 1991a). This means that the adequate deciphering of social and epistemological signs in the mathematics classroom remains a very difficult task (cf. Walkerdine 1990: 187).

Symbols, symbolic relationships, and introduction to the use and reading of symbols are essential aspects for the formation of every culture (Wagner 1981, 1986). Mathematics deals per se with symbols, signs, symbolic connections, abstract diagrams, and relations. The use of symbols in the culture of mathematics teaching is constituted in a specific way, giving social and communicative meaning to letters, signs, and diagrams in the course of ritualized procedures of negotiation. Social interaction constitutes a specific teaching culture based on school-mathematical symbols that are interpreted according to particular social conventions and methodical rules. This ritualized perception of mathematical symbols seems to be insufficient according to the analysis of critical observers, because it does not advance to the genuine essence of the symbol but remains on the level of pseudorecognition.

The openness of the intentional references of epistemological and social signs and symbols in interactive mathematical processes is at the heart of the problems with the concept of understanding mathematics. This openness is responsible for the contradictory statements made about understanding. Finding the "correct" intentional references for social and epistemological signs always depends on individual and common insights, on personal activity, and on acceptance, as well as on construction of completely new relations and use of already known referential interpretations of mathematical concepts. It may provide complete, sudden understanding of a problem and, at the same time, give an idea of how to develop an increasingly deeper understanding.

In the following we shall relate our notion of mathematical understanding to conceptions discussed in the educational literature. The following two positions that emphasize the linking of the learning subject with the mathematical content reflect important aspects important for our formation of the concept of understanding. The first definition points to the external and corresponding internal representation of mathematical knowledge:

A mathematical idea or procedure or fact is understood if it is part of an internal network. More specifically, the mathematics is understood if its mental representation is part of a network of representations. The degree of understanding is determined by the number and the strength of the connections. A mathematical idea, procedure or fact is understood thoroughly if it is linked to existing networks with stronger or more numerous connections. (Hiebert and Carpenter 1992: 67)

The second description tries to relate epistemological constraints of mathematical knowledge to the active role of the learner:

It is proposed to conceive of understanding as an act (of grasping the meaning) and not as a process or way of knowing. (Sierpinska 1990a: 24)

An act of understanding happens only in the attentive mind, who is willing to identify objects, to discriminate between them, to perceive generality in the particular and the particular in the general, to synthesize large domains of thought and experience. Our minds are not being passively "imprinted with ideas of things without." . . . It needs active construction to even see what everybody seems to see in an effortless and natural way. (Sierpinska 1994: 101)

The first definition claims the integration of the new mathematical knowledge into existing networks of mental representations; the second description interprets understanding as an "active construction," an "act of grasping the new meaning," and in this way points to the limits of integrating the new knowledge into existing networks of representations. From our epistemological perspective of understanding as the deciphering of social and mathematical signs referring to a variable intentional reference context, some modifications have to be considered with regard to these conceptions of understanding. Signs do not only refer to fixed elements in networks of representation, but essentially to *relations* to be constructed in the reference fields (this makes the network of representations and the process of integration more complicated and may even lead to radical changes and reconstruction of networks of representations where the true symbol function of the sign is imperative; cf. Steinbring to appear). The active construction of meaning and overcoming of epistemological obstacles is necessarily embedded within the framework of the classroom culture of conventions, generalizations, rules, methodological procedures, and socially accumulated and accepted knowledge that is at everybody's disposal.

In order to analyze how mathematical understanding develops in the classroom culture, we propose to expand the perspective of investigation in terms of the following two dimensions:

Mathematical knowledge:

        social aspects ↔ epistemological aspects

Participating persons:

        students' understanding ↔ teacher's understanding
        (of mathematics and of students' and teacher's intentions)

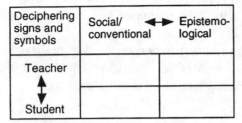

Figure 13.1. The matrix of relations between the social dimension and the knowledge dimension.

In this way, conceiving of mathematical understanding as the deciphering of social and epistemological signs and symbols necessarily leads to a reciprocal process of understanding between teachers and students, everyone interactively trying to decipher social and epistemological signs communicated by other persons. The analytic distinction between social/conventional and epistemological aspects will help to clarify the inherent dependencies (see Figure 13.1).

The following section deals with the relationship between social/conventional and epistemological constraints on mathematical knowledge. The third section will analyze the problem of understanding the development of mathematical knowledge embedded within the classroom culture. The investigation is based on specific examples of understanding mathematical knowledge and on exemplary cases of classroom episodes.

Mathematical knowledge and understanding

When trying to decipher mathematical signs, one has to construct a relationship between those mathematical signs (and their network of representation) that have been constructed already and the new signs that still have to be understood. What kind of relationship is this? Is it a reduction, an equilibrium, or a dependency? And what kind of epistemological constraints and social conventions are involved in this relationship?

In our first example, we shall analyze the understanding of written symbols with the help of procedures describing the correct mathematical operations. In most cases, the understanding of decimal fractions is explained by describing the correct algorithms for the elementary mathematical operations of addition, subtraction, multiplication, and division. For this description, one may find the use of the *rule of shifting the point:* The multiplication (division) of a decimal fraction by 10, 100, 1,000, . . . , is done by shifting the point one, two, three, . . . , places to the right (left).

This rule describes the transformation of the decimal fractions into natural numbers. It is used to "define" the mathematical operations, for instance, addition and multiplication of decimal fractions:

Rule of addition and subtraction:

Example:

$$2.743 + 3.85 \ = 6.59$$

$$\downarrow .1000 \quad \downarrow .1000 \quad \uparrow \div 100$$

$$2743 \ + 3850 = 6593$$

Both terms of the sum are multiplied by a power of 10 (here 1,000) that produces natural numbers that can then be added. Afterward, the enlargement is undone by division by 1,000. Here, as well, the law of distributivity is used.

Rule of multiplication:

Example:

$$3.45 \cdot 2.3 \ = \ 7.935$$

$$\downarrow .100 \quad \downarrow .10 \quad \uparrow \div 1000$$

$$345 \ \cdot 23 \ = \ 7935$$

Here, the theorem is used: If a factor is multiplied, the result is also multiplied accordingly (Postel 1991: 19).

This way of introducing the arithmetical operations for decimal fractions clearly shows that the fractions are first transformed into natural numbers. In this domain, the known arithmetical operations are performed, and afterward the result is retransformed into a decimal fraction. This is declared as the result of the new operation. For certain, the rule of shifting the point should contain the new conceptual knowledge of decimal fractions. In most cases, however, this rule (or a similar version) is made conceptually vain by taking it as a technical device for counting the correct place of the point by reading this off from the surface of the written symbols.

This way of explaining the decimal fraction by technically explaining the arithmetical rules is also inherent in the usual manner of lining up the two decimal fractions by putting the points in a line and then performing the already known operation.

Suppose the students have represented the procedure in such a way that they have connected the mechanics of aligning digits with the combining of quantities measured with same unit (ones, tens, hundreds, and so on). When these students encounter addition and subtraction with decimal fractions, they are in a good position to connect the frequently taught procedure – line up the decimal points – with combining quantities measured with the same unit. If they build the connection, the addition and subtraction procedure becomes part of the existing network, the network becomes enriched, and adding and subtracting decimals is understood. (Hiebert and Carpenter 1992: 69)

This reductive explanation of the arithmetical operations of decimal fractions with the help of the known algorithms of written procedures for the arithmetical operations displays two important aspects:

First, the new concept of decimal fractions is not used itself; it is transformed into natural numbers with which students are already able to operate.

Second, the rule of transforming decimal fractions into natural numbers that should conceptually regulate the use of the new sign of the point (and in this way clarify an important conceptual aspect of the decimal fraction) is devaluated to a formal counting scheme that can be used without any conceptual idea of the number concept.

When using the technique of lining up the decimals according to the point, the counting of the correct position of the point is even superfluous for addition and subtraction, thus really suggesting that decimal fractions are some kind of natural numbers and not something conceptually new. This kind of understanding of decimal fractions as natural numbers is known from analyzing students' errors. A widely used implicit idea of decimal fractions is the understanding that the point separates the given number into two natural numbers (cf. Wellenreuther and Zech 1990). For instance, the following types of transformations and arithmetical operations could be observed according to this perspective:

$$5.3 + 2.42 \rightarrow 7.45$$
$$18.27 \div 9 \rightarrow 2.3$$
$$0.2 \cdot 0.4 \rightarrow 0.8$$

This way of understanding and of manipulating decimal fractions can, in principle, be justified on the same basis of conventional rules described:

First, the decimal fractions are transformed into natural numbers in order to perform the required operations.

Second, the rule of transforming decimal fractions into natural numbers is an algorithmic scheme without conceptual connections: There is one natural number to the left and one to the right of the point.

The analysis shows that both ways of operating with decimal fractions, that is, the students' procedure and the "official" one, are justified on the same grounds; when reducing the concept of decimal fractions to natural numbers and their operations and at the same time intending to make the rule of transformation as simple as possible (an intention that necessarily leads to an evacuation of conceptual relations that were initially contained in the meaning of the rule), then there is no possibility of explaining why the students' rule is incorrect. The correctness of rules to be used and applied is reduced to external social conventions. It is the teacher's authority that decides which rule is correct.

This first example of introducing decimal numbers shows that understanding a new mathematical concept cannot be done by reducing it completely to concepts that are already known. When trying to enhance understanding by describing operational rules in such a reductive process, there is a great danger of changing the status of rules from mathematical operations referring to conceptual aspects into formalized conventional recipes. Understanding is not simply complete integration into the existing network of representations. For the concept of decimal fractions, it is the other way around: Decimal fraction symbols do not have to be deciphered as natural number symbols, but the natural number symbols have to be reconsidered as a kind of decimal fraction symbols (cf. Steinbring 1989, 1991b, 1992).

A second example analyzes the understanding of written symbols and mathematical concepts with the help of introducing referential objects to provide mathematical meaning to new symbols. Similar to the understanding of decimals, fractions are often introduced by transforming the new symbols into natural numbers and performing the elementary arithmetical operations in this number domain. For example, the following operation with fractions $4/5 \div 2/15$ is explained according to the recipe, "Fractions are divided by multiplying with the reciprocal value of the second fraction. Fractions are multiplied by multiplying the denominators and multiplying the numerators." In this way, one obtains arithmetical operations that the students already know: $4 \cdot 15$ and $5 \cdot 2$ leading to the result: $60/10$ or 6. This kind of procedural reduction of new symbols to known symbols has been explained. The present focus is on the construction of referential meaning for the new symbols.

Consider the following problem from a textbook for sixth-grade students (Figure 13.2). This problem deals with the division of fractions

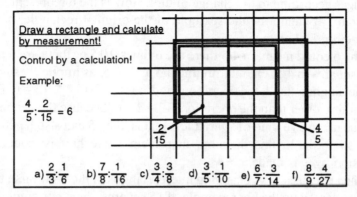

Figure 13.2. Visualizing the division of fraction.

and tries to use a graphic diagram to mediate in a direct way the meaning of fraction division. This contrast between formula and graphic diagram is suitable to clarify some epistemological aspects of sign and object (or referent) in school mathematics. On the one side, there are numerical signs connected by some operational symbols, functioning as a little system: On the other side, there is a geometric reference context, intended to enhance the understanding of the signs and operations. The diagram should support the process of constructing a meaning for the formula. The *relational structures* in the geometric diagram and the formula are the important aspects and not the signs themselves.

How can this diagram give meaning to the formula? Is it possible to use it to deduce the idea of the division of fractions? Is it adequate to conceive of the elements in this diagram as concrete objects for directly showing the meaning of division?

First of all, one observes that all problems to be tackled have denominators that are a multiple of the denominator of the other fraction. Consequently, the intended explanation with the help of the diagram cannot be universal. A certain type of fraction seems to be presupposed, indicating a first reciprocal interplay between diagram and formula. There are more indications of this interplay: In this representation, a variable comprehension of 1 or the unit is necessary. At one time, the big rectangle with the 15 squares is the unit, used to visualize the proportions of 4/5 and 2/15 as four rectangles (with 3 squares each) and as a rectangle of 2 squares, respectively. The composition of three squares to a rectangle represents a new unit or 1. When interpreting the operation $4/5 \div 2/15 = 6$, the epistemological meaning of the result "6" changes according to the change of unit. How is the 6 represented in the diagram? It cannot be the sextuple of the original rectangle, hence no pure empirical element.

The 6 could mean, in 4/5 there are 6 times 2/15 or there are 6 pairs of two squares in 4/5. Or, interpreting 4/5 as 12/15, as implicitly suggested in the diagram itself, the operation modifies to 12/15 ÷ 2/15 = 6. But this is nothing other than the operation: 12 ÷ 2 = 6, because the denominator can be taken as a kind of "variable"; that is, the 15 can also be 20, or 27, and so forth. In this division, in principle, the half is calculated; a division by 2 is made.

The analysis shows changing interpretations of the unit: First, the unit is represented by the big rectangle of 15 squares; then, one single square also represents the unit. The epistemological reason is that a fraction like 12/15 is not exclusively the relation of the two concrete numbers 12 and 15, but a single representative of a lot of such relations: 4/5, 8/10, 16/20, 20/25, ... What is defined as the unit in the diagram is partly arbitrary and made by some convention. Furthermore, the constraints of the geometric diagram and the given numerical sign structure partly determine the choice of the unit. For instance, for this arithmetical problem, it would not be an adequate choice to take the rectangle of $5 \cdot 7$ squares as the unit, whereas a rectangle of $6 \cdot 10$ squares, or a subdivision of the squares into quarters, would be valid.

The intentional variability implicit in the numerical structure of a fraction is partly destroyed in the geometric diagram used to represent the fraction; this variability has to be restored in the diagram by means of flexibly changing the unit. The concrete single diagram, with its parameters once chosen, has to be conceived of as a "general" diagram.

The meaning of the new symbols cannot be provided directly by the given reference context in its customary perception. When starting with visual representations of units and parts for displaying the idea of fractions as a means to distribute something, there is then a sudden shift in this relation between the symbols and their reference context when dealing with the division of fractions, and this can no longer be explained within the pregiven frame of empirical objects and their concrete distribution. The symbol system 4/5 ÷ 2/15 does not simply show a greater complexity, but also drastically changes its referential relationship: The former network of representations has changed in such a way that new *relations* in this network have to be constructed. The diagram no longer displays empirical objects (rectangles of different shapes) but relations between objects (the relation 4/5 or the "unit" relation in all possible combinations of geometric objects). Understanding new math-

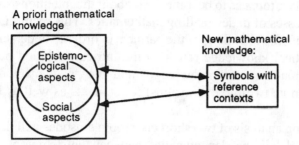

Figure 13.3. The relation between familiar and new mathematical knowledge.

ematical symbols, being able to decipher epistemological and social signs, requires the construction of a relationship between knowledge that is already known and the new concept (the symbol with its new meaning) as is depicted in Figure 13.3.

This relation between new and old knowledge cannot be a complete reduction of new symbols and operations to old symbols and operations. When the intention is to base understanding exclusively on the procedural links between new and old symbols, there is a danger of converting mathematical rules into a conventionalized scheme of actions that becomes void of mathematical meaning. When relating new concepts to the referential objects already used for the known concepts, there also is the danger of simplifying the meaning of the new concepts by turning the symbols into names for objects (cf. Steinbring 1988). The new mathematical concept has to be conceived of as a new, proper symbol with a changed reference context: The new intentional references of the symbol that still has to be deciphered force the transformation and change of the old reference context. This perspective of looking from the new concepts to the old concepts is only possible if there is an interplay between the epistemological and social/conventional aspects of mathematical knowledge.

Classroom interaction and understanding

When trying to investigate processes of understanding mathematics in the everyday classroom culture, it becomes obvious that every person's inclination to understand always means to discover and to compensate the different intentional meanings of the epistemological and the social, conventionalized signs. Thus, for students, it is important to figure out the teacher's intentions with regard to the school mathematical knowledge. The teacher, too, has to become clear about the intentions students follow in processes of understanding mathematics. This reciprocal way of understanding the intentions of the partners in order to understand the new mathematical knowledge relies on the intricate relation between the "a priori mathematical knowledge" and the "new mathematical knowledge" (in its twofold dimensions) for students as well as for the teacher.

The following analysis of two short classroom episodes will demonstrate in more detail the dependencies between epistemological and social/conventional aspects of the knowledge in question and how the students' intention to understand relates to the understanding the mathematics teacher has of their understanding. The first example concentrates on how the students are pushed to follow the a priori understanding of the teacher. This leads to an acceptance of the conventionalized aspects of the knowledge without real understanding of the epistemological point of the concept dealt with. The focus of the second example is on how a student understands an epistemological relation in the new mathematical knowledge that is not accepted by the mathematics teacher because she strictly adheres to her fixed a priori mathematical knowledge.

During the *first episode* dealing with the topic "What is relative frequency?" (see appendix 1), the teacher tries to recall the concept of relative frequency. She starts with the question, "What do we understand by relative frequency, Markus?" (1). The way of formulating "what do *we* understand" indicates that some accepted, conventionalized form of noting the concept of relative frequency exists already. The first student, Markus, collects nearly all the necessary elements, that is, "the number of cases observed," "the number of trials," and he points out in a joking way that these elements have to be combined.

Then the student Klaus proceeds in another direction, saying, "Relative frequency means, for example, often, it is, hem, a medium value" (6). The teacher rejects his proposal, and the student Frank continues with the old idea, trying out various possibilities for combining the elements already identified: "The trials are divided with the cases observed, I think, or multiplied" (9, 10). This is strongly approved by the teacher, who states, "Markus already put it quite correctly, it's just the decisive word that's missing" (11). This now clarifies the accepted framework for the students to handle the question: The way of searching for the needed pieces of a puzzle is expected, the two pieces "the number of cases observed" and "the number of trials" are already found, but another decisive piece (the mathematical combination of the two others) is still missing. The very quick proposals "Subtract" and "Take it minus" provoke a severe rejection by the teacher: "That's incredible!" (15), leading to an alternative approach by asking how this should be written down.

The phase (17–42) opened by the question "You should ask yourselves how will you write it down" (17) introduces the context of fraction and fractional calculus into the discussion. Contrary to her expectations, the key words "fraction" and "fractional calculus" (24, 26) evoke student contributions such as "subtracted in the fractional calculus" (31), "one as denominator and the other as numerator" (35, 36), "To calculate a fraction!" (37), and "To reduce to the common denominator" (42). The teacher simply intended to point to the link between fraction and division, a seemingly simple transfer that, toward the end of this phase, could only be achieved by using a funnellike pattern (Bauersfeld 1978).

The teacher's question "What kind of calculation is it then, if you write a fraction?" (38, 39) forces the "correct" answer: "Dividing." The last piece of the puzzle is found. What now remains is to combine correctly all three pieces (phase 43–48). The teacher's first proposal, "relative frequency is . . . the number of cases divided by?" (43, 44), does not use the detailed vocabulary. Consequently, one student's answer does not fit: "the number of cases observed" (45). Finally, the correct formulation is initiated by the repeated, complete teacher question, "No, the cases observed divided by? Aha!" (46). At last, relative frequency is also formulated by means of fractional calculus.

The official understanding of the concept of relative frequency codified before is represented in the statement "Relative frequency is the number of cases observed divided by the number of trials." This verbal description is visualized often as a fraction in the following manner:

$$\text{relative frequency} = \frac{\text{number of cases observed}}{\text{number of trials}}$$

This representation in the form of a fraction simply should express the operation of division; no further relations to fractional calculus are implied. The teacher starts the repetition of the concept of relative frequency with its accepted "definition." This seems to represent her a priori mathematical knowledge that has to be reconstructed.

Her formulation "What do *we* understand by relative frequency?" (1) is a signal to the fact that nothing completely new has to be discovered, but rather something already introduced and codified. Nevertheless, the student Klaus presents something that is very open and could lead to an epistemological and conceptual consideration of this concept. But the teacher strictly rejects this orientation, and she adheres to the course already taken, that is, searching for the matching pieces of the conventionalized definition. The two main ingredients are already detected, and the teacher signals that only "the decisive word" is still missing. This sign can be interpreted only in part by the students: They know they are on the right track, and they have to look for a mathematical operation, but which one?

This becomes a guessing game that is empty of mathematical meaning, and the teacher gives another signal demanding that the students remember how the relative frequency is *written down*. This produces the framing of a fraction, evoking all sorts of descriptions connected with fractions and fractional calculus. The signals "fraction" and "fractional calculus" mislead the students in some way, and they are not able simply to decipher the analogy of the fraction bar with the operation of division that the teacher is aiming at. This becomes possible only after some discussion and with a further direct signal: "What kind of calculation is it then, if you write a fraction?" (38, 39). At the end of this episode, the teacher feels forced to make her signal very explicit in order to obtain the correct, expected answers.

The signals given by the teacher during the course of this episode are designed to orient the students to the conventionally agreed "definition" of the concept of relative frequency. This intention is understood very quickly by most students, and the rejection of Klaus's proposal as leading away from this orientation reinforces the understanding of this intention. The more the accepted answers are restricted to the social/conventional side, the more the students are deprived of some epistemological means of justification. They are only able to make proposals to be approved or rejected by the teacher. The epistemological helplessness of the students becomes obvious when the signal "fraction" is understood too extensively instead of simply reading off the operation of division from the fraction bar.

The a priori knowledge of the teacher dominates the course of this episode: First she starts with her fixed understanding of the concept of relative frequency as already "defined" in the class as a special fraction; second, she presupposes that the students have already arrived at a similar "definition" to her own. In this way, the process of understanding is simply reduced to something like the refinding of a conventionalized description for a concept already given. All the signals the students give during this discussion are judged by the teacher solely according to how they ensure that the intended goal will be reached as directly as possible. The restriction of the interaction on the social/conventional level turns the understanding of a piece of mathematical knowledge that is already judged as definitely given and fixed a priori into a process of negotiating the right words, names, and rules, and in most cases its validity can be delivered only by the teacher's authority (cf. Steinbring 1991a).

During the *second episode* coping with "the area formula for the trapezoid" (see appendix 2), the students are asked to explore a new area formula in two different ways. They already have some experience, because they have discussed the formula in a similar way for the parallelogram. With her first long statement (1–8), the teacher gives some introductory hints. She signals positively and negatively how the two expected methods of geometrical construction and argumentation might function. For the first solution, she indicates to enlarge the trapezoid and to construct a parallelogram. With reference to the fabrication of the area formula for the parallelogram done the day before, she explicitly points to the way to cut off angles and add them on other sides in order to gain the second solution; here, it is not possible to cut off a whole angle, as in the case of the parallelogram, but one still has to try to construct a rectangle.

The students perform the expected first solution; now the discussion of a possible second solution starts. A student makes a proposal that cannot be understood yet in all its details: "Could one not, this line here, the height . . . make parallel to, to this other line, and then one would have such a quadrangle and on the other side a triangle" (11, 13, 14). The student points to a drawing (Figure 13.4).

In line with her hints given at the beginning, the teacher rejects this proposal, because she understands that a whole angle, that is, the big triangle, should be cut off, and this is not possible because she believes "the problem is, that one knows only the whole baseline g_1 and not just this little piece there. That's totally unknown. Well, when you really have concrete lengths, you could draw and measure it. But, in general, I can't do it. How should I then say, g_1 minus some small piece, but how long is this little piece at all? Hence, that does not work!" (17–21). By looking at only one triangle to be cut off from the trapezoid, the first impression is that one cannot arrive at a formula, because one cannot determine explicitly the baseline of this triangle. (This approach can lead to a solution, if the other triangle on the left side is also cut off. Then the baselines of both triangles together will be $g_1 - g_2$. This information is sufficient to deduce a formula for the trapezoid.)

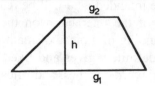

Figure 13.4. The area of trapezium.

The student is not impressed by the teacher's argument; he again takes up his idea, which seems to be different from the idea rejected by the teacher. "Yes, if we now, ehm the longer side . . . well, take g_1 minus g_2 . . . and then, ehm, then one has, there would be a remainder; let's say g_1 would be 5, hence 8 minus 5, giving 3 and then, if one would now push g_2 backwards, in a way giving a right angle, then something would remain there on the right side" (23, 26–29). The teacher does not want to follow his idea, but the student continues: "I do mean something else, there you take 5 times h, then, what is left here, there remains 3 centimeters, that is a slope, yes, then, ehm, you also take the slope . . . the remainder is 3 times the height then, divided by two" (32–34, 35). The teacher interjects: "I really have understood this" (35) when the student develops his line of argumentation, but she has not understood the student's idea. Instead, she intends to explain that she really has understood the apparent incorrectness of his proposal; with her last contribution, the teacher begins to present what she thinks is a correct second solution (37–40).

Probably, the student's proposal could be explained by Figure 13.5; this was not shown in this way by the student; he simply expressed his idea verbally. One change the student makes is to replace the variables g_1 and g_2 by concrete numbers, but still he is able to use these numbers to develop, in principle, a correct universal formula. The student has transformed the trapezoid into a rectangle together with a triangle. By what he has called "if one would push now g_2 backwards," the student constructs a rectangle and a rectangular triangle allowing for the calculation of the whole area. Then he determines the areas of both surfaces: $5 \cdot h$ for the rectangle and $3 \cdot h/2$ as the area for the triangle. If the teacher could accept this, the concrete values 5, 3, and 8 could be retranslated into the corresponding variables, providing a general formula for the area of the trapezoid:

$$A = g_2 \cdot h + \frac{(g_1 - g_2) \cdot h}{2} = \frac{2 \cdot g_2 \cdot h + (g_1 - g_2) \cdot h}{2} = \frac{(g_1 + g_2) \cdot h}{2}$$

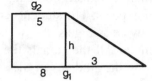

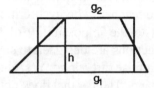

Figure 13.5. Transforming the trapezium and keeping the area constant.

Figure 13.6. Giving the area of the trapezium by the area of a rectangle.

This is a perfect solution, but unfortunately not one within the teacher's scope. She is unable or unwilling really to uncover the student's argument. She has a fixed a priori geometric construction in mind that prevents her from understanding his solution. In her last contribution, the teacher explains what she expects. The students should have drawn the medium line, in agreement with the hint given earlier not to cut off a whole angle but only parts in order to construct a rectangle (see Figure 13.6).

When developing this construction of the medium line, the same problems might arise as those that the teacher classified as generally unsolvable, that is, to determine the length of the baseline of the small triangles to be cut off. From the final complete drawing with the medium line, the length of the medium line can generally be determined as

$$m = \frac{(g_1 + g_2)}{2}$$

What sorts of understanding can be observed in this short episode, both on the side of the teacher and on the side of this student? Similarly to the episode discussed before, here too, the teacher starts with her definite a priori knowledge about the admitted geometric constructions and linked deductions of the area formula for the two solutions sought. At the beginning, the teacher in some way informs the students of her intentions and her a priori ideas about the mathematical problem. The first solution is presented just as the teacher expects: The students construct the correct parallelogram with an area twice as big as that of the trapezoid, thus developing the correct formula. All the signals the teacher had given before could be deciphered by the students in the expected way.

When discussing the second solution, this frame of trying to decipher correctly the signals given before and keeping to the further hints, rejections, and reinforcements of the teacher changes. One student seems to be very sure about his solution. He keeps on presenting it despite serious counterremarks by the teacher, who even criticizes the mathematical correctness and points to the impossibility of the proposed solution. First, the teacher's reactions are designed to signal that the student, as are others, is pursuing a false direction: "Yes, ehm, some others also made this proposal" (15), and "Hence, that does not work!" (21), or later, "Pay attention" (30). But because the student is insensitive to the teacher's hints, she now directly rejects the student's proposal, "I really have understood this" (35), expressing in this way that the student should now definitely stop presenting his unacceptable argumentation. Finally, the teacher introduces her correct mathematical deduction of the second solution with a loud and distinct "No!"

At no time is the student willing to abandon his argument and to enter the intended frame of discussion by trying to uncover the a priori fixed geometrical construction the teacher has in mind. No remark of the teacher hinders the student from finishing his argument, which is always in conflict with the solution the teacher is aiming at. The discussion of the first solution develops according to the expected context; the students refer to the teacher's ideas and try to match their own contributions to the social conventions and expectations. Now there is a rupture in this former implicit didactic contract (Brousseau 1984). One student presents a completely new and unforeseen solution; the teacher is unable to decipher the student's proposal. At the same time, this is prevented by the fact that she always has in mind her own a priori idea, which she wants to push forward, and therefore rejects the student's proposal.

During this short episode, the teacher is neither able to move the interaction into the planned frame nor to reduce the mathematical discussion to the social/conventional level; the student in some sense violates the conventions of how to proceed in classroom discussion when developing new knowledge in agreement with the teacher's expectations, requirements, and definition of the accepted context and manners of justifying new knowledge (mostly by way of reducing it to the teacher's implicit or explicit ideas and demands).

The persistent attitude of the student, based on his conviction of having found a mathematical solution that is consistently in conflict with the teacher's expectation, is a main reason that here the process of understanding focuses on epistemological aspects of the new knowledge. The new knowledge (i.e., the area formula developed in the student's argument) provides a true change in perspective on the a priori knowledge that can lead to new insights for the old knowledge. The teacher, however, remains unaware of this; she is not willing to understand that this student has developed an advanced mathematical understanding of the area formula for the trapezoid. He has not simply discovered the already existing solution of the teacher by reducing all his ideas and proposals to this ready-made old knowledge and by entering the question–answer game between students and teacher on the social/conventional level. The teacher's critical attitude toward the student's proposal also shows the frame of methodical constraints and expectations within which the teacher has to work; she cannot, and she is not willing to, spend too much time on a particular and perhaps misleading solution. She wants to stress a "simple" solution that can be understood by other students as well.

The necessary equilibrium between social and epistemological aspects in interactive processes of understanding – consequences

We have started our reflection on the problem of understanding mathematics in interactive processes with the description that understanding is the deciphering of social and epistemological signs and symbols. An important aspect inherent in this idea is the fact that signs and symbols possess an intentionality; that is, they are referring to something else that has to be detected and constructed, in epistemological as well as in social/communicative terms. The intentionality of signs and symbols displays an openness that, at the same time, cannot be arbitrary but has to be framed by epistemological constraints and by social conventions.

The framing of signs and symbols in an interactive mathematical culture is manifold. From elementary algebra it is known that, for instance, the variables x, y, and a, b, and so forth, evoke specific interpretations (often methodically desired by the teacher); students also have learned to read correctly the implicit communicative signals the teacher is sending (examples can be found in our two short episodes). In this way, the negotiation of meaning in classroom interaction is simultaneously a process of mutually framing the intentionality of the signs and symbols that are communicated (cf. Bauersfeld 1983, 1988; Voigt 1984a, 1985). The feedback one gets on the remarks made configures in some instances the intentionality of the signs and symbols under discussion.

When trying to organize the understanding of new mathematical knowledge during classroom interaction, one cannot observe only the reciprocal framing and interpreting of signs and symbols expressed by the persons participating in the mathematical classroom culture. This interactive negotiation is also based on a discursive frame that structures and legitimizes the process of understanding and the accepted forms of understanding. Thus, understanding a mathematical concept or problem during classroom teaching is not simply a direct, correct determination of the intentionality of the mathematical symbols in question. What could be the right and adequate interpretation also depends on the social and conventional rules and patterns of what is an accepted understanding (cf. Maier and Voigt 1989; Voigt 1984b).

Accordingly, the understanding of mathematical knowledge is a reciprocal relationship of understanding the new mathematical knowledge in its own right (i.e., integrating the old knowledge structure into the future, new knowledge structure, as in the example of the negative numbers), and, at the same time, of organizing and formulating this understanding of new knowledge in the frame of the conventional, legitimized, and accepted patterns of social knowledge justification (or of demonstrating the necessity of modifying the social, conventional framework). The student has to understand the new mathematics while being able to express his or her mathematical understanding within the cultural setting of the conventional social patterns, descriptions, and metaphors. Both requirements have to be fulfilled before it can be "verified" that the student really has understood.

This socially conventionalized discursive frame of understanding is indispensable, but it cannot be conceived of as a fixed organizational scheme according to which understanding mathematics could be defined correctly or even be deduced. It describes, for instance, what are legitimate patterns of mathematical argumentation, what are accepted forms of proof, what are accepted analogies, what degree of rigor is needed, whether an example is sufficient as a general argument, and so forth. In this way, the conventionalized discursive frame of understanding is open to change and modification occurring during the course of further mathematical development.

The reciprocal relationship between directly understanding the new mathematical knowledge and organizing it in a socially convention-alized discursive frame of understanding is based on a balanced inter-play between social and epistemological aspects of mathematical knowledge. Certain intentions of mathematical signs and symbols have to be agreed upon socially, generalizations of mathematical concepts growing in the course of knowledge development have to be socially sanctioned, and the relational structure of this expanding new knowl-edge has to be controlled epistemologically. Epistemological changes and reorganizations of mathematical knowledge can be performed only if they are in agreement with the socially conventionalized patterns of argumentation and understanding, which themselves depend on the epistemological character of mathematics in that their modifications and changes are subject to evolving new epistemological constraints.

The balanced interplay between understanding new mathematical knowledge and organizing it in a socially conventionalized discursive frame of understanding is to some extent destroyed in everyday mathe-matics culture. The discursive pattern becomes an interactively con-stituted methodical ceremonial for generating common understanding of school mathematical knowledge. The essence of this methodical ceremonial is to collect the necessary elements (i.e., descriptions and operational signs) for the knowledge under discussion and to combine them in order to obtain the expected methodical metaphor for the math-ematical knowledge (cf. Steinbring 1991c). In our first episode, this was the concept of relative frequency, described by the statement "Relative frequency is the number of cases observed divided by the number of trials," written in the shape of a fraction:

$$\text{relative frequency} = \frac{\text{number of cases observed}}{\text{number of trials}}$$

In the second episode, the teacher expected the construction of the geometrical methodical metaphors of a parallelogram (made of two trapezoids) and a rectangle for developing the area formula; the students are expected to collect all necessary elements, to combine them, and to reconstruct what the teacher already knows.

The main reason for the dominance of this methodical ceremonial is the central objective of mathematics teaching to make mathematical understanding (directly) possible, and this simultaneously requires that the teacher has already understood herself. On the basis of his or her understanding, that is, his or her fixed a priori knowledge, the teacher now organizes the methodical course of teaching, which, in many cases during everyday teacher–student interaction, causes the total replacement of epistemological aspects of knowledge by conventionalized communicative strategies. Since the teacher has already understood the problem and integrates all knowledge elements into a network of representation, the only task that remains for the students is that of correctly discovering these elements. This search is guided and facilitated by the teacher's social signs and hints indicating for the students whether they are on the right track and how far away from the goal they still are.

A dilemma arises for real interactive processes of mathematical understanding: the more and better the teacher already has understood the new mathematical knowledge in question (and, of course, the teacher has the strong obligation to understand all the mathematics beforehand), the greater the danger that the organization of processes of understanding in mathematics teaching degenerates into the described methodical ceremonial for generating common understanding of school mathematical knowledge, which, in its essence, means a prevention of true mathematical understanding. The first episode shows the degeneration of the concept of relative frequency (indeed some kind of medium value) to a written (and spoken) form of fraction (numerator and denominator): The epistemological relation of this mathematical concept is in no way modeled operationally. In the second episode, the teacher cannot understand (i.e., integrate into her a priori knowledge) the epistemological reflections the student is developing for finding the area formula.

A reestablishment of the balanced interplay between understanding new mathematical knowledge and organizing it in a socially conventionalized discursive frame of understanding is only possible in everyday mathematics interaction when the teacher becomes aware of the need to understand something in this process deeply for himself or herself, and that not everything is already understood beforehand. In most cases, this understanding the teacher has to construct concerns the problems students have when actually going through an interactive process of understanding, and even the understanding of new *mathematical* knowledge may be required by the teacher, too. In the first episode, the consideration of the student's remark "Relative frequency means for example often, it is, hem, a medium value" (6) could have led to a better mathematical understanding of this concept by starting to construct a new conceptual metaphor instead of reducing it to the fraction concept.

The second episode clearly shows that there really is something – even mathematically – for the teacher to understand that she did not know before. However, she refuses this understanding because of the obvious differences to her a priori knowledge, on which she wanted to base her teaching process methodically. Indeed, the argument of the student is not easy to understand, and it is not just the collection of already existing elements and their operational combination. It is the actual, developmental construction of new mathematical relations, simultaneously using some social conventions. For the student, the central idea of understanding his own conceptual metaphor is the functional change of the trapezoid into a geometric surface fulfilling different requirements: Keeping the area unchanged and producing geometric figures for which the area formulae are already known. At the same time, the student uses a convention, widely accepted by students, namely, taking concrete numbers instead of variables, but using them without any further restriction: a real mathematical convention, whose acceptability might be negotiated.

The analysis of the student's process of understanding shows the interrelation of social, conventional, and epistemological constraints that has to be developed and related in order to produce real understanding, that is, actively to grasp the new meaning. Here, the teacher herself could really gain much new understanding instead of trying to force the student to enter her fixed methodical frame of understanding. In this way, the teacher trying to understand the student's process of understanding could then support the integration of this kind of understanding into the conventional, legitimized, and accepted patterns of social knowledge justification for the other students. This would allow questioning and modifying the social conventions of codifying the understanding of school mathematical knowledge according to new, unforeseen epistemological insights.

Appendix 1: Transcript 1, grade 7: "What is relative frequency?"

1. T.: What do we understand by relative frequency, Markus?
2. S.: Relative frequency is, if you take the number of the cases observed
3.   together with the number of trials, well you throw all into one pot, and stir it up.
4. S.: Hahaha, good appetite!
5. T.: Klaus?
6. S.: Relative frequency means for example often, it is, hem, a medium value.
7. T.: Yes, what . . . ? Then it's better to say nothing!
8. T.: Relative frequency, Frank?
9. S.: Hem, relative frequency does mean, when the trials are divided with

10.   the cases observed, I think, or multiplied.
11.   T.: Markus did already say it quite correct, just the decisive word did
        miss,
12.   T.: . . . but now you do know it, Markus?
13.   S.: Subtract.
14.   S.: Take it minus.
15.   T.: That's incredible! Silvia?
16.   S.: In a chance experiment, the number of cases observed, when you . . .
17.   T.: Hem, Markus, you should ask how you will write it down?
18.   S.: Oh yes, OK, well, relative frequency is, when you, the number of
19.   trials with the number of cases observed, well, if you then . . .
20.   T.: But how you write it down? How do you write it down?
21.   T.: Come on, write it on the blackboard!
22.   S.: Shortly writing down.
23.   T.: Thank you, Ulli.
24.   S.: Writing as a fraction.
25.   T.: Aha, as a fraction, so, what is it then, what kind of calculus?
26.   S.: Fractional calculus.
27.   T.: Well, a complete sentence, Markus!
28.   S.: Hem, well relative frequency is . . .
29.   T.: Let it aside.
30.   S.: . . . the number of, hem, the number of cases observed with the
31.   number of trials, yes, subtracted in the fractional calculus.
32.   T.: Subtracted in the fractional calculus!!! Markus, that's incredible!
33.   T.: Ulli, formulate it reasonably!
34.   S.: If you take that, the number of the, the cases observed, yes, and the
35.   number of trials, hem, well, if you then, one as denominator and the
36.   other as numerator, hem, I don't know what you mean.
37.   S.: To calculate a fraction!
38.   T.: Either to write as a fraction, or what kind of calculation is it then, if
39.   you write a fraction?
40.   S.: Dividing.
41.   T.: Yes, that's what I think too. OK, yes. Nobody doesn't know it any
        more?!
42.   S.: To reduce to the common denominator.
43.   T.: Well, relative frequency is, and Markus has said it correctly, the
44.   number of cases divided by . . . ?
45.   S.: . . . the number of cases observed.
46.   T.: No, the cases observed divided by . . . ? Aha! Or, if you formulate it as
47.   Ulli said, the cases observed as numerator and the number of trials as
        . . . ?
48.   S.: Denominator.

Appendix 2: Transcript 2, grade 8, "The area formula for the trapezoid"

Some properties of the trapezoid are discussed in the class and compared with those of the parallelogram and the rectangle. The students have been asked by the teacher to explore the area formula of the trapezoid in partner work. She gives the following hints:

1. T.: The problem is to find a formula for the area . . . there are two
2. possibilities to come to a formula for the area . . . you can, eh, enlarge this trapezoid . . . eh . . .
3. but one can also draw it in some other way, or extend it, that one gets a
4. parallelogram afterwards . . . the other possibility is, that you indeed try,
5. as we have done it yesterday with the parallelogram, that
6. somehow one divides it, cutting off angles and adding them on the other side, so that one gets a
7. rectangle at the end . . . No, not a whole angle, that does not work . . . only parts!
8. You should try it in your groups!

Subsequent to the group work of the students, the proposals for the solutions are discussed. The first proposal is totally in accordance with the expectations of the teacher. Some students have turned a copy of the trapezoid upside down and have put it alongside the old trapezoid, now having a parallelogram of which they know the area formula; they use it correctly for developing the area formula of the trapezoid, getting $A = [(g_1 + g_2) \cdot h/2]$. In the following the second proposal for obtaining the formula is discussed.

9. T.: . . . OK, the second possibility, even if I would like to stop teaching here . . .
10. Ehm, the second possibility . . .
11. S.: . . . could one not, this line here, the height . . .
12. S.: . . . could you please speak a bit louder . . .
13. S.: . . . make parallel to, to this other line, and then one would have
14. such a quadrangle and on the other side a triangle . . .
15. T.: . . . yes, ehm, some others also made this proposal . . . only the
16. problem is if I would draw a line of height here, yes, and this triangle
17. is cut off there, with all these proposals and possibilities the problem is, that one only knows the
18. whole basis line g_1 and not only this little piece there. That's totally
19. unknown. Well, when you really have concrete lengths, you could
20. draw and measure it. But in general I can't do it. How should I then
21. say, g_1 minus some small piece, but how long is this little piece at all?
22. Hence, that does not work! . . . Hem, I would like to have another possibility . . . Jochen!

23. S.: Yes, if we now, ehm the longer side . . . well, take g_1 minus g_2
24. . . .
25. T.: . . . aha . . .
26. S.: . . . and then, ehm, then one has, there would be a remainder; let's say g_1
27. would be 5, hence 8 minus 5, giving 3 and then, if one would push now g_2
28. backwards, in a way giving a right angle, then something would remain
29. there on the right side . . .
30. T.: . . . pay attention, now you are again coming up with pushing back-
31. wards g_2, later we will deal with it . . .
32. S.: . . . I do mean something else, there you take 5 times h, then, what is
33. left here, there remains 3 centimeters, that is a slope, yes, then, ehm, you also take
34. the slope . . .
35. T.: . . . I really have understood this . . .
36. S.: . . . the remainder is 3 times the height then, divided by two . . .
37. T.: . . . no!! A second possibility, ehm, one tries to construct a rectangle,
38. and I have seen how some of you have used the drawing
39. triangle and they have moved it in this way . . . and at this one point . . . . . I will draw here, in this trapezoid something like a medium line . . . . but first of all one has to get this idea, OK!

References

Bauersfeld, H. (1978). Kommunikationsmuster im Mathematikunterricht – Eine Analyse am Beispiel der Handlungsverengung durch Antworterwartung. In H. Bauersfeld (Ed.), *Fallstudien und Analysen zum Mathematikunterricht* (pp. 158–170). Hannover: Schroedel.

Bauersfeld, H. (1983). Subjektive Erfahrungsbereiche als Grundlage einer Interaktionstheorie des Mathematiklernens und -lehrens. In H. Bauersfeld et al. (Eds.), *Lernen und Lehren von Mathematik* (pp. 1–56). Köln: Aulis.

Bauersfeld, H. (1988). Interaction, construction and knowledge: Alternative perspectives for mathematics education. In D.A. Grouws, T.J. Cooney, and D. Jones (Eds.), *Effective mathematics teaching* (pp. 27–46). Reston, VA: NCTM & Lawrence Erlbaum.

Brousseau, G. (1984). The crucial role of the didactical contract in the analysis and construction of situations in teaching and learning. In H.G. Steiner (Ed.), *Theory of Mathematics Education. IDM – Occasional Paper Nr. 54.* (pp. 110–119). Bielefeld: Institut für Didaktik der Mathematik.

Herscovics, N., and Bergeron, J.C. (1983). Models of understanding. *Zentralblatt für Didaktik* der Mathematik, 15(2), 75–83.

Hiebert, J., and Carpenter, T.P. (1992). Learning and teaching with understanding. In D. Grouws (Ed), *Handbook of research in teaching and learning mathematics* (pp. 65–97). New York: Macmillan.

Maier, H. (1988). "Verstehen" im Mathematikunterricht – Explikationsversuch zu einem vielverwendeten Begriff. In P. Bender (Ed.), *Mathematikdidaktik – Theorie und Praxis, Festschrift für Heinrich Winter* (pp. 131–142). Berlin: Cornelsen.

Maier, H., and Voigt, J. (1989). Die entwickelnde Lehrerfrage im Mathematikunterricht (Teil 1 u. 2). *mathematica didactica*, 12, 23–55, 87–94.

Pirie, S.E.B. (1988). Understanding: Instrumental, relational, intuitive, constructed, formalised . . . ? How can we know? *For the Learning of Mathematics*, 8(3), 2–6.

Pirie, S.E.B., and Kieren, T. (1989). A recursive theory of mathematical understanding. *For the Learning of Mathematics*, 9(3), 7–11.

Postel, H. (1991). Konzeptionen zur Behandlung der Dezimalbruchrechnung in der Bundesrepublik Deutschland. *Der Mathematikunterricht*, 37(2), 5–21.

Saxe, G.B. (1991). *Culture and cognitive development: Studies in mathematical understanding*. Hillsdale, NJ: Lawrence Erlbaum.

Schroeder, T.L. (1987). Students' understanding of mathematics: A review and synthesis of recent research. In *Proceedings of Psychology of Mathematics Education, XI* (pp. 332–337).

Sierpinska, A. (1990a). Some remarks on understanding in mathematics. *For the Learning of Mathematics*, 10(3), 24–36.

Sierpinska, A. (1990b). Epistemological obstacles and understanding: Two useful categories of thought for research into the learning and teaching of mathematics. In *Proceedings of the 2nd Bratislava International Symposium on Mathematics Education. August 1990*. Bratislava: University of Bratislava.

Sierpinska, A. (1994). *Understanding in Mathematics*. London: The Falmer Press.

Skemp, R.R. (1976). Relational understanding and instrumental understanding. *Mathematics Teaching*, 77, 20–26.

Steinbring, H. (1988). Nature du Savoir Mathématique dans la Pratique de l'Enseignant. In C. Laborde (Ed.), *Actes du Premier Colloque Franco-Allemand, de Didactique des Mathématiques et de l'Informatique* (pp. 307–316). Grenoble: La Pensée Sauvage.

Steinbring, H. (1989). Routine and meaning in the mathematics classroom. *For the Learning of Mathematics*, 9(1), 24–33.

Steinbring, H. (1991a). Mathematics in teaching processes: The disparity between teacher and student knowledge. *Recherches en Didactique des Mathématiques*, 11(1), 65–107.

Steinbring, H. (1991b). Eine andere Epistemologie der Schulmathematik — Kann der Lehrer von seinen Schülern lernen? *mathematica didactica,* 14(2/3), 69–99.

Steinbring, H. (1991c). The concept of chance in everyday teaching: Aspects of a social epistemology of mathematical knowledge. *Educational Studies in Mathematics,* 22, 503–522.

Steinbring, H. (1992). *The relation between social and conceptual conventions in everyday mathematics teaching (Manuscript).* Bielefeld: IDM.

Steinbring, H. (1994). Symbole, Referenzkontexte und die Konstruktion mathematischer Bedeutung – am Beispiel der negativen Zahlen im Unterricht. *Journal für Mathematik-Didaktik,* 15, (3/4), 277–309.

Steinbring, H. (to appear). Epistemology of mathematical knowledge and teacher–learner interaction. *The Journal of Mathematical Behavior.*

Voigt, J. (1984a). *Interaktionsmuster und Routinen im Mathematikunterricht – Theoretische Grundlagen und mikroethnographische Falluntersuchungen.* Weinheim: Beltz.

Voigt, J. (1984b). Der kurztaktige, fragend-entwickelnde Mathematikunterricht. *mathematica didactica,* 7, 161–186.

Voigt, J. (1985). Patterns and routines in classroom interaction. *Recherches en Didactique des Mathématiques,* 6(1), 69–118.

Vollrath, H.-J. (1993). Paradoxien des Verstehens von Mathematik. *Journal für Mathematik-Didaktik,* 14(1), 35–58.

Wagner, R. (1981). *The invention of culture.* Chicago: University of Chicago Press.

Wagner, R. (1986). *Symbols that stand for themselves.* Chicago: University of Chicago Press.

Walkerdine, V. (1990). *The mastery of reason.* London: Routledge.

Wellenreuther, M., and Zech, F. (1990). Kenntnisstand und Verständnis in der Dezimalbruchrechnung am Ende des 6. Schuljahres. *mathematica didactica,* 13(3/4).

Part IV

Outlook

---

14 About the notion of culture in mathematics education

HEINRICH BAUERSFELD

It is one of the worst concepts ever formulated.
N. Luhmann[1]

Culture has become a household word, if not a catchword, as in "the culture of the (mathematics) classroom." We run into difficulties as soon as we interpret the term "culture" in more general terms as the constructing *of a human-made reality.* Reality for whom? For all, for learners, or for members only? Constructing from what or with what? Are there means, prerequisites, or (other) prior realities? Though *constructing* indicates a process, the obvious aim is the *evolving reality,* a product. Is it both, process and product? If it is a changing process, and, for that reason, an issue with a history, what is its identity? There is no end to this.

This brings to mind Gregory Bateson's famous metalogues,[2] discussions between father and daughter about problematic themes. The last and richest one, "What is an instinct?" (Bateson 1969), defines *instinct* as an "explanatory principle" and ascertains that an explanatory principle explains "anything you want it to explain," and thus it "really explains nothing. It's a sort of conventional agreement between scientists to stop trying to explain things at a certain point" (1969: 12).

This, perhaps, may serve as a promising starting point. Culture, like instinct, context, intelligence, learning, and many other terms in frequent use, serves as an explanatory principle.[3] It would be worthwhile for more than one doctoral dissertation to analyze the many different uses in education, and in mathematics education in particular. An application of Derrida's (1977) method of "deconstruction" would promise helpful insights into many contradictions, inconsistencies, cloudy extensions, and even suppressed meanings and distortions (see Merz and Knorr-Cetina 1997).

Generally, every attempt to arrive at the precise definition of any concept ends in hopelessness or circularity, in $\alpha\pi\text{o}\rho\iota\alpha$, as we know from Plato's Dialogues. Indeed, the history of the sciences can be read as an uninterrupted chain of such failures. A visitor from another galaxy might be tempted to state that theories seem to be the more successful, the more open their key concepts are. Weakly defined concepts or related loose nets of concepts leave ample space for the users to fill in individual interpretations. We identify such difficulties more easily in theories about the mind, though the problem also exists in principle in the natural sciences and even mathematics (Goedel 1931 cited in Nagel and Newman 1958; Lakatos 1976; Davis and Hersh 1980). Accordingly, although we face impressive progress and development in terms of technology, we still seem to face the very old problems of people living together, trying to understand each other, coping with both their individual and societal difficulties, and educating their children. The only things that are new are the changing historical versions and facets in which these appear.

The adoption of natural science methods in psychology, in order to arrive at a comparable precision of results, has been only a limited success, as evidenced by, for instance, a century of endeavors to obtain a blanket definition of "intelligence." The often heard defense, "Intelligence means what the test measures," may serve as a revealing caricature of the problem. In return, the critique of the alternative descriptive or soft methods exaggerates the same type of argument: rich descriptions with poor meanings.[4]

It is not by chance that the increasing use of the term "culture" is accompanied by the opening up of educational research and theory, an opening up for softer methods, for relativistic perspectives, accompanied by a growing sensitivity about the use of language. French postmodernism teaches us that there is no simple way out. All we can do is to tell "stories"[5] to each other (Derrida 1992), and theories are likewise stories in this sense. Nonetheless, we still have a modest chance of comparing and analyzing such stories and thus obtaining *relatively* better and more motivating "stories."

What can be offered here is only to point out several areas that become more obscured rather than clarified through this well-worn notion of culture, and not a systematic analysis or critical comparison of the different uses, denotations, and connotations. Thus, the following paragraphs aim at presenting a few constructive hints at issues that usually receive little attention – where, in Bateson's sense, scientists stop too early with their attempts at explaining culture. Since the term "culture of the classroom" is used preferentially for the early school years, I shall limit my stories about neglected but important facets to this domain.

The unity of mind and body

The biologists worked hard to unmind the body; and the philosophers disembodied the mind.

Gregory Bateson *(1991: XVII)*

Recent research contributions to the learning of language, the mother tongue, have begun to reveal the extent and intensity of the development that precedes the speaking of the first word (e.g., Forrester 1996; Armstrong, Stokoe, and Wilcox 1995; Garton 1992; Trevarthen 1980). The most convincing example for me is Hanus Papousek's (1989, 1992) finding that, in the majority of early mother–infant interactions, it is the baby who opens and elicits the interaction, though mothers believe in their guiding function and think that they themselves stimulated the activity. Through body language and prelinguistic sound production, the infant can successfully open, maintain, and close an interaction with the mother, as well as with other persons when they can "read" the child's communicative attempts and can join the "attunement" (Bruner 1996: 175).

This means that babies are already in command of fairly developed fundamental prerequisites for communication long before they personally begin to speak in the strict verbal sense. Moreover, the influence as well as the further development of this fundamental nonverbal formation of the mind continue to be active and powerful in later years as well, although spoken language usually becomes dominant (Lakoff 1990). As Johnson (1987) has summed up:

There is meaning that comes through bodily experience and figurative processes of ordering, all of which are ignored by dominant objectivist treatments of language, meaning, understanding, and reasoning. In particular, there is the functioning of preconceptually meaningful structures of experience, schematic patterns, and figurative projections by which our experience achieves meaningful organisation and connection, such that we can both comprehend and reason about it. (Johnson 1987: 17)

By observing animal communication, Gregory Bateson (1991; in his "Last Lecture," September 1979) gained the insight that "It seems to me important . . . that we accept very firmly that body and mind are one." He announced to use these notions "as if they were already obsolete, where I hope they will arrive soon." "For the formal separation," says Bateson, "we could perhaps blame Descartes in the seventeenth century" (1991: 308). Stephen Toulmin has discussed this point in detail in "Cosmopolis" (1990).

Related to processes of "enculturation," the important part of this perspective is the typically *subconscious* functioning of related activities. One can never say how one has learned how to ride a bicycle, how to swim, how to speak, or how to submerge oneself in reflection. We cannot describe sufficiently how the necessary coordination of motor, sensory, and cognitive activities has been achieved. We speak of *imitation, intuition,* or *doing what is similar.* But these notions are just as weak and just as much "dormitive principles" as the notion of culture itself.

In adult communication as well, we permanently take into account our partner's mood and feelings. We "read" the partner's body language in parallel to the ongoing verbal part of our communication, and we permanently shape our own actions and reactions according to it. These fundamental abilities develop in mother–infant interactions. No wonder that children in the early school years are experts at reading their teacher's body language (I am convinced that teachers are manipulated in this manner by their students more often than they realize). The teacher's face can tell them whether their answer is taken as right or wrong before they hear the teacher's voice.[6] Children learn quite reliably about how the teacher assesses mathematics, about the teacher's likes and dislikes regarding the students' task performance, and much more about motives, reservations, and other issues behind the overt activities.

Thus a necessarily subjective framework of associations, evaluations, and relations emerges subconsciously with each of the students, a rich set spanning the whole range of our senses. These individual frameworks influence the students' future activities and decisions, also quite reliably and subconsciously. And in many aspects they are contrafunctional to the teacher's intentions and aims. In general, we underestimate the influence of mathematics teaching and of the students' mutual experiences in the classroom on these covert processes from writing in a straight line to the steps and strategies to be taken for the solution of a problem; from the values attached to an activity to the formation of the students' school mathematical habitus.

Bateson (1991: 75) has stated that "learning is always a colearning." From my perspective, this is to say that with every activity and usually in parallel to the officially planned or intended actions, we actually learn much more than just the subject matter thematized. The (more political) discussion of the "hidden curriculum" many years ago (e.g., Stodolsky 1988: 117) and more recently the investigations of "indirect learning" (e.g., Putnam et al. 1990) relate to this perspective. Colearning marks the insight that certainly the majority of what we learn until about the age of 10 is learned as a powerful by-product; it is learned casually and indirectly.

The claimed unity of body and mind has found much support from recent analyses of brain functioning (e.g., Bauersfeld 1993, 1996; Ellis and Young 1996). There is no separate processing of a pure mind. Within the large network of our brain, practically all domains are inter-connected. The body, feelings, and all senses are more or less present in all activations, and their intimate connections, therefore, emerge through the person's active participation in a culture.

The crucial function of human interaction and cooperation

If we have wrong ideas of how our abstractions are built – if, in a word, we have poor epistemological habits – we shall be in trouble – and we are.

Gregory Bateson (1991: 233)

Summarizing a comparative analysis of 46 research reports related to the cooperation/competition controversy from 1929 to 1993, Quin, Johnson, and Johnson (1995) report:

Members of cooperative teams outperformed individuals competing with each other on all 4 types of problem solving. These results held for all individuals of all ages and for studies of high, medium, and low quality. (1995: 129)

The authors propose the following possible reasons for this:

The exchange of information and insights among cooperators, the generation of a variety of strategies to solve the problem, increased ability to translate the problem statement into equations, and the development of a shared cognitive representation of the problem. (1995: 139)

This is a strong vote for the promotion of cooperation in the classroom.

From a different perspective, though with similar consequences to Bateson, Bruner has pointed to the need for cooperation, for "active negotiation," if individual development is not to be doomed to failure or to severe retardation. As investigations of the development of human-reared chimpanzees in comparison with chimp-mother-reared chimpanzees demonstrate, "The more a chimpanzee is exposed to human treatment, treated *as if* he were human, the more likely he is to act in a human-like way" (Bruner 1996: 179). The "human-reared chimpanzees outperformed all the others, human and ape alike." For instance, when told, "Do what I do!" the human-reared apes performed even better than human children of the same age.

Kanzi [the most gifted of the chimpanzees] and some other "enculturated" apes are different. Why? For precisely the reasons you articulate for children: from an early age Kanzi has spent his ontogeny building up a shared world with humans – much of the time through active negotiation. An essential element in this process is undoubtedly the behavior of other beings. . . . *Apes in the wild have no one who engages them in this way.* (M. Tomasello, cited in Bruner 1996: 181)

These findings illustrate the discrimination between biologically "primary" and "secondary" psychological dispositions (Geary 1995). The primaries develop "principally in response to evolutionary demands" (Bruner 1996: 171), they come naturally, and such dispositions can be found in many animal species. The secondaries require conscious effort and interaction with peers of the same species, through education or cooperation. They do not develop quasi-automatically, and are found only in a few specially trained primates and in human beings.

Trevarthen (1980) has used the similar terms "primary intersubjectivity" and "secondary intersubjectivity" for the related discrimination in mother–infant interactions. According to his observations, motives for primary intersubjectivity are (among others) "to seek proximity and face-to-face confrontation with persons" and "to express clear signs of confusion or distress if the actions of the partner become incomprehensible or threatening" (Trevarthen 1980: 326–327). Secondary intersubjectivity emerges when infants begin to see themselves as "causing the mother to modify her actions and her speech." Now, games become possible that are much more sophisticated than the primary "peek-a-boo" game, because the child, in reciprocity, tries to take over the mother's role in the games and no longer just wants "to slavishly imitate a model act" (Trevarthen 1980: 330–331).

Bruner infers

that the humanoid mind/brain complex does not simply "grow up" biologically according to a predestined timetable but, rather, that is opportunistic to nurturing in a humanlike environment. (1996: 183)

The remarkably superior quality of the human-reared chimpanzee's activities demonstrates a degree of development of the secondaries that does not appear in chimpanzees in their natural environment. If this is already the case in primates, what does it mean then, to treat or educate a human being in a *humanlike* way?

In the early school years, consequently, we shall have to engage much more in interactions than in arranging for a set of tasks to be solved by the single child in competitive isolation. As teachers, we shall have to act much more carefully in all classroom interactions, taking into account that our children actually learn along more fundamental paths, and actually learn deeper lessons than those that we think we are teaching them. Stressing the term "culture" in this relation may mean that, as teachers, we shall have to open ourselves up to a more differentiated exchange and more sensitive interaction with our students, trying to understand their tentative inventions and to tolerate their ways of presenting subject matter, through offering our descriptive means (but neither by imposing them nor by misusing them as an instrument of evaluating control or selection), and, last but not least, through making our own motives and reflections transparent.

Language, self, and reflection

The mystery is what we call conscious.
Gregory Bateson (1991: 173)

"The human being can be the object of his own actions" (Blumer 1969: 79, referring to Mead). This is because the human has a self. And this dealing with oneself, as far as it is done consciously, proceeds mainly, though not only, in language.

In small group interaction we often find scenes in which one child stops working and speaks to himself or herself quite openly with lowered voice and the eyes turned inward.[7] These disruptions of direct action indicate hesitation, an escaping from overt interaction and diving away from the environment, an opening of reflection because of suddenly arising difficulties, doubts, resistance, or uncertainties. Despite all the problems we might have with interpreting the related meanings, the point is that means and ways of performing such reflections emerge from the interaction with others, with peers – that is, through culture. And these tentative early steps toward reflection and an understanding of oneself require promotion and encouragement.

Such support appears to be necessary from another perspective as well. Every kind of reflective process proceeds within a different "language game" (in Wittgenstein's sense) than the language game used for the description of the direct action that is under consideration. And, again, these language games for reflection have a chance to be learned in classroom cultures only through richly differentiated and transparent reflective practices. One might name them *language games of a second order.*

Through one particular finding, the comparative analysis of Quin, Johnson, and Johnson (1995) quoted previously provides indirect support for this view: "The superiority of cooperation, however, was greater on non-linguistic than on linguistic problems" (1995: 129). The authors hypothesize that one possible cause of the reported difference may be that "the need for linguistic skills may reduce performance on linguistic problems." We can take this difference as an indicator of a deficit of education in the early school years, not least because "older students performed better on linguistic problems than did younger students" (1995: 136–137). We cannot blame natural individual development (the "primaries") for this deficit, because "where humans are concerned there is no natural mind," as "Geertz would say" (Bruner 1996: 172). The deficit, as I would interpret it, refers more likely to the need for an intensive exchange and negotiation in elementary education about both the teacher's and the students' verbal presentations and descriptions of all processes. And this requires the teacher's attention in small-group work as well as in whole-class discussions. Usually, there is not enough opportunity for the individual student actively to develop new language games to improve her or his verbal flexibility in dealing with mathematical tasks (for the development of argumentation in classroom discussions, see Krummheuer 1995).

Are teachers under the pressures of the complex classroom processes sufficiently aware of their personal impact and how they function as models with regard to these crucial reflections? Do they, as teachers, exploit every opportunity to expose and make transparent their own related reflections, how they deal with their own mistakes, their decisions for different strategies and approaches when faced with tasks, their faith and ability to "carry on regardless" in case of momentary and even continued failure, and their personal ways of maintaining self-control and showing fairness to everybody?

Quite common is the final confession of severe theoretical deficits in the Quin, Johnson, and Johnson research report.[8] The authors find "still unclear: the conditions under which competition may facilitate problem solving, and the internal dynamics determining how effectively cooperative groups approach problems and teach their members how to solve them." The enthusiastic dissemination of the constructivist principle and the ease that has brought about the term "social constructivism" should not disguise the theoretical deficits. Although we have many psychological theories of learning and teaching, we have only first approaches to sociological theories of learning that deal with the deficits mentioned (Krummheuer 1992; Miller 1986).

About the teacher's role

We can only know by virtue of difference. This means that our entire mental life is one degree more abstract than the physical world around us.

Gregory Bateson (1991: 309)

Much earlier it was Albert Einstein (1946: 287) who expressed this point even more incisively: "The concepts which arise in our thought and in our linguistic expressions are all – when viewed logically – the free creations of thought which can not inductively be gained from sense-experiences." The conviction "according to which the concepts arise from experience by way of 'abstraction,' i.e. through omission of a part of its content" appears to him to be a "fatal conception." When we identify a difference with Bateson (1991: 162) "as an elementary idea," the question arises of how and where these ideas and constructions arise in the classroom. Since there is no direct teaching, we are, again, led back to the facets of classroom cultures discussed, that is, the support from encouraging interaction and cooperation, language games, and reflection, and a more sophisticated understanding of the unity of mind and body and of the totality of our experiences.

Working in different research projects has made me aware of a fatal swing in the changing model of teaching that participating teachers sometimes show, especially those very interested in and willing to take over the aims of the projected reform. These teachers easily accept that didactic teaching (in the traditional sense), telling the students, and playing the "recitation game" (Hoetker and Ahlbrand 1969) can have devastating effects on the students. Convinced that there is no transmitting of knowledge, they favor the belief that nothing should be taught that the students can find themselves. This swing from one extreme to another produces a manner of teaching that comes very near to the questioning technique of Joe Weizenbaum's famous computer program "Eliza": They just take up the students' doubted or incorrect utterance, and react by using the same words and shaping them into a question. Sometimes they just wait or give neutral hints as stimuli. The effects are sometimes no less demotivating than the teaching methods abandoned before. Obviously, it is very difficult to "live" an encouraging culture, when one has had no previous opportunity to gain vivid experiences as a learning member in a vivid culture of the same or, at least, a similar type (see Chatterley and Peck 1995). This leads us to the problem of initial teacher education.

Like a searchlight focusing on the problems under discussion, the research findings of Wisniewski and Medin (1994) demonstrate how sensitive human interaction is. Their study of children in a "rule-learning" experiment shows that "a simple manipulation (varying the label associated with a category) dramatically influences rule learning" and thereby "undermines a basic assumption of many concept learning models that learning operates over a space of well defined features" (1994: 277). They conclude:

At least in some domains, learning is not a process in which prior knowledge first exerts its influence on data and then empirical learning takes over afterward. Rather, it is a process in which prior knowledge and experience are closely interwoven.

I would prefer to say, a process in which prior domains of experiences are activated and (re)combined for the recent purposes, closely merging with the formation of new experience. The authors hold the view that learning is "a highly interactive process" (1994: 278).

Last not least: The agents

What is disastrous is to claim an objectivity for which we are untrained and then project upon an external world premises that are either idiosyncratic or culturally limited.

Bateson (1991: 77)

Culture differs in meaning when we speak of the "French culture" and the "Japanese culture" or when we speak of the "culture of the mathematics classroom." A classroom culture clearly cannot exist independently of the culture of the society that maintains this institution. The notion *subculture* would be more adequate then. But another difference is much more important: Cultures have no *curriculum* although an observer can interpret functions, structures, and development. And one can neither *create* nor *reform* a culture, although cultures change (not only but also) through the inventions of their members. Cultures, especially, are not formed through agents.[9] Subcultures also emerge, but they also depend on the influence of explicitly trained agents, particularly in educational settings.

The tragedy is that our university departments train these agents to become experts in their related special subject matter, but not in the subculture wanted in the classroom. One becomes a member of a culture through active participation only, and this takes time. Living in a culture is an adaptive as well as a constructive process. Through imitation, intuition, and "tuning in," the future member learns to do things as *one does them in this culture.* And through doing this, necessarily in each person's individual way, the member contributes to the maintenance and the development of this culture. Also, and consciously, the future member takes over regulations and learns to obey norms for everyday life. Along the way, these formations, emerging with the individual, become habitual and subconscious (if their parts have ever been formed consciously), and this to the extent that the member can act adequately and in accordance with this culture even in situations never experienced before. Bourdieu (1977, 1980) has named this formation "habitus."

Without doubt, 4 or 5 years of initial teacher education do have an impact on the student teacher's habitus. But because the universities concentrate mainly on their own development, on the generation of qualified researchers and subject matter experts, their institutional culture is quite different from the needs, say, of elementary education and the related subcultures. As a consequence of this young teachers pass through a painful process of reorientation when encountering the realities of the new culture, and, in the end, they adapt themselves to the pragmatic needs of school practice: a sad process when we think of the children, of the wasted efforts and endeavors; and often it is more a process of subjection and surrender rather than one of active influence on and change of the system.

The universities will have to develop their teaching toward more compatibility with the wanted culture in school settings. This cannot mean that university professors should act like elementary teachers. But reform is necessary in order to prevent the contrafunctional effects of the initial training. As long as the teacher education institutions do not specialize in the needs of their student teachers, there is not much hope for an effective reform of the elementary school.

These tentative remarks may close with a statement from the final page of the book that inspired me to these considerations, Jerome Bruner's *The culture of education* (1996: 184):

If psychology is to get ahead in understanding human nature and the human condition, it must learn to understand the subtle interplay of biology and culture. Culture is probably biology's last great evolutionary trick.

Acknowledgment

I am grateful to Falk Seeger for constructive discussions during the preparation of this paper.

Notes

1. Verbal communication.
2. Several of these are published in *A Review of General Semantics,* vol. X, 1953, 127–130, 213–217, 311–315, and vol. XI, 1954, 59–63; another in *Impulse, Annual of Contemporary Dance,* 1954, 23–26. For readers of German access is easier, because all the metalogues (including one not published before) have been printed at the beginning of the German translation of *Oekologie des Geistes* ("Steps to an ecology of mind"), Frankfurt/Main: Suhrkamp 1981.
3. At other places Bateson also uses the notion *dormitive principle* (Bateson 1991: 170f.), adopting Molière's proposition for the "learned doctor's" explanation of "the physiological effects of opium" (1991: 135).
4. In a research community that has become more sensitive to the specificity of theoretical bases and to related methods, both paradigms can serve reasonable purposes beyond the usual antagonism.
5. According to Bateson (1969: 12) it was Isaac Newton who already "thought that all hypotheses were just *made up* like stories."
6. Helga Jungwirth has investigated the different effects on boys and girls in related teacher–student interactions in the early secondary grades (Jungwirth 1991).
7. Since the early 1970s my research group at the Institut für Didaktik der Mathematik (IDM) at Bielefeld University, namely, Wolfgang Herrlitz, and later Goetz Krummheuer and Jörg Voigt, has produced many transcripts from videotaped classroom lessons and related analyses through adopting and extending social interactionist, ethnomethodological, and sociolinguistic views and methods (see, e.g., Bauersfeld 1978; Herrlitz, Dauk, and Heinl 1979; Krummheuer 1992; Voigt 1984, 1995; and the many books in the series "Untersuchungen zum Mathematikunterricht," Köln: Aulis Verlag.
8. Franz Emanuel Weinert, former director of the Max Planck Institute for Psychological Research in Munich, used to caricature the common final conclusions of research reports by saying, "Two things can be taken for sure in educational research: All elements are connected to each other and further research is needed!" (personal communication).
9. The enormous efforts of the French to promote their language all over the world can be interpreted as an attempt to promote their culture. The limited success puts these efforts more in line with the efforts to reform school cultures.

References

Armstrong, D.F., Stokow, W.C., and Wilcox, S.E. (1994). *Gesture and the nature of language.* Cambridge: Cambridge University Press.

Bateson, G. (1969). Metalogue: What is an instinct? In T.A. Sebeok and A. Ramsay (Eds.), *Approaches to animal communication* (pp. 11–30). Paris: Mouton.

Bateson, G. (1991). Sacred unity: Further steps to an ecology of mind. (R.E. Donaldson, Ed.). New York: Harper Collins.

Bauersfeld, H. (1978). Kommunikationsmuster im Mathematikunterricht – Eine Analyse am Beispiel der Handlungsverengung durch Antworterwartung. In H. Bauersfeld et al. (Eds.), *Fallstudien und Analysen zum Mathematikunterricht* (pp. 158–170). Hannover: Schroedel.

Bauersfeld, H. (1993). Theoretical perspectives on interaction in the mathematical classroom. In R. Biehler, R.W. Scholz, R. Sträßer, and B. Winkelmann (Eds.), *Didactics of mathematics as a scientific discipline* (pp. 133–146). Dordrecht: Kluwer.

Bauersfeld, H. (1996). Wahrnehmen – Vorstellen – Lernen. Bemerkungen zu den neurophysiologischen Grundlagen im Anschluß an G. Roth. In P. Fauser and E. Madelung (Eds.), *Vorstellungen bilden: Beiträge zum imaginativen Lernen* (pp. 143–163). Velber: Friedrich Verlag.

Bauersfeld, H., and Cobb, P. (Eds.) (1995). *The emergence of mathematical meaning: Interaction in classroom cultures.* Hillsdale, NJ: Lawrence Erlbaum.

Bauersfeld, H., Kampmann, B., Krummheuer, G., Rach, W., and Voigt, J. (1982). *Transkripte aus Mathematikunterricht und Kleingruppenarbeit.* University of Bielefeld: IDM.

Blumer, H. (1969). *Symbolic interactionism: Perspective and method.* Englewood Cliffs, NJ: Prentice-Hall.

Bourdieu, P. (1977). *Outline of a theory of practice.* Cambridge: Cambridge University Press.

Bourdieu, P. (1980). *Le sens pratique.* Paris: Les Éditions de Minuit.

Bruner, J.S. (1983). *Child's talk: Learning to use language.* Oxford: Oxford University Press.

Bruner, J.S. (1996). *The culture of education.* Cambridge, MA: Harvard University Press.

Chatterley, L.J., and Peck, D.M. (1995). We're crippling our kids with kindness! *Journal of Mathematical Behavior,* 14, No. 4, 429–436.

Davis, P.J., and Hersh, R. (1980). *The mathematical experience.* Basel: Birkhäuser.

Derrida, J. (1973). *Speech and phenomena and other essays on Husserl's theory of signs.* Evanston, IL: Northwestern University Press.

Derrida, J. (1977). *Of grammatology.* Baltimore, MD: Johns Hopkins University Press.

Derrida, J. (1978). *Writing and difference.* London: Routledge and Kegan Paul.

Derrida, J. (1992). "Differance." In A. Easthope and K. McGowan (Eds.), *A critical and cultural studies reader.* London: Open University Press.

Donald, M. (1991). *Origins of the modern mind: Three stages in the evolution of culture and cognition.* Cambridge, MA: Harvard University Press.

Einstein, A. (1946). Remarks on Bertrand Russel's theory of knowledge. In P.A. Schilpp (Ed.), *The philosophy of Bertrand Russell* (pp. 279–291). New York: Tudor Publishing.

Ellis, A.W. and Young, A.W. (Eds.) (1996). *Human cognitive neuropsychology.* Hove: Psychology Press.

Forrester, M.A. (1996). *Psychology of language: A critical introduction.* London: Sage Publications.

Garton, A.F. (1992). *Social interaction and the development of language and cognition.* Hillsdale, NJ: Lawrence Erlbaum.

Geary, D.C. (1995). Reflections of evolution and culture in children's cognition: Implications for mathematical development and instruction. *American Psychologist, 50,* No. 1, 24–37.

Herrlitz, W., Dauk, E., and Heinl, G. (1979). Zur mathematischen Kommunikation in der Schülergruppe (DIMO – project report, 2 volumes). University of Bielefeld: IDM.

Hoetker, J. and Ahlbrand, W.P. (1969). The persistence of the recitation. *American Educational Research Journal, 6,* No. 2, 145–167.

Johnson, M.J. (1987). *The body in the mind: The bodily basis of meaning, imagination, and reason.* Chicago: University of Chicago Press.

Jungwirth, H. (1991). Interaction and gender: Findings of a microethnographical approach to classroom discourse. *Educational Studies in Mathematics, 22,* No. 3, June, 263–284.

Krummheuer, G. (1992). *Lernen mit "Format" – Elemente einer interaktionistischen Lerntheorie.* Weinheim: Deutscher Studien Verlag.

Krummheuer, G. (1995). The ethnography of argumentation. In P. Cobb and H. Bauersfeld (Eds.), *The emergence of mathematical meaning: Interaction in classroom cultures* (pp. 229–269). Hillsdale, NJ: Lawrence Erlbaum.

Lakatos, I. (1976). *Proofs and refutations: The logic of mathematical discovery.* Cambridge: Cambridge University Press.

Lakoff, G. (1990). *Women, fire, and dangerous things: What categories reveal about the mind.* Chicago: University of Chicago Press.

Merz, M. and Knorr-Cetina, K. (1997). Deconstruction in a "thinking" science: Theoretical physicists at work. *Social Studies of Science, 27,* 73–111.

Miller, M. (1986). *Kollektive Lernprozesse – Studien zur Grundlegung einer soziologischen Lerntheorie.* Frankfurt/Main: Suhrkamp.

Nagel, E. and Newman, J.R. (1958). *Goedel's proof.* New York: New York University Press.

Papousek, H. (1989). Patterns of rhythmic stimulation by mothers with three-month-olds: A cross-modal comparison. *International Journal of Behavioral Development,* 12, No. 2, June, 201–210.

Papousek, H., Jürgens, U., and Papousek, M. (Eds.). (1992). *Nonverbal vocal communication.* Cambridge: Cambridge University Press

Putnam, R.T., Lampert, M., and Peterson, P.L. (1990). Alternative perspectives on knowing mathematics in elementary schools. In C.B. Cazden (Ed.), *Review of Research in Education* (Vol. 16, pp. 57–150). Washington, DC: AERA.

Quin, Z., Johnson, D.W., and Johnson, R.T. (1995). Cooperative versus competitive efforts and problem solving. *Review of Educational Research,* 65, No. 2, 129–143.

Stodolsky, S.S. (1988). *The subject matters: Classroom activity in math and social sciences.* Chicago: University of Chicago Press.

Toulmin, S.E. (1990). *Cosmopolis: The hidden agenda of modernity.* New York: The Free Press/Macmillan.

Trevarthen, C. (1980). The foundation of intersubjectivity: Development of interpersonal and cooperative understanding in infants. In D.R. Olson (Ed.), *The social foundation of language and thought* (pp. 316–342). New York: Norton.

Trevarthen, C. (1986). Form, significance, and psychological potential of hand gestures of infants. In J.-L. Nespoulos, P. Perron, and A.R. Lecours (Eds.), *The biological foundation of gestures: Motor and semiotic aspects* (pp. 149–202). Hillsdale, NJ: Lawrence Erlbaum.

Untersuchungen zum Mathematikunterricht (1981–1992). 8 volumes: IDM – Band 3, 5, 6, 10, 16, 18, 19, and 22, Köln: Aulis Verlag Deubner.

Voigt, J. (1984). *Interaktionsmuster und Routinen im Mathematikunterricht.* Weinheim: Beltz.

Voigt, J. (1995). Thematic pattern of interaction and sociomathematical norms. In P. Cobb and H. Bauersfeld (Eds.), *The emergence of mathematical meaning: Interaction in classroom cultures* (pp. 163–201). Hillsdale, NJ: Lawrence Erlbaum.

Walkerdine, V. (1988). *The mastery of reason.* London: Routledge.

Wisniewski, E.J. and Medin, D.L. (1994). On the interaction of theory and data in concept learning. *Cognitive Science,* 18, 221–281.

Author index

Subject index